IEE TELECOMMUNICATIONS SERIES 51

Series Editors: Professor J. O'Reilly
Professor G. White

Security for mobility

Other volumes in this series:

Security for mobility

Edited by
Chris J. Mitchell

The Institution of Electrical Engineers

Published by: The Institution of Electrical Engineers, London,
United Kingdom

The Institution of Electrical Engineers,
Michael Faraday House,
Six Hills Way, Stevenage,
Herts., SG1 2AY, United Kingdom

British Library Cataloguing in Publication Data

Security for mobility
    1. Computer security 2. Mobile communication systems
    I. Mitchell, Chris, 1953- II. Institution of Electrical Engineers
    005.8

ISBN 0 86341 337 4

Typeset in India by Newgen Imaging Systems
Printed in the UK by MPG Books Limited, Bodmin, Cornwall

Contents

Chris J. Mitchell and Ralf Schaffelhofer

Preface

Over the last 10–15 years, mobile telecommunications has grown from being a small niche technology to a massive industry. Mobile telephones are now ubiquitous, and the divisions between PCs, PDAs, mobile phones and other mobile devices are becoming increasingly blurred. Against this background, the security of information handled by these devices, and of the mobile devices themselves, becomes ever more important. This is the focus of this book.

Each chapter of the book has been contributed by a different collaboration of authors, including many of the leading European experts currently working within the field. Thus, each chapter is self-contained and can be read independently, although there are many relationships between the various chapters.

The main text of this book is divided into five parts, with the following headings and contents.

- Part I: *Underlying technologies* contains five chapters, all concerned with technologies underlying the provision of security in mobile networks. First, a very brief introduction to cryptography is provided, focussing on the techniques of particular importance to the mobile environment. The next two chapters examine aspects of Public Key Infrastructures, that is, infrastructures for the distribution and management of public keys, in the context of mobile networks. Two further chapters consider current and future uses of smartcards and other smart tokens.
- Part II: *Network security* contains a total of six chapters. The first chapter provides a review of existing network access security for UMTS. This is followed by two further chapters looking at possible future techniques for secure mobile access. The remaining three chapters consider three separate mobile network security issues, namely security in the Personal Area Network, routing security in ad hoc networks and security issues for MobileIPv6.
- Part III: *Mobile code issues* is made up of two chapters. The main issue dealt with here is that of software that can migrate around a mobile network. The issue of mobile code is something of growing importance across both mobile and fixed networks, and there is growing awareness of both the potential and threat of mobile code, for example, in the form of viruses and applets. There are two major mobile code issues of particular relevance to mobile networks, namely mobile

agents and downloaded code for personal mobile devices, and these two topics are the focus of this part.

- Part IV: *Application security* contains a further three chapters looking at two major applications of mobile networks, and the particular security issues these two applications bring. The first chapter considers secure mobile commerce, that is, the conduct of e-business over mobile networks. Whilst there are clearly large overlaps with more conventional e-commerce security issues, the use of a mobile platform, typically with limited processing and/or storage capabilities, poses its own special set of challenges and opportunities. The other application issue, considered in the other two chapters, is the secure delivery of proprietary content (e.g. music, video or textual material) via mobile networks. Again there are major business opportunities, but also major hazards, including risks of piracy.
- Part V: *The future* is made up a single chapter containing a roadmap for future research in mobile security and privacy.

Acknowledgements

First and foremost, I would like to thank the authors of the chapters of this book. Without their enthusiasm for the project, and willingness to spend long hours preparing their respective contributions, this book would clearly never have existed.

Second, I must also thank a number of present and former staff members at the IEE for enabling this book to be created. In this context, I must mention Paul Staniforth, who suggested the creation of a special issue of the IEE *Electronics and Communications Engineering Journal*, which appeared in October 2002, and the articles from which form the core of this book. Roland Harwood then suggested this book, and encouraged me to start work on this project. His initial encouragement was followed by vital help and support from Robin Mellors-Bourne and Sarah Kramer.

Third, the vast majority of the chapters in this book have come about as a result of work within a small number of major collaborative research projects. The EU 5th framework projects SHAMAN and PAMPAS are primarily responsible for 11 of the 17 chapters. The UK Mobile VCE Consortium, through the Core II programme of research, has supported the writing of a further two of the chapters. I would like to thank all the consortium members in these projects for supporting the writing of this material.

Last, but by no means least, I must thank my wife Sue, for allowing me the time and space to pursue projects such as this, while she gets on with running the house, the family and much else besides.

Contributors

Derek Babb
Advanced Technology and Standards
Samsung Electronics Research Institute
Communications House
South Street
Staines
Middlesex TW18 4QE
UK
Email: derek.babb@seri.co.uk

Craig Bishop
Advanced Technology and Standards
Samsung Electronics Research Institute
Communications House
South Street
Staines
Middlesex TW18 4QE
UK
Email: craig.bishop@seri.co.uk

Krister Boman
Ericsson AB
Lindholmspiren 11
SE-41756 Gothenburg
Sweden
Email: krister.boman@erv.ericsson.se

Niklas Borselius
Information Security Group
Royal Holloway, University of London
Egham
Surrey TW20 0EX
UK
Email: Niklas@Borselius.com

Joris Claessens
European Microsoft Innovation Center
Ritterstrasse 23
52072 Aachen
Germany
Email: jorisc@microsoft.com
(Work performed while at COSIC,
Katholieke Universiteit Leuven,
Belgium)

Jozef Dankers
Siemens Atea n.v.
Atealaan 34
B-2200 Herentals
Belgium
Email: jozef.dankers@siemens.com

Terence E. Dodgson
Advanced Technology and Standards
Samsung Electronics Research Institute
Communications House
South Street
Staines
Middlesex TW18 4QE
UK
Email: tdodgson@seri.co.uk

James Edwards
Room 3.39
University of Bristol
Merchant Venturers Building
Woodland Road
Bristol BS8 1UB
UK
Email: james.edwards@bristol.ac.uk

Hubert Ertl
Giesecke& Devrient GmbH
Prinzregentenstrasse 159
D-81607 Munich
Germany
Email: Hubert.Ertl@de.gi-de.com

Theo Garefalakis
Information Security Group
Royal Holloway, University of London
Egham
Surrey TW20 0EX
UK
Email: theo@comm.utoronto.ca

Christian Gehrmann
Ericsson Mobile Platforms AB
Nya Vattentornet
SE-22183 Lund
Sweden
Email:
Christian.Gehrmann@ericsson.com

Christian Günther
Corporate Technology
Siemens AG
Otto-Hahn-Ring 6
D-81739 Munich
Germany
Email:
christian.guenther@siemens.com

Silke Holtmanns
Ericsson Eurolab Deutschland GmbH
Ericsson Allee 1
D-52134 Herzogenrath
Germany
Email: Silke.Holtmanns@ericsson.com

Günther Horn
Corporate Technology
Siemens AG
Otto-Hahn-Ring 6
D-81739 Munich
Germany
Email: guenther.horn@siemens.com

Peter Howard
Vodafone Group Services Ltd
Vodafone House
The Connection
Newbury
Berkshire RG14 2EN
UK
Email: peter.howard@vodafone.com

Keith Howker
Vodafone Group Services Ltd
Vodafone House
The Connection
Newbury
Berkshire
RG14 2EN
UK
Email: keith.howker@vodafone.com

Robert Hulsebosch
Telematica Instituut
P.O. Box 589
7500 AN Enschede
The Netherlands
Email: Bob.Hulsebosch@telin.nl

Glyn Jones
Thales Research and
Technology (UK) Ltd
Worton Drive

Worton Grange
Reading
Berkshire RG2 0SB
UK
Email: Glyn.Jones@thalesgroup.com

Dritan Kaleshi
Department of Electrical and
Electronic Engineering
University of Bristol
Merchant Venturers Building
Woodland Road
Bristol BS8 1UB
UK
Email: dritan.kaleshi@bristol.ac.uk

Heiko Knospe
T-Systems
Systems Integration
Am Kavalleriesand 3
D-64295 Darmstadt
Germany
Email: heiko.knospe@t-systems.com

Bruce MacFarlane
Pedagog Ltd.
Tredomen Business Centre
Hengoed CF82 7FN
UK
Email: bruce@pedagog.com

Robert Maier
Dept. Electrical Engineering-ESAT
SCD/COSIC
Katholieke Universiteit Leuven
Kasteelpark Arenberg 10
B-3001 Leuven-Heverlee
Belgium
Email:
Robert.Maier@esat.kuleuven.ac.be

Chris J. Mitchell
Information Security Group
Royal Holloway, University of London

Egham
Surrey TW20 0EX
UK
Email: C.Mitchell@rhul.ac.uk

Alistair Munro
Department of Electrical and
Electronic Engineering
University of Bristol
Merchant Venturers Building
Woodland Road
Bristol BS8 1UB
UK
Email: alistair.munro@bristol.ac.uk

Valtteri Niemi
Nokia Research Center
P.O. Box 407
FIN-00045
Nokia Group
Finland
Email: valtteri.niemi@nokia.com

Kaisa Nyberg
Nokia Research Center
P.O. Box 407
FIN-00045
Nokia Group
Finland
Email: kaisa.nyberg@nokia.com

Kenneth G. Paterson
Information Security Group
Royal Holloway, University of London
Egham
Surrey TW20 0EX
UK
Email: Kenny.Paterson@rhul.ac.uk

Bart Preneel
Dept. Electrical Engineering-
ESAT-SCD/COSIC
Katholieke Universiteit Leuven

Kasteelpark Arenberg 10
B-3001 Leuven-Heverlee
Belgium
Email:
Bart.Preneel@esat.kuleuven.ac.de

Ralf Schaffelhofer
T-Systems
Systems Integration
Am Kavalleriesand 3
D-64295 Darmstadt
Germany
Email:
ralf.schaffelhofer@t-systems.com

Dirk Scheuermann
Fraunhofer Institute for Secure
Telecooperation (SIT)
Rheinstrasse 75
D-64295 Darmstadt
Germany
Email:
Dirk.Scheuermann@sit.fraunhofer.de

Marko Schuba
Ericsson Eurolab Deutschland GmbH
Ericsson Allee 1
D-52134 Herzogenrath
Germany
Email: Marko.Schuba@ericsson.com

Scarlet Schwiderski-Grosche
Information Security Group
Royal Holloway, University of London
Egham
Surrey TW20 0EX
UK
Email:
scarlet.schwiderski-grosche@rhul.ac.uk

Vaia Sdralia
Advanced Technology and Standards
Samsung Electronics Research Institute

Communications House
South Street
Staines
Middlesex TW18 4QE
UK
Email: vaia.sdralia@samsung.com

Hannes Tschofenig
Corporate Technology
Siemens AG
Otto-Hahn-Ring 6
D-81739 Munich
Germany
Email: hannes.tschofenig@siemens.com

Adrian O. Waller
Thales Research and
Technology (UK) Ltd
Worton Drive
Worton Grange
Reading
Berkshire RG2 0SB
UK
Email: Adrian.Waller@
thalesgroup.com

Toby Whitley
Centre for Communications Research
Department of Electrical and Electronic
Engineering
University of Bristol
Merchant Venturers Building
Woodland Road
Bristol BS8 1UB
UK
Email: toby.whitley@bristol.ac.uk

Angus Wood
Inspired Broadcast Networks
1-7 Livonia Street
London W1F 8AD
UK
Email: gus@shady.org

Timothy Wright
Vodafone Group Services Limited
Vodafone House
The Connection
Newbury
Berkshire RG14 2EN
UK
Email: timothy.wright@vodafone.com

Po-Wah Yau
Mobile VCE Research Group
Information Security Group
Royal Holloway, University of London
Egham
Surrey TW20 0EX
UK
Email: p.yau@rhul.ac.uk

Part I

Underlying technologies

This first part is concerned with providing an understanding of the main technologies on which security for both current and future mobile systems will be built, and how these technologies will develop. Chapter 1, *Cryptography for mobile security*, provides a brief summary of the cryptographic background to the remainder of the book. This chapter can be skipped by those readers already familiar with cryptographic concepts.

Until now, the cryptographic techniques employed for mobile telecommunications security have primarily been of symmetric or 'secret key' type. That is, the schemes employed rely on the use of pre-established shared secret keys. However, the situation is likely to change in the future heterogeneous computing and communications environment, where UMTS and GSM technologies will be just two amongst many communications techniques. In such environments, public key cryptography is likely to be of increasing importance. The use of public key cryptography requires the establishment of a Public Key Infrastructure (PKI) and issues associated with the management and use of a PKI in a mobile environment are the focus of Chapter 2, *PKI in mobile systems*.

The theme of PKI is continued in Chapter 3, *The personal PKI*. This describes the notion of a PKI introduced specifically to support security within a Personal Area Network. Whilst managing a large multi-purpose PKI is a highly non-trivial activity, the Personal Area Network is likely to be under the control of one, or a small number of individuals, and many of the most challenging issues for large-scale PKIs no longer apply.

Security for GSM and UMTS mobile devices relies on an internal smartcard for the secure storage of cryptographic keys (and other security parameters). The use of such a portable and removable security token is likely to be vital in the future provision of security services for a wide range of mobile devices. Chapter 4, *The smartcard as a mobile security device*, describes the evolving smartcard technology that makes this possible, and also outlines future possible applications of these devices. This theme is continued in Chapter 5, *Secure mobile tokens – the future*, which considers the likely future development of smartcard technology, and the ways in which such devices might be used in future mobile networks.

Chapter 1

Cryptography for mobile security

Chris J. Mitchell

Cryptography underlies the provision of security for just about every kind of communications network, and mobile networks are no exception. Just about every chapter in this volume makes use of cryptographic techniques and terminology, with which the reader is assumed to have a basic familiarity.

The role of this preliminary chapter is to provide a very brief introduction to cryptography. Whilst reading this chapter is no substitute for a more detailed study of the subject, the main terms are introduced, with the goal of enabling much of the remainder of the book to be understood.

There are many books on cryptography, and the interested reader is strongly encouraged to consult these books to gain a better grasp of the topic. Two books that can be recommended are Piper and Murphy's recent brief introduction to cryptography [1], which provides a basic introduction to the main concepts of cryptography, and Menezes, van Oorschot and Vanstone's encyclopaedic work on cryptography [2], which, despite now being some six years old, remains an enormously reliable source of information on all aspects of cryptography.

1.1 Security services

When discussing the provision of security for communicated data, it is important to establish a terminology for the objectives of security mechanisms. That is, when specifying the security requirements for a mobile system, it is important to have a way of expressing the requirements in a way that is mechanism-independent. We use the term *security service* for this concept, sometimes also referred to as a *security feature*.

There are four main security services of relevance to this book, namely:

- authentication,
- data integrity,
- data confidentiality, and
- non-repudiation.

Authentication can be divided into two somewhat different services. *Entity authentication* is the verification of the identity of a communicating party at an instant in time, whereas *(data) origin authentication* provides evidence of the origin of received data. To use a simple analogy, origin authentication is obtained for a physical letter by examining the postmark on the envelope and checking the sender's signature; entity authentication in a telephone conversation would typically be obtained by recognising the voice of the other party.

Data integrity refers to the protection of data against change. In a communications context, it is often not possible to prevent changes to data by an 'active' attacker on a communications channel. In such a context, data integrity does not mean prevention of change – instead it refers to the ability of the receiver of data to reliably detect whether changes have been made or not. Manipulated data can then, for example, be rejected, and, where necessary, re-transmission can be requested.

Confidentiality of data means that unauthorised parties are unable to read the data. There are some variants of confidentiality, including *traffic confidentiality*, where an interceptor of traffic is prevented from learning the volumes of traffic intended for particular recipients (and where the traffic comes from).

There are a number of variants of non-repudiation services. *Non-repudiation of the origin of data* is a service provided to the recipient of data; it provides the recipient with evidence that the data were genuinely sent by the originator. In particular, the evidence should not be deniable by the data originator and hence should be of value in the event of a dispute about whether data were, or were not, actually sent. *Non-repudiation of the receipt of data* is a service provided to the originator of data, providing evidence that data were received by its intended recipient. This evidence should not be deniable by the recipient, and hence should again be of value in the event of a dispute – in this case about whether data were, or were not, actually received.

It is important to note that all these services can be provided by a variety of different techniques, not just cryptographic means. This is one reason why it is important to distinguish between cryptographic techniques, designed to provide services, and the services themselves. Identifying which security services are needed comes from a requirements analysis of a system – deciding which cryptographic techniques should be employed to provide the services, and how they should be managed, is an implementation decision.

1.2   Symmetric and asymmetric cryptography

Regardless of which cryptographic methods are employed there is a need to generate and manage cryptographic keys – what is known as *key management*. It is normally assumed that the opponent knows exactly how the cryptographic algorithms in use work – the only things known to the legitimate participants that are not known by the opponents are the values of certain secret or private keys.

Cryptographic techniques can be divided into two main classes, known as secret key and public key techniques (or symmetric and asymmetric techniques), depending on the nature of the keys used. In secret key (symmetric) cryptography, all cryptographic algorithms use shared secret keys. That is, the sender and the receiver of encrypted or integrity-protected messages will share a secret key, which is used both by the sender to encrypt or integrity-protect the message, and by the recipient to decrypt or verify the message. This is the 'classical' model for cryptography, and, until the 1970s, essentially the only model.

However, in the 1970s, Diffie and Hellman devised public key cryptography, in which keys typically belong to individual entities, and are generated in matching pairs. One half of the pair is called the public key, and can be made widely known, and the other half of the pair is the private key, which is normally known only to the key owner. The uses made of the key pair depend on the type of public key cryptographic algorithm – we briefly review the uses made of key pairs by two types of such algorithm.

- For public key encryption algorithms, the public key is used to encrypt the data (i.e. to obtain the *ciphertext* from the *plaintext*), and the corresponding private key is used to decrypt the ciphertext to recover the correct plaintext. This means that anyone who has access to a user's public key can encrypt data to generate ciphertext that only the user (using his/her private key) can decrypt.
- In a *digital signature algorithm*, the private key is used by the sender of a message to generate a *digital signature* (or just a *signature*) on the message. This signature will typically be in the form of a check value sent with the message. Anyone with the public key can then verify the signature, and thereby verify the origin and integrity of the message.

1.3   Secret key techniques

Three main classes of secret key cryptographic techniques are now briefly discussed, namely encryption, Message Authentication Codes and hash-functions. One major security technique that incorporates the use of cryptographic algorithms is also very briefly introduced, namely the notion of an *authentication protocol*.

1.3.1   Symmetric encryption

Symmetric encryption techniques come in a variety of different types. Probably the most widely known type of symmetric cipher is the *block cipher*. In a block cipher, data are processed in blocks, for example, of 64 or 128 bits. A block cipher algorithm possesses two related operations – an encryption operation, which will take as input a block of plaintext and a secret key and output a block of ciphertext, and a decryption operation, which, when given the same secret key, will always map a ciphertext block back to the correct plaintext block.

In fact, there are several different ways to use a block cipher to encrypt a long data string, but the simplest is known as the *Electronic Code Book* (*ECB*) mode of operation. To encrypt a message using ECB mode, the data must first be divided into blocks of length equal to the block length of the cipher. Each data block is then input to the block cipher (along with the secret key) to obtain a ciphertext block. The encrypted message consists simply of all the encrypted blocks. Decryption then can work in a directly analogous 'block by block' way.

There are a number of widely known and widely used block ciphers, of which the best known is almost certainly the Data Encryption Standard (DES) algorithm. This algorithm, dating back to the mid-1970s, has been a *de facto* standard for certain industries, notably in banking, for over 20 years. However, DES secret keys only contain 56 bits, which means that, with modern technology, it is possible to search through all possible keys (i.e. all 2^{56} of them) until the correct one has been found. Indeed, to demonstrate the inadequacy of DES, recently a machine was developed specifically to do this. As a result, DES is decreasingly often used, at least in its basic form – however, the use of DES in a compound form known as 'triple DES', with two or three different DES keys, has given the algorithm a new lease of life.

Another block cipher of increasing importance is the so called Advanced Encryption Standard (AES) algorithm, which was developed in the late 1990s as a replacement for DES. This algorithm uses much longer keys than DES (of at least 128 bits) and also has a 128-bit block length, as opposed to the 64-bit blocks used by DES. One other block cipher algorithm of importance in a mobile context is KASUMI, an algorithm incorporated into the 3GPP specifications (see Chapter 6).

Another important class of symmetric encryption algorithm is the stream cipher, as exemplified by the A5 algorithm used in GSM (see Chapter 6). A stream cipher is different from a block cipher in ECB mode in that data are encrypted 'bit by bit'. The major component of a stream cipher algorithm is a sequence generator (or *keystream generator*), which takes a secret key as input and generates a pseudo-random sequence of bits as output. This sequence (known as the *keystream*) is bit-wise EXORed with the plaintext bit sequence to derive the ciphertext. Decryption is exactly the same process as encryption.

The A5 cipher used in GSM employs a specifically designed keystream generator – in fact, there are two versions known as A5/1 and A5/2. However, a different approach has been adopted in 3GPP. It is possible to use a block cipher to construct a keystream generator – indeed, there are two well-known and standardised ways of achieving this using the Output Feedback (OFB) and the Counter (CTR) modes of operation. The CTR mode combined with KASUMI is used in 3GPP.

Finally, note that the principle function of encryption is to provide the confidentiality service for transmitted or stored data. However, it is possible to provide other services if encryption is used in appropriate ways.

1.3.2 *Message authentication codes (MACs)*

The second major cryptographic scheme of the symmetric (secret key) type is the MAC. A MAC algorithm is a cryptographic function that takes as input a message and a secret key and outputs a short, fixed length, block of bits known as the MAC. This MAC is then sent or stored with the message, and acts to protect its integrity and guarantee its origin.

If the recipient of a MAC is equipped with the correct secret key, then the key can be used with the received message to re-compute the MAC value. If this re-computed value agrees with the MAC value sent or stored with the message, then the recipient knows that the message has not been changed and that it must have been sent by someone who knows the secret key (presumably the legitimate originator).

To be effective it must be infeasible to compute a MAC for a message without knowing the secret key, and it must also be infeasible to compute the secret key by observing valid MACs. The most widely used method for generating MACs is almost certainly the 'CBC-MAC'. This is obtained by using a block cipher in a mode known as the Cipher Block Chaining (CBC) mode. DES-based CBC-MACs have been very widely used for many years, although in many applications they are likely to be replaced by other functions for security reasons. Another method for generating MACs is to use a cryptographic hash-function (described below) in a special way.

Finally, note that MACs can be used to provide both data integrity and data origin authentication services. However, unless used in a special way, MACs cannot provide non-repudiation services. This is because the process of validating a MAC is essentially the same as the process of generating a MAC. That is, when verifying a MAC, it is necessary to re-compute it – hence, the sender of a MAC-protected message can potentially repudiate it and claim that the recipient 'forged' the MAC using the shared secret key. That is, the MAC is of no long-term value as evidence that a message was genuinely sent. However, the situation is different for digital signatures, as discussed below.

1.3.3   Hash-functions

The third type of cryptographic scheme we consider is the (cryptographic) hash-function. Hash-functions are somewhat different to the functions we have considered so far in that they do not use keys. A hash-function takes as input an arbitrary data string and gives as output a short, fixed-length, value that is a function of the entire input; this output is known as a *hash-code* or *hash-value*. Hash-functions must have the one-way property, that is, they must be designed so that they are simple and efficient to compute, but also so that given an arbitrary output, it is computationally infeasible to find an input that gives the chosen output. One well-known and widely used example of a hash-function is the SHA-1 function (Secure Hash Algorithm revision 1), which gives a 20-byte output.

If such a one-way hash-function is available then it can be used for a variety of security purposes. First, if it is necessary to protect the integrity of a large amount of data, the entire set of data can be input to a hash-function, and the output stored securely (it can even be written down). At some later stage the same data can be input to the hash-function, and the output compared with the stored value. If they agree, then the user can be confident that the data have not changed. Second, a hash-function is a key component of most practical digital signature schemes (discussed below). Third, using a special construction involving a secret key (e.g. a function known as HMAC) a hash-function can be used to generate a MAC.

1.3.4   Authentication protocols

An authentication protocol is a defined exchange of messages between two (or possibly more) parties, with the objective of providing one or both parties with an entity authentication service. That is, the objective is for one or both of the parties to verify the identity of who it is they are exchanging messages with, and that the other party is 'live', that is, the messages are not replayed versions of 'old' messages.

Such authentication protocols make use of cryptographic techniques to protect the origin and integrity of individual messages. One common approach is to employ MACs for this purpose. As an alternative to the use of MACs to protect the protocol messages, it is also possible to use digital signatures (discussed below).

1.4   Public key techniques

Analogously to the three types of symmetric (secret key) crypto-techniques, we now discuss three main classes of asymmetric (public key) algorithms, namely encryption, digital signatures and key establishment.

1.4.1   Asymmetric encryption

Public key encryption algorithms are somewhat similar to block ciphers, in that they possess an encryption operation that transforms blocks of plaintext into ciphertext blocks, and a decryption operation that reverses this process. However, the main difference, as outlined above, is that different keys are used for the two operations – the public key of the intended recipient for encryption and the recipient's private key for decryption. Many algorithms have been proposed, but one public key encryption scheme is by far the most widely known, that is, the RSA scheme, named after its inventors, Rivest, Shamir and Adleman. Implementing this algorithm requires the implementation of multi-precision arithmetic, involving exact arithmetic calculations with numbers of size 2^{1024} or more. As a result, public key encryption with RSA (and also with other algorithms) tends to be far more computationally intensive, and hence slower to compute, than secret key encryption algorithms such as DES or AES.

1.4.2   Digital signatures

As briefly discussed above, a digital signature is computed using the signer's private key, and can then be verified by anyone equipped with the signer's public key. When computing a signature it is almost always the case that a hash-function is applied to the message being signed. The most common form of a signature gives a value that, much like a MAC, is sent or stored with the message it is protecting.

One key difference from a MAC is the way in which signatures are verified. As mentioned above, verifying a MAC essentially involves re-computing it. However, this is not the case for a digital signature; verifying a digital signature uses a special verification function that takes as input the signature, the message and the public verification key, and gives as output an indication as to whether the signature is valid or not. Thus, just because an entity can verify the correctness of a signature, it does not mean that it is possible to forge a signature. Thus, as well as being able to provide data integrity and data origin authentication functions, a digital signature can also provide non-repudiation services. The disadvantage is that digital signature functions are generally significantly more complex to compute than MAC functions.

Examples of signature functions include those based on the RSA scheme, and also variants of the El Gamal scheme, such as the Digital Signature Algorithm.

1.4.3   Key establishment

Key establishment schemes are a third type of asymmetric cryptographic schemes. In such schemes, two parties (who know each other's public keys) can exchange messages in such a way that they agree on a shared secret key based on the messages exchanged and their respective private keys, and yet anyone observing the messages cannot compute the secret key. There are a variety of such schemes, although the best known are based on the Diffie–Hellman scheme.

1.5   Key management

We conclude this brief introduction to cryptography by considering some widely used methods for distributing keys. We separate the discussions of managing keys for symmetric and asymmetric algorithms.

1.5.1   Managing shared secret keys

The key management problem for symmetric algorithms is that of establishing shared secret keys between any pair of parties needing to exchange secret or integrity-protected messages. There are a variety of ways this can be achieved. We briefly outline two such approaches.

In one scheme, a single entity (or 'Trusted Third Party'), which must be permanently online, acts to provide keys whenever they are required. This entity, known as a Key Distribution Centre (KDC), shares a distinct secret key with each entity in the system. Whenever one entity wishes to establish a secret key with another entity, a message is sent to the KDC, which generates a secret key for this specific pair of entities. This secret key is then encrypted twice, once under each of the secret keys the KDC shares with the two entities. This encrypted key is then distributed to the two parties, who are now able to interact securely. Such an approach may sound rather cumbersome, but it is widely used in practice and can be very effective in closed environments.

A second approach is to use asymmetric cryptography to establish the desired secret key. Suppose every entity within a network possesses a key pair for an asymmetric key establishment scheme (as discussed above). Then, an asymmetric key establishment mechanism can be used to establish a shared secret between any two parties. Such an approach has the advantage that no online third party is required.

However, public keys do need to be reliably distributed to support the asymmetric scheme (see below), and so the advantages of such schemes are not always so clear-cut in practice.

1.5.2   Managing public keys and public key infrastructures (PKIs)

The key management problem for public key cryptographic schemes is rather different to that for symmetric schemes. In a public key scheme, the keys being distributed do not need to be kept secret. That is, one entity will generate its own key pair, and will then distribute its public key to all parties who may wish to use it – there is essentially no risk in the public key becoming known, and hence key distribution can be done using public channels – sometimes such keys are even published on the web.

However, whilst this may sound as if there is no key management problem at all, it is still necessary for the recipient of a public key to know that it is genuine, and not the public key of some third party masquerading as the expected entity. In an informal setting this can be achieved in a variety of ways, for example, by manually exchanging hash-codes of public keys or by phoning the other party and asking them to confirm randomly selected digits of the public key. However, in a large network where key verification must be an automatic process, this type of approach will simply not work.

The most widely discussed solution to this problem is the use of *public key certificates*. A public key certificate is a copy of a user's public key, concatenated with other data including the user name and the expiry date, all digitally signed by a trusted third party. This third party is known as a Certification Authority (CA). If an entity possesses a trusted copy of the CA's public key, this can be used to verify the signature in the certificate, and hence obtain a reliable copy of the public key for the entity to whom the certificate belongs.

The collection of CAs within a domain and the certificates they generate is usually referred to collectively as the PKI. There are many issues to be covered when installing and operating a CA, including how to interoperate with other CAs, and how to deal with 'revoked' certificates, that is, certificates for users who are no longer entitled to be part of the system or whose private keys have been compromised. These issues all form part of the PKI management problem, discussed in Chapter 2.

References

1 PIPER, F. C. and MURPHY, S.: 'Cryptography: a very short introduction' (Oxford University Press, Oxford, 2002)
2 MENEZES, A. J., VAN OORSCHOT, P. C. and VANSTONE, S. A.: 'Handbook of applied cryptography' (CRC Press, Boca Raton, FL, 1996)

Chapter 2

PKI in mobile systems[1]

Jozef Dankers, Theo Garefalakis, Ralf Schaffelhofer and Timothy Wright

In current mobile systems, some applications already use public key techniques and an underlying Public Key Infrastructure (PKI) to provide security, and such use is widely expected to grow. This chapter provides an overview of the basic techniques and the entities that are involved in a PKI and describes how they are used in current mobile systems. The chapter also highlights the envisaged use of PKI in future mobile systems and the accompanying challenges it brings, drawing on recent results of the European Union's SHAMAN project.[2]

2.1 Introduction

Mobile systems for communication have spread all over the world at a rapid pace, so that today, in many countries, most people are equipped with mobile devices. In the past, mobile phones were used mainly for telephony services; nowadays, however, other services, such as the delivery of information, are gaining increasing importance. Not only are new services coming into the mobile world, but also new kinds of devices are being introduced. These new devices differ in that they provide an extended range of features and they may access networks in new ways.

Along with this evolution, PKI, which is used in the fixed network to support end-to-end security, has been seen as the enabling technology for the provision of the security required for the applications and services and access means offered by mobile systems. This chapter provides an overview of PKI in Section 2.2, highlights its use in current mobile systems in Section 2.3 and its possible use in future mobile systems in Section 2.4.

2.2   PKI overview

All security mechanisms deployed today are based either on symmetric/secret key or asymmetric/public key cryptography or sometimes a combination of the two. We introduce the basic aspects of secret key and public key techniques and compare their main characteristics. A detailed description of cryptographic mechanisms and their application can be found in Reference 1. Next, the most important elements and procedures that constitute a public key infrastructure on which public key techniques rely on are briefly explained. A comprehensive description of a PKI can be found in Reference 2.

2.2.1   Secret key techniques

Secret key techniques are based on the fact that the sender and recipient share a secret key prior to communicating in a secure manner. This shared secret key is used to perform various cryptographic operations, such as encryption and decryption of messages or creation and verification of message authentication data. With symmetric techniques, there is a need for a separate out-of-band exchange/generation of the shared secret prior to the intended communication. As an example, in GSM the secret key that is shared between the mobile subscriber and the home operator is installed on the SIM owned by the mobile subscriber and administered in the home operator database of the subscriber. The need for secret key exchange, prior to the intended communication, complicates the provision of security between entities that do not have a pre-established trust relationship.

Authentication is achieved proving the possession of the pre-shared secret key to each other. A widely used method to prove possession of the pre-shared secret key is the challenge-and-response method. A challenge is sent to the challenged node, which calculates a response using the challenge and the secret key as inputs to an algorithm. This response is sent to the challenger, which performs the same operation and compares the result with the received response.

The administration and management of secret keys, including generation, distribution, renewal and tamper-resistant storage, can become very complicated, as the number of keys grows quadratically: for each pair of entities a secret key has to be created and distributed. For a group of n entities communicating with each other, this means $n * (n - 1)/2$ keys.

Due to the need for pre-shared secret keys, secret key-based solutions have low scalability. A major advantage of secret key techniques is that they are computationally very fast compared to public key techniques. This is the main reason why many protocols today still use secret key mechanisms for authentication.

2.2.2   Public key techniques

Public key techniques are based on the use of asymmetric key pairs, composed of a 'public key' and a 'private key'. Each user possesses one or more key pairs. One of the keys (the public key) is made publicly available, while the corresponding key (the private key) is kept private. Because one of the keys is publicly available there is no need for a secure out-of-band key exchange, but there is a need for an infrastructure to distribute the public key authentically. Because there is no need for pre-shared secrets prior to a communication, public key techniques are ideal to support security between parties previously unknown to each other.

Authentication is performed by proving possession of the private key. One mechanism to prove possession of the private key is a digital signature. A digital signature is generated with the private key and verified using the corresponding public key, that is, by the public key bound to the entity generating the signature.

Public key techniques make it possible to dynamically establish secret session keys. A simplified procedure is that one entity calculates the secret session key and sends it encrypted with the public key of the entity with which it wants to initiate a session to that entity. The receiving entity obtains the secret key by decrypting the received information with its private key.

As the public key of the key pair is usually published in a directory, the overhead associated with the distribution of keying material to communicating parties is reduced significantly when compared with solutions based solely on secret key techniques. The number of keys in the system is also less than with symmetric systems – for a group of n entities communicating with each other, only n key pairs are required.

A drawback of public key techniques is that they are computationally very intensive, which makes them less suitable for limited devices, such as, mobile phones.

The advantages of public key techniques, as described above, do not come for free. The cost includes additional organisational measures and more sophisticated client logic. Furthermore, this additional overhead could lead to extended user-interaction, caused by, for instance, the handling of certificates by the user.

2.2.3   Certificates

A key element in the use of public key techniques is the certificate, a data structure that binds a public key to an entity in an authentic way. The data structure is signed by an independent third party, which has to be trusted by the entities using certificates issued by this trusted party. The certificate guarantees that the public key is bound to the entity that is stated in the certificate. The assurance of the correct binding and, therefore, the assurance that the identification of the certified party was done thoroughly is the crucial requirement for public key techniques. The certificate contains, among other information:

- A certificate number, which is a unique number relative to the certificate issuer.
- The name of the issuer of the certificate.
- The name of the certificate owner.
- The public key of the owner.
- The algorithm used to calculate the signature.
- A validity period, which specifies the period in which the certificate is valid. A limited validation period increases security, as if it is found, for example, that

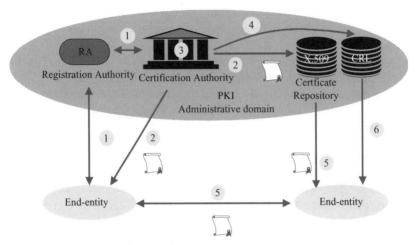

Registration Authority: registers the end-entity
Certification Authority: trusted party certifying the public key
Certificate Repository: storage system for issued certificates
End-entity: mobile user, ASP,SM, ME, NAP
CRL Repository: storage system for a signed list of serial
numbers of revoked certificates

ASP: Application service provider
SM: Smart module (e.g. USIM)
ME: Mobile equipment
NAP: Network access point
CRL: Certificate revocation list

Figure 2.1   Basic concepts of a PKI

the certificate owner has given a false name, the (false) certificate cannot be used permanently.

- Extensions, which are optional. An example extension would be the extension that refers to the policy or rules according to which the certificate is issued in a particular administrative domain. Mainly due to the use of extensions, there exist different formats of certificates profiled for specific uses and each format may have different versions. This may cause major interoperability problems between different administrative domains, each having their own policies on extension usage.

2.2.4   PKI

The management of certificates in an administrative domain, during their lifecycle, requires an infrastructure, which is called the PKI. The steps in the description below refer to the steps in Figure 2.1, which contains a representation of an elementary PKI.

The core component of a PKI is the Certification Authority (CA). This authority is trusted by the end-entities in its administrative domain and is responsible for the status of the certificates.

The main steps of the certificate lifecycle management are as follows.

1. *Registration and key pair generation*. Before end-entities can use the services supported by the PKI, they must register with the PKI. During registration the

identity of the end-entity is established and verified according to the policy of the administrative domain. The registration procedure varies depending on the entity that generates the key pair:

- If the CA generates the key pair the private key is securely passed, via out-of-band mechanisms, to the registering end-entity.
- If the end-entity generates the key pair, then the public key is passed to the CA, which checks whether the registering end-entity really possesses the corresponding private key (by means of proof of possession mechanisms).

The CA may offload certain registration functions to a Registration Authority to enhance scalability and decrease operational cost. The issuance of certificates and certificate revocation lists (CRLs) rests solely with the CA, however.

2. *Certificate generation and distribution.* Once the end-entity is verified and the key pair is generated, the certificate is issued and distributed to the end-entity and to a certificate repository.
3. *Certificate expiration.* The CA has to renew certificates when they expire. The CA is informed by the end-entity of expired certificates.
4. *Certificate revocation.* Another task of the CA is to revoke certificates, for example, when the corresponding private key is compromised.
5. *Certificate retrieval.* End-entities retrieve certificates from a certificate repository or may exchange certificates, depending on the security protocol.
6. *Certificate validation.* To validate certificates the end-entities need to retrieve CRLs from the CRL repository or may make use of online certificate status check protocols (such as OCSP).

2.2.5 Standardisation activities

Standardisation activities with relevance to PKI includes the definition and formalisation of PKI concepts, and the specification of certificate formats and processing rules. With regard to the specification of certificate formats a distinction has to be made between groups that define certificate formats and groups that profile defined certificate formats for specific environments and uses. Groups defining certificate formats include X.509, WAP Forum (now included in the Open Mobile Alliance, OMA), SPKI, Open PGP and EDIFACT. Those profiling certificate formats, primarily X.509v3 certificates, include PKIX, TC68 and the OMA. Certificate and CRL repository issues are connected to the X.500 and LDAP efforts.

The most important and widely accepted certificate format is the X.509 ITU-T recommendation [3], which is also published as ISO/IEC International Standard 9594-8 [4].

One of the most important standards activities related to PKI today takes place in the PKIX working group of the IETF[3] [22]. As mentioned above, PKIX deals with the definition of so-called 'X.509-profiles'. Since the original ITU-X.509 certificate standard leaves many options for the contents of a certificate, an X.509-profile [5] exactly defines the fields for Internet certificates in order to allow for better interoperability.

Furthermore, PKIX standardises a variety of other formats and protocols required to manage and operate a PKI. Another series of widely used specifications, known as 'Public-Key Cryptography Standards' (PKCS), has been issued by RSA Security Inc.[4] It deals with data structures and algorithm use for basic applications of asymmetric cryptography.

2.2.6   PKI in fixed networks

Nowadays, public key techniques and the supporting PKI are used in the fixed network by a number of security protocols[5] to support the establishment of session keys required by the protocol to provide confidentiality and integrity, combined with authentication of the parties involved in the initiation of the session.

Public key techniques are also used to support the provision of secure execution environments by signing downloadable code.

The most important application for public key techniques has been their ability to provide end-to-end security between two unknown parties, first in closed environments and later on in the Internet.

2.3   PKI in current mobile systems

2.3.1   Characteristics of current mobile systems

In today's mobile communication systems, access to services is granted based on a subscription the user has with a service provider. Even if access is granted based on a pre-paid method, there is a contractual arrangement between the service provider and the subscriber. The concept of pre-established security relationships offers the possibility of integrating security with mechanisms based on symmetric cryptography. The provider of second- or third-generation mobile networks delivers smartcards with pre-installed symmetric keys, which are used by the network to authenticate the mobile device and, in the case of third-generation networks, also by the mobile device to authenticate the access network. The authentication method is based on the trust relationship that exists between the access network provider and the service provider, via a roaming agreement, and between the user and the service provider, via the service subscription. The symmetric session keys for confidentiality and integrity protection of data sent over the air are derived during the authentication phase.

Confidentiality and integrity extending over the whole path between the two parties actually communicating, that is, end-to-end security, is not provided by the access network security of second- and third- generation systems and, therefore, has to be provided on one of the network, transport or application levels.

Public key mechanisms and the supporting PKI are not used in current mobile environments to provide network access security because:

- a pre-shared secret key between the mobile node (e.g. a smartcard) and the service provider can be installed relatively easily, as part of the subscriber subscription procedure in current mobile systems;

- non-repudiation is not a stringent requirement for network access; and
- symmetric cryptography provides much higher performance compared to public key cryptography.

2.3.2 Recent usage of PKI in mobile environments

PKI is used in mobile environments by a number of security protocols and/or security schemes, in the same way as it is used in the fixed network, but with a PKI that is adapted to cope with the limitations of mobile environments.

These protocols and/or security schemes, which are described in Section 2.3.4, may be used by applications to provide end-to-end security.

One such application is the Wireless Application Protocol (WAP), specified by the OMA[6] (formally the WAP forum), which defines specifications for access to Internet data and services by wireless devices.

WAP 1.x (WAP 1.0, WAP 1.1, WAP 1.2 and the 'June 2000' release will all be referred to as WAP 1.x) uses the Wireless Transport Layer Security (WTLS) protocol (see Section 2.3.4.1) to protect messages in the wireless network part and some way into the wired network, that is, between the wireless device and a 'WAP Gateway'. The WAP gateway transforms the WAP 1.x stack to/from the wired internet (http/TCP/IP) stack,[7] relays the data between the wireless and the wired network and communicates with the Web Server that the mobile device is accessing. Because WTLS and SSL/TLS are incompatible, content must be decrypted and re-encrypted as it passes through the WAP gateway. This means that client and server both have to trust the WAP gateway not to eavesdrop on the data as it appears in plaintext form in the WAP gateway. As decryption and encryption has to be done in the WAP gateway, WAP 1.x does not provide a mechanism for performing true end-to-end security between the client and the web server and this defect delays and reduces uptake of WAP for some m-commerce services, for example, WAP banking (see [20]).

WAP 2.0 was the release of WAP specifications published in June 2001. WAP 2.0 does not require the WAP gateway, as such, as the mobile WAP 2.0 browser supports standard Internet network and transport protocols, that is, TCP/IP. This means that TLS can be supported by the handset and can run end to end from the handset to a web server supporting TLS, without the need for decryption at the gateway.

The OMA defined a profile of TLS suitable for the mobile environment. This specifies cipher suites within the large number specified in TLS that must be supported by the handset and server in order to guarantee interoperability of ciphering, integrity protection and authentication algorithms. In addition, as TLS uses the X.509v3 certificate format as profiled in RFC 2459[8] (X.509-PKIX), the WAP forum also specified a profile of the X.509v3 certificate format, the X.509-WAPcert [5]. The X.509-WAPcert profile is optimised for the handset but still should allow processing by the handset of most X.509-PKIX certificates in existing use for TLS and SSL.

A further use of public key cryptography within the OMA specification is the signText WMLScript function, which is a function that a server can use to request a mobile device to produce a digital signature using the mobile device's private key. This function is described in [18].

As the standards by which Internet data can be moved to and from wireless devices are defined by WAP, it may be considered as the enabler of m-commerce. Security for m-commerce is described in [20].

2.3.3   PKI on hand-held PCs and smart phones

Over recent years, hand-held PCs and smart phones have gained increasing importance in the mobile world. Due to recent technological developments we are now dealing with powerful mobile devices with regard to computing power and storage capabilities. Different operating systems are used on these devices. The currently most widely spread operating system is Pocket PC from Microsoft.

Pocket PC in its most recent version (Pocket PC 2002) only includes minor PKI capabilities. The Pocket Internet Explorer is able to perform TLS operations with server authentication. However, dedicated certificate management is not included in the current software. This means that there is neither a certificate store as known from the desktop operating systems from Microsoft nor integration into applications, such as secure e-mail. Furthermore, only a few third party providers offer PKI-based applications. In general, the support of PKI-related functionality should be possible, as the architecture of Pocket PC relies on the Crypto API 2.0 concept as introduced in Windows 2000. At least Microsoft Windows CE 3.0, which forms the basis for the derivation Pocket PC 2002, includes all the Crypto API functionality necessary for certificate handling.

Besides Pocket PC there are two more promising candidates, namely Palm OS 5.0 and Symbian v7, the latter used on smart phones. The Symbian OS architecture includes a security module, offering a component for certificate management. Interfaces for the storage of certificates and private keys are specified and a user interface for certificate management is available. Furthermore, a certificate management module is provided for authentication of third parties to the user of the phone and for authentication of the phone's user. Even the checking of revocation information via CRLs and online checking (OCSP) is mentioned as a current feature in a white paper on the Symbian website [6].

Palm OS 5.0 includes a security API, which supports encryption/decryption, hashing and signature verification. Furthermore, the Palm Web-Browser 2.0, which runs on Palm OS 5.1, also supports SSL/TLS. It is not clear if, or to what extent, a certificate management component has been implemented.

2.3.4   Protocols and security schemes

The following subsections are an overview of the current and emerging use of public key cryptography and the underlying PKI in mobile telecommunications, identified by WP3[9] of the SHAMAN project. A more detailed description is available in Reference 7.

2.3.4.1 WTLS

The WTLS protocol [8] is a PKI-enabled security protocol, designed for securing communications and transactions over wireless networks. It is used with the WAP transport protocols to provide security on the transport layer between the WAP client in the mobile device and the WAP server in the WAP gateway. The security services that are provided by the WTLS protocol are authentication, data confidentiality and data integrity. Applications are able to selectively enable or disable WTLS security services depending on their security requirements and the characteristics of the underlying network (e.g. an application may disable the WTLS confidentiality service on networks already providing this service at a lower layer). WTLS provides functionality similar to the Internet transport layer security systems TLS and SSL, and has been largely based on TLS, but has been optimised for use over narrow-band communications and incorporates datagram support. The fact that WTLS was designed to support datagram transport, that is, where packets may be lost or arrive out of order, led to it being incompatible with TLS and SSL, which are designed for use over reliable connection-orientated transport such as TCP. WTLS is implemented in most major micro-browsers and WAP servers.

The WTLS protocol consists of two layers. The lower layer is the record layer protocol and the upper layer is primarily the handshake protocol (although the alert protocol, used to indicate error conditions, is also part of the upper layer). The handshake protocol is used to initiate a secure session. It allows the WAP client in the mobile equipment and the WAP server in the WAP gateway to agree on a protocol version, to select cryptographic algorithms, to optionally authenticate each other and to generate a shared secret. The shared secret is used by the record layer protocol, to provide data integrity and confidentiality. To provide the authentication service and to generate the shared secret, the WTLS handshake protocol may use public key cryptographic techniques and a PKI designed for wireless environments (WPKI), which is described in Section 2.3.4.2. The possible use of uncertified public keys for the generation of the shared secret is also provided. However, this implies that neither the client nor the server is authenticated, which makes the system vulnerable to man-in-the-middle attacks.

To guarantee optimum security, a tamper-resistant device should be used to store the sensitive data (permanent private keys) and perform the security functions using these sensitive data (e.g. cryptographic operations). A specification for an interface into a tamper resistant storage, the Wireless Identity Module (WIM) was also developed within the WAP Forum (see [19]). Certificates are integrity protected by the signature of the issuing party, so they can be exposed without danger.

A new certificate format was devised for WTLS, the 'WTLS' certificate format. This was designed to have a reduced size and to reduce the complexity of certificate processing on the client. It is very similar to version 1 of the X.509 certificate format, having no support for extensions.

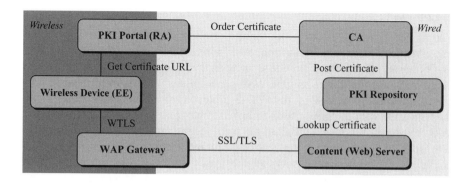

Figure 2.2   Major components and operational flow of WPKI

2.3.4.2 WPKI

Just as WML and WTLS are optimised versions of HTML and TLS with respect to mobile environments, WPKI is an optimised extension of a traditional PKI for the wireless environment. It is concerned with the requirements on a PKI imposed by WTLS and the signText function. The WPKI specification [9] was designed to be used in wireless networks just as the IETF PKIX standards are used in wired networks. In reality, as the number of mobile devices with their own private keys is actually very small, the WPKI specification has not been implemented or used a great deal.

2.3.4.2.1 WPKI architecture for WAP1.x

As shown in Figure 2.2 WPKI requires the same components as a traditional PKI: an End-Entity Application (EE), a Registration Authority (RA), a Certification Authority (CA) and a PKI Repository.

However, in WPKI, the end entities (EE) and the registration authority (RA) are implemented differently, and a new entity, referred to as the PKI Portal, is introduced.

The EE in WPKI runs on the WAP device. It is responsible for the same functions as the EE in a traditional PKI.

The PKI Portal can be a dual-networked system, like a WAP gateway. It typically functions as the RA and is responsible for translating requests of WAP clients to the RA and CA. The PKI Portal interoperates with WAP devices on the wireless and with the CA on the wired network.

An extensive description of the WPKI is contained in Reference 9.

2.3.4.2.2 Wireless-specific adaptations in WPKI and other OMA specifications

WPKI includes optimised (with respect to mobile environments) PKI protocols, certificate profiles and cryptographic algorithms.

- Traditional PKI service request encoding is optimised to use protocols such as WTLS and signText that the mobile device may already support and so avoid the support of relatively complex protocols such as PKIX CMP.

- A new certificate profile, which is reduced in size compared to a standard X.509 certificate, is defined for certificates that need to be sent over the air: the X.509-WAPCert format, a X.509v3 format profiled for WAP, for client certificates (see [5]). The sending of client certificates over the air is avoided as far as possible. Instead, X.509-PKIX certificate formats, X.509v3 formats as profiled in RFC2459, are stored in a PKI directory from which the server can retrieve them. Another possibility is that the WAP client presents to the server the location (URL) of its X.509-PKIX certificate, to reduce storage and transmission bandwidth.
- Traditional RSA- and DSA-based signature schemes are supported. Also, the specifications support the use of Elliptic Curve Cryptography (ECC) based schemes. These were seen as beneficial because of their shorter key lengths and more efficient signature computation. In practice, though, the actual use and implementation of ECC according to the WAP specifications is rare.

The use of certificate status validation mechanisms, like CRLs and OCSP for the wireless client, are not yet specified for WPKI. To provide a work-around for the lack of client-side status validation facilities, short-lived server certificates were introduced to obviate the need for a separate revocation check. The CA authenticates a server typically for 1 year. The CA issues a new short-lived certificate, with a lifetime of typically 48 h, every day of that year. For revocation, the CA simply ceases issuing further short-lived certificates. The client requires sufficiently accurate time awareness.

2.3.4.3   SIM application toolkit

The SIM Application Toolkit (SAT) or simply 'SIM toolkit' is a specification of SIM and terminal functionality that allows the SIM to take control of the terminal (the SIM is usually a 'slave' to the 'master' terminal) for certain functions. There is a specific SAT specification describing how SMS messages can be exchanged encrypted between the SIM and an external server, transparently to the terminal.

Symmetric methods are presently used to secure these messages, but it would also be possible to secure these methods using public key methods. For example, a set of SMS messages to be sent from the server to the SIM could be encrypted using a symmetric key that was itself encrypted using a public key corresponding to a private key on the SIM. The SIM would then be the only entity that could decrypt the SMS set. The SMS set could also or alternatively contain a signature, generated with the private key of the server that could be verified using the corresponding public key present on the SIM. Similarly, in the reverse direction, the SIM could send out encrypted and signed SMS sets.

However, it should be noted that the increased length of signatures using public key methods as compared to 'signatures' (e.g. Message Authentication Codes) using symmetric methods is very significant in the SMS environment, as SMS are only 160 bytes long. A single SMS would be required for the signature alone if RSA keys with 1024 bit moduli were being used.

In defence of the use of public key methods for SAT secure messaging it might be said that the use of public key techniques would allow easier key management than if only secret key methods were used if multiple parties wanted to communicate securely with the SIM. Against this though, it could be said that the operator would not want an open model for secure communication with the SIM (such communication could have access to privileged functions of the SIM and the terminal) but would want to be fully in control of which parties could communicate with the SIM in this way.

The successor of the SAT for UMTS is the USIM Application toolkit (USAT). Apart from SMS, USSD (Unstructured Supplementary Service Data) may also be used as bearer [5].

2.3.4.4 mSign approach

mSign, a consortium of companies related to mobile security (www.msign.org), published the mSign protocol specification that is intended as a standard for interoperable mobile signatures. Message formats, message types as well as security levels are specified within this publication.

The major goal of the specification was to provide a framework for the introduction of mobile signatures for devices with different cryptographic capabilities. The new idea that was introduced in the mSign protocol is that the signature is generated in accordance with the crypto-capabilities of the client device. That means, a signature may be generated on the client device or may be generated by a trusted party on behalf of the original user. A powerful device may perform the cryptographic computations itself, other devices may generate a signature that does not conform to standards and is then 'translated' by the mobile operator. A third possibility that was proposed is to delegate the generation of the signature completely.

In summary, the mSign protocol may be used to provide digital signatures by means of PKI but offers alternative ways for signature generation at the client side.

Currently, there is little activity in the mSign consortium. The standard, however, has been introduced to the ETSI M-COMM working group (www.etsi.org), which is currently working on an adaptation of the standard's core ideas.

2.3.4.5 IKE

Internet Key Exchange (IKE) is an authentication and key management protocol [10] used by IPSec to establish security associations (SAs) between parties communicating over the Internet protocol. Security associations (essentially a shared secret key) can be established either manually or dynamically. IKE is a protocol for authenticated key exchange, thus providing dynamic key management.

The operation of the IKE protocol is divided into two phases, mainly for performance reasons. In phase 1, the communicating parties establish a common secret by means of an authenticated Diffie–Hellman key exchange mechanism. The authentication of the communicating peers can be done by means of asymmetric key pairs or by means of pre-shared secret keys. In the case of asymmetric key pairs, RSA signed or encrypted nonces may be used. RSA signed nonces require a PKI, while RSA encrypted nonces require only a public key, which may be established manually or dynamically (e.g. derived from a previous SA established by means of RSA signed nonces). After the successful completion of phase 1, a so-called IKE-SA is established. The IPSec SA, which is needed by IPSec protocols such as AH (Authentication Header) and ESP (Encapsulating Security Payload), is created in phase 2, and is based on the IKE-SA established in phase 1.

As explained in the previous paragraph, with the RSA signed nonces method certified public keys are used (during phase 1 of IKE). There is, therefore, a need for a public key infrastructure. IKE specifies the use of the PKIX profile of X.509v3 certificates described in References 11 and 12. As many implementations do not adhere completely to the specified format, an Internet draft [13] has been produced to achieve interoperability in the Internet market.

The IKE protocol has a number of deficiencies. The three most important are that the number of rounds is high, causing high latency for the setup of an IPsec SA, that it is vulnerable to denial-of-service attacks and, lastly, the complexity of its specification. Therefore, an IKE-replacement protocol is being developed at the IETF. From a number of initial proposals, IKEv2 [14] and JFK [15] (Just Fast Keying) were discussed as possible successors to IKE. Finally, in October 2002, JFK and IKEv2 were merged into a single scheme known as IKEv2 [21], which includes significant elements from JFK.

2.3.4.6   IKEv2

For IKE phase 1, four methods for peer authentication are defined and each of these methods can be used in two operational modes: normal and aggressive. Only the normal operation mode, consisting of six messages, provides identity protection. The aggressive mode consists of three messages. This means that eight exchanges are possible to establish an IKE-SA.

IKEv2 [14] greatly simplifies IKE by replacing the eight possible phase 1 exchanges with a single exchange based on either public signature keys or shared secret keys. The single exchange, which consists of two request/response pairs, provides identity hiding. The IKEv2 phase 2 exchange to set up an IPsec SA is one request/response pair, and can be used to create or delete an IPsec SA, re-key or delete the IKE-SA, or give information such as error conditions. Moreover, the phase 2 exchange can be piggybacked on the initial phase 1 exchange, still further reducing the latency of setup of an IPsec SA, compared with IKE.

2.3.4.7   JFK

One of the design goals of the JFK [15] protocol is User Identity Confidentiality. Several protocol variants are considered: first, the protection can cover the initiator, or the responder or both, and second, the protection can be valid either against active attackers or alternatively only against passive eavesdroppers. Two variants of the protocol, denoted JFKi and JFKr, are specified. These variants are very similar in many respects, apart from two main differences: JFKi provides active identity protection for the initiator and no identity protection for the responder, whereas JFKr provides active identity protection for the responder and passive identity protection for the initiator. In addition, JFKi contains an additional (amortisable) signature.

2.3.4.8 XML-related protocols

XML, extensible mark-up language, a universal format for structured documents and data on the Web as specified by the World Wide Web Consortium (W3C, www.w3c.org), has been extended with PKI-based functionality over the last few years. A Key-Management-Framework was introduced and standards for encryption and generation of signatures have been written.

The philosophy of key certificate management in XML is to outsource some of the functionality that is usually done by the client, to a trusted party.

The XML key management specification (XKMS) comprises two parts: the XML key information service specification (X-KISS) and the XML key registration service specification (X-KRSS).

X-KISS allows a client to delegate part or all of the tasks required to process XML signature elements to a trusted party. A key objective of the protocol design is to minimise the complexity of applications using XML signature. So the application is relieved of the complexity and syntax of the underlying PKI and is independent of the specification this PKI is based upon.

A good example for the concept of delegating functionality to a trusted party is that XKMS would provide functions similar to those provided by OCSP (Online Certificate Status Protocol) for requesting the current status of a certain certificate without the client needing explicit OCSP functionality.

The X-KRSS specification defines a protocol for a web service providing registration of public key information. The protocol provides authentication of the applicant and Proof of Possession (PoP) of the private key in the case that the client has generated the key pair.

Besides the key management system an XML compliant syntax for signatures (XML Signature) as well as for encrypted content and information for the recipient to decrypt this content (XML Encryption) was specified. The latter specification defines, in addition, processes to encrypt/decrypt digital content. As the general XML syntax allows the assembly of different items of information so that they can be looked at as one document, these different parts may have to be handled in different ways. Therefore, it must be possible to apply the security functions, as mentioned above, on these parts separately, making the syntax more complex. One example is a customer record, where different people are allowed to view different parts of the encrypted document.

2.3.4.9 MExE and MIDP 2.0

Mobile Execution Environment (MExE) provides a specification of a security framework for execution environments for second- and third-generation mobile devices. The specification [16] was initially produced by ETSI and then handed over to 3GPP. A major goal was to provide a security (and capabilities negotiation) framework that was independent of the actual execution environment of the mobile device. MExE makes use of PKI to verify the origin and the integrity of a downloaded application and grant/deny certain rights after checking.

This is done with the following mechanism: the MExE environment receives an executable signed by the providing party. The signature is verified and the application is matched to one of the following security domains: 'Manufacturer', 'Operator' or 'Third party'. The domain that a signed executable is assigned to is determined by the root certificate that the application's certificate can be verified with. If the application's certificate is ultimately verified by a root certificate assigned to the Operator domain (say) then the application is assigned to the Operator domain.

Executables belonging to one of the above Security Domains are implicitly regarded as 'trusted'. MExE executables that are not signed, or signed with a signature key that cannot be verified by the mobile device, are designated 'untrusted'. Untrusted executables operate within a 'sandbox', an environment with very limited access to the device's functionality and services. The owner of the device has the possibility to specify a policy on the rights an application belonging to a certain security domain is granted.

No MExE mobile devices have actually been implemented as commercial products, but many of the principles behind MExE were used in the development of the Mobile Information Device Profile (MIDP) Version 2.0 (see [17]). MIDP 2.0 is a specification of a security framework for Java applications designed to run within the MIDP Java environment. MIDP uses a Java Virtual Machine of reduced complexity designed specifically for mobile devices. MIDP 2.0 specifies how a signed Java application can be verified to belong to a domain defined by a root certificate and an associated policy file. The policy file specifies the capabilities of Java applications within that domain. The Recommended Practice document for MIDP 2.0 specifies the policy files that should be associated with Operator, Manufacturer and Third Party domains. MIDP 2.0 is due to be implemented on handsets being released in 2003.

2.4 PKI in future mobile systems

2.4.1 Characteristics of future mobile systems

It is commonly believed that mobile systems beyond the third generation will be characterised by a variety of wireless access networks connected to an all-IP-based core network, enabling global roaming using the radio technique best suited to support the requested service, and by terminals that may consist of several different components forming personal area networks (PANs) by themselves.

Security provision for these future mobile systems requires extensions to the symmetric cryptography based security methods available for third-generation mobile systems. For various areas, considered in Section 2.4.2, public key based security methods and the underlying PKI offer the means to provide the required security.

Although PKI is a mature concept for providing security in homogeneous environments like enterprise networks, a number of challenges have to be overcome to use it in these future mobile systems, which are characterised by heterogeneity. Section 2.4.3 highlights the challenging issues.

2.4.2   Areas to use PKI in future mobile systems

2.4.2.1   Access to mobile networks

Access to future mobile networks may be subscription based as in today's public mobile communications, but access may also be granted on the basis of online payment methods, such as, for example, credit cards.

In the case of access granted by online payment the user does not have a subscription with the provider of the requested service, which means that there is no pre-established relationship between them. A public key based mechanism for authenticating the access network can be used. Although public key mechanisms are not used for network access security in current mobile systems, use is possible in future mobile systems because they avoid the involvement of the home network. Public key based authentication and key agreement protocols use a certificate-based trust infrastructure. The certificates are issued by an entity trusted by the mobile node and the access network. The common trust entity may, for example, be a third party or the service provider of the mobile node. SHAMAN (http://www.ist-shaman.org/) has evaluated the public key based authentication and key agreement protocols IKEv2 [14] and JFK [15] (both successors of IKE) against sets of cryptographic requirements that are relevant in two different network access scenarios. The first scenario is the subscription-based access where the user has a long-term trust relationship with a home network. The second scenario is the alternative access where no home network is available, and hence there is no pre-established relationship.

The assessment for the subscription case yielded that JFKi fulfils all requirements for network access. However, JFKr and IKEv2 do not protect the user from active attacks from bogus access networks and, therefore, cannot be recommended for the network access scenario.

In the alternative access case, the set of requirements differs because the mobile node may use preliminary authentication or even remain anonymous initially. This implies that 'authentication of the mobile node towards the access network' and 'mobile node confidentiality' must be considered differently to the subscription case. However, the mobile node requires the access network to authenticate itself towards the mobile node. The assessment of the public key protocols for the alternative access case yielded that JFKi, JFKr and IKEv2 are all suitable for the network access scenario.

2.4.2.2   Communication between mobile core networks

To enable global mobile communication for the user there is a need for communication between mobile core networks of different operators. The use of each other's network to support global roaming is based on roaming agreements between the operators. In public mobile systems, like GSM, security for communication between the mobile core networks is not included in the present specifications. The GSM architecture was introduced in the late 1980s. At that time only few providers were able to offer services at all and these providers trusted each other. Nowadays, it is far easier for companies to provide mobile services so that implicit trust between all the providers cannot be assumed anymore. To secure the communication between these core networks, PKI may be the right instrument.

2.4.2.3 Intra-PAN communication

It is becoming quite common for users to possess several mobile devices (e.g. mobile phone, PDA, notebook computer, earpieces, etc.) that are able to communicate over wireless interfaces. Roughly speaking, the devices of a single user, which are physically close, define a PAN. For a rigorous definition of the notion we refer to Chapter 9. As radio communications are, in general, easier to intercept and/or change wired communication, the internal PAN communications should be secured. That is, source authentication, data integrity and data confidentiality should be provided. As the characteristics of the PAN are quite different from those of a large-scale communications network (e.g. the devices are few, have limited capabilities and are owned by a single person), some interesting problems arise. We describe some of those problems, in Section 2.4.3.5.

2.4.2.4 Inter-PAN communication

An interaction between networks that is quite different to that between core networks is the communication between PAN networks that was described in Section 2.4.2.3. As PANs are local structures and are managed internally – which means the role of the certifying party is probably within the PAN – new requirements arise. The structure of communicating PANs can be different, too. On the one hand there might be a master component in each PAN acting as a gateway; on the other hand, every device of one PAN might be able to communicate with every component from the other PAN. Of course, the first alternative is more complex to handle internally, whereas the second has to deal with trust hierarchies and is, therefore, more complex between the PAN networks. The most important issue to solve in this context is how the two PANs get an authentic initial connection. This means a kind of cross-certification process has to be performed beforehand. Certainly, this mechanism is easier for the first architecture described above, as only a trust relationship between two dedicated devices has to be established, whereas in the second approach 'trust domains' have to be 'linked'.

2.4.3 Challenging issues

2.4.3.1 Managing the complexity of a PKI for limited devices

Mobile devices have limited processor capacity and memory storage: a trade-off between power, battery life and computing ability must usually be made. Therefore, we should start from the assumption that resources are scarce. Public key techniques and a PKI, however, demand high processor capacity and memory storage. However, many mobile devices now use dedicated hardware to perform public key operations, which reduces the complexity of the problem considerably.

A high processing capacity is required for performing public key operations and for the construction and validation of certificate chains. The processing of certificate chains may be a very complicated and time intensive operation, dependent on the length of the certificate chain and the possibility of cross-certification. Cross-certification is required when users from different PKIs need to trust each other's certificates. Also the use of public key cryptography in the generation of a shared secret session key consumes processor resources.

There is also a high demand for storage capacity to store the certificates and certificate revocation lists. These are stored/cached for performance reasons and to save bandwidth on the wireless link.

A possible way to cope with limited devices is to outsource some activities to a server, and this is what XKMS is designed to enable. Other protocols that do this are the Delegated Path Validation (DPV) protocol, which allows delegating all path validation to an OCSP server, and the Delegated Path Discovery (DPD) protocol, which allows delegating path construction to an OCSP server. Another example is the Simple Certificate Validation Protocol (SCVP), which allows the outsourcing of certificate handling to a server. Note, however, that DPV, DPD and SCVP are not being progressed within the PKIX group.

Also important is the selection of an appropriate public key algorithm. Guidelines are available from the SHAMAN results (http://www.ist-shaman.org/).

2.4.3.2   Managing the complexity of a PKI for limited bandwidth

Another challenge is managing the complexity of a PKI for a limited bandwidth environment. Bandwidth constraints may arise due to limited channel capacity between the security module (SM) and the module device hosting the security module. The radio link will probably not be a problem in the future: fourth-generation mobile networks will offer broadband capacities that will allow a fast and effective transmission of PKI protocol messages. It still may not be possible to download large CRLs, however, as for a commercial CA these can be very large. This problem gets even worse if the device has no capabilities to store CRLs locally, as it is necessary to download the CRL each time a PKI-based action performed. Possible solutions include the use of delta revocation lists, of single certificate status protocols such as OCSP and the delegation of activities to a network server.

2.4.3.3   Interoperability issues

Several interoperability issues have to be addressed if a PKI will be used in future mobile systems. These issues need to be investigated in detail and finding appropriate solutions is a major challenge.

One interoperability issue is the existence of different certificate formats. These different formats have come about for several reasons. A first reason is that the format is tailored to the environment in which it is used. An example nowadays is WTLS, which uses certificates with a limited number of parameters for use in a wireless environment. The most widely used format is the ITU format X.509, which is profiled by the IETF to be used on the Internet. Nevertheless, the format may be interpreted in different ways, so that different profiles have been specified, for example, the PKIX profile. Even the profiles offer the possibility to specify private extensions, which may lead to problems with some applications, especially when the extensions are set as 'critical'. If this is done, a certificate is only considered valid when the application can evaluate the extension.

When the same format is used, interoperability problems may arise due to the certificate extensions that are defined. These extensions may have standardised or proprietary values. When an entity receives a certificate with an extension marked as critical, and it does not understand the extension, then the certificate is rejected, complicating the security functions.

Interoperability can be increased:

- by restricting the use of extensions (especially proprietary extensions) and the use of the criticality flag;
- CAs strictly following profiles such as RFC 3280 [12] when issuing certificates.

Even when the certificate conforms to the standard profiles, the applications using the certificates may be incompatible, for example, due to implementation errors caused by misinterpretation of the specifications.

Another source of interoperability problems may be the use of directory services based on different protocols.

Different PKI authorities issuing certificates may use different policies. This means that certificates issued in one security domain may not be acceptable in another security domain with possibly more restrictive policies.

It becomes even worse if major suppliers of CAs use non-standard certificate features and/or do not follow standardized profiles.

2.4.3.4 Organisational issues

Some issues to be solved are organisational ones. A number of different scenarios for the generation of key pairs to be used in Public Key Infrastructures exist.

The key pair can actually be generated in the mobile device, or on the smartcard, or centralised by the manufacture of the device or the owner of the smart card and put on the device/smartcard during manufacture. Another scenario is the central generation of the key pair by a service provider at the request of a mobile user and the secure distribution of the key pair to the mobile device. For each of the scenarios the following points need to be considered in order to select a particular scenario: whether a cryptographically good key can be assured, whether the public key can be certified in a secure way and whether the private key can be securely (very securely) transmitted to the mobile device.

Another issue to be considered is the storage of the private key. Possible locations are the Security Module issued by the network operator (e.g. the SIM in GSM), an additional Security Module issued by a third party service provider (for instance a WIM) or put on the device by the device manufacturer, or a file on a device (software private key).

To develop a PKI for future mobile systems, a trust model needs to be defined. This is achieved by assigning functions or roles to the PKI-involved parties and by identifying the required trust relations between them. The role of a CA, the basic element of a PKI, could be taken by the home network provider, the access network provider or by a trusted third party.

A PKI may be more complex when the access network provider takes the role of CA, especially when the access network providers do not cooperate. Issuance of certificates by the home network provider provides the subscriber with greater identity privacy towards the access network provider.

In defining the trust model, minimisation of the trust relations must be pursued, as this increases the security level. Another challenge is to minimise the user interaction, because the user is not acquainted with the procedures behind security and it is inconvenient for him/her.

2.4.3.5   Personal PKI

As mentioned in Section 2.4.2.4, the internal communications of a PAN need to be secured. Symmetric cryptographic techniques can, in principle, provide a solution. Public key techniques, however, offer certain advantages that make them preferable. For instance, they provide the devices with digital signature capabilities. Furthermore, if symmetric techniques are employed, then any two devices have to agree on a common secret key before they can communicate securely. This will imply several initialisations (imprinting devices with secret keys) for each device if there are a number of devices in the PAN. In this section, we consider some challenging problems regarding what we call the Personal PKI, that is, the public key infrastructure employed in the PAN. The assumption is that the PAN contains at least one device that is capable of functioning as a CA and is known to every device in the PAN (This topic is considered in more detail in Chapter 3).

2.4.3.5.1   Initialisation

Before a device can communicate securely, it needs to create a key pair and communicate the public part authentically to the CA. In return, it receives a public key certificate signed by the CA. One possible way to initialise the devices that has been considered within SHAMAN is by imprinting. This is based on weak password authenticated data exchange. Roughly speaking, the device and the CA exchange information over an insecure channel. Subsequently, the integrity of the received data is checked by means of a short MAC that is computed using the weak password, and is readily verified by the user. In scenarios where both devices are equipped with displays and keypads, the MAC is never transmitted, but rather displayed, so it cannot be intercepted, and thus the password is secure against offline dictionary attacks. Challenging issues arise when the device under initialisation is very limited, for example, has no display, no numerical keypad, etc.

2.4.3.5.2   Multiple CAs

PAN are necessarily of a more ad hoc nature than fixed networks, as devices may enter and leave the PAN at any time. This causes complications when devices with certain functionalities are not present in the PAN. With respect to security, the device with the most important functionality is the CA. In order to avoid a single point of failure, when the CA is unavailable, one may consider the possibility of having several devices in the PAN that can potentially function as CAs. In this 'multiple CA' environment, several interesting problems arise. For instance, if the acting CA leaves the PAN, another 'secondary' CA takes over. This transition should be as transparent to the other PAN devices as possible. Therefore, the secondary CA should keep a state practically indistinguishable from the primary CA, raising synchronisation issues. Furthermore, an issue usually not present in conventional PKIs is that of the security of the CA. That is, a CA may become unavailable because it has been compromised (e.g. stolen). In this case, more drastic measures have to be taken, as the signature of the compromised CA is not to be trusted any more.

2.5   Conclusions

Nowadays, PKI, tailored for wireless environments, is used by a number of security protocols to enable (end-to-end) security for services and applications, such as WAP and m-commerce in current mobile systems. These systems do not use PKI for securing network access, because there always exists a pre-established relation between the service provider and the mobile subscriber, allowing the use of secret key based methods, which are more efficient.

Future concepts for mobile devices accessing networks (such as those based on Bluetooth or IEEE 802.11b) may be based on spontaneous networking, that is, gaining access to a network without a prior relationship to the network provider. This means that the communicating parties have to provide credentials for authentication without knowing each other from prior sessions. In this case, authentication must be based on certificates and a common trusted third party. A PKI is needed for certificate management through their lifecycle. For the same reason, securing the communication between PAN components may require a local PKI. However, enhancements to current PKI procedures, being investigated in the SHAMAN project, are needed before PKI can be used in new scenarios in future mobile networks.

Notes

1   The work described in this chapter has been supported by the European Commission through the IST Programme under Contract IST-2000–25350. The information in this document is provided as is, and no guarantee or warranty is given or implied that the information is fit for any particular purpose. The user thereof uses the information at his/her sole risk and liability.

2   SHAMAN (Security for Heterogeneous Access in Applications and Networks) is an IST project that aims to provide security solutions for heterogeneous access and distributed dynamically configurable terminals, and investigates the use of PKI to support the selected security solutions.

3   Charter of the PKIX working group: http://www.ietf.org/html.charters/pkix-charter.html

4   PKCS standards and PKI related information: http://www.rsasecurity.com/rsalabs/pkcs/index.html

5   Internet Key Exchange (IKE), Transport Layer Security (TLS), e-mail security using Secure Multi-purpose Internet Mail Exchange (S/MIME), and e-commerce applications like Secure Electronic Transactions (SET).
6   Open Mobile Alliance: http:.//www.openmobilealliance.org
7   WDP to/from TCP/IP, WTLS to/from SSL/TLS, WTP/WSP to/from HTTP and binary WML to/from WML (XML).
8   RFC 2459 has now been superceded by RFC 3280, but the WAP Certificate profile was based on RFC 2459.
9   WP3 is the working party within SHAMAN investigating PKI issues to support the security mechanisms for future mobile systems, specified by the working parties WP1 and WP2.

References

1   MENEZES, A., VAN OORSCHOT, P. and VANSTONE, S.: 'Handbook of applied cryptography' (CRC Press, Boca Raton, FL, 1996), www.cacr.math. uwaterloo.ca/hac
2   ADAMS, C. and LLOYD, S.: 'Understanding public-key infrastructure, concepts, standards, and deployment considerations' (Macmillan Technical Publishing, 1999)
3   ITU-T Recommendation X.509 (03/2000): 'Information technology – Open systems interconnection – The directory: public key and attribute certificate frameworks'
4   ISO/IEC 9594–8: 2001 (4th edn.): 'Information technology – Open systems interconnection – The directory: public key and attribute certificate frameworks'
5   WAP-211-WAPCert (Version 22 May 2001); http://www1.wapforum.org/tech/ documents/WAP-211-WAPCert-20010522-a.pdf
6   White paper: 'Symbian OS Version 7 Functional description'; http://www. symbian.com/technology/symbos-v7x-det.html
7   SHAMAN, D04: 'Initial report on PKI requirements for heterogeneous roaming and distributed terminals'; http://www.ist-shaman.org/
8   Wireless Transport Layer Security (Version 6-apr-2001): http://www1.wapforum. org/tech/documents/WAP-261-WTLS-20010406-a.pdf
9   WPKI WAP-217-WPKI (Version 24-apr-2001): http://www1.wapforum.org/ tech/documents/WAP-217-WPKI-20010424-a.pdf
10 Internet RFC 2409: 'The internet key exchange', November 1998
11 HOUSLEY, R., FORD, W., POLK, W. and SOLO, D.: 'Internet RFC 2459, Internet X.509 public key infrastructure – certificate and CRL profile', January 1999; http://www.ietf.org/rfc/rfc2459.txt
12 Internet RFC 3280: 'Internet X.509 public key infrastructure – Certificate and CRL profile', April 2002; http://www.ietf.org/rfc/rfc3280.txt
13 Internet draft: 'A PKIX profile for IKE'; http://community.roxen.com/ developers/idocs/drafts/draft-ietf-ipsec-pki-req-05.html
14 Proposal for the IKEv2 Protocol: 'draft-ietf-ipsec-ikeve-02', April 2002

15 Just Fast Keying (JFK): 'draft-ietf-ipsec-jfk-04'

16 3GPP TS 23.057 v4.1.0 (2001–03): 'Mobile Execution Environment (MExE)';
    Functional Description Stage 2 (Release 4)

17 JSR-000118 Mobile Information Device Profile, 2.0 (Final Release);
    http://jcp.org/aboutJava/communityprocess/final/jsr118/index.html

18 WAP-161-WMLScriptCrypto (Version 20 June 2001); http://www1.wapforum.
    org/tech/documents/WAP-161-WMLScriptCrypto-20010620-a.pdf

19 WAP-260-WIM (Version 12 July 2001); http://www1.wapforum.org/tech/
    documents/WAP-260-WIM-20010712-a.pdf

20 Secure M-commerce. This volume

21 IKEv2 Protocol: 'draft-ietf-ipsec-ikev2–04' (January 2003)

22 http://www.ietf.org/html.charters/pkix-charter.html

Chapter 3

The personal PKI[1]

Chris J. Mitchell and Ralf Schaffelhofer

The term personal Public Key Infrastructure (PKI) was devised within the SHAMAN project to describe a public key infrastructure specifically designed to support the distribution of public keys in a Personal Area Network (PAN). In this chapter, a variety of issues relating to the operation and management of a personal PKI are discussed. Following a general discussion of requirements for personal PKIs, the main topics covered are: the operation of personal Certification Authorities, device initialisation, proof of possession and revocation.

3.1   Introduction

This chapter is concerned with methods for the deployment of PKI techniques to support secure communications between devices in a PAN. One major issue dealt with in this chapter is the development of methods for two PAN components to securely exchange their public keys, as required to support the pairing of 'second party components' (as defined in the PAN reference model within Annex 2 of Reference 1). It is assumed that the two devices cannot rely on either existing symmetric shared keys or connection to a global PKI that both devices trust.

The term 'Personal PKI' is used throughout for a PKI deployed to support communications in a PAN. The idea is that by deploying a PKI in such a limited environment, many of the problems associated with PKI deployment in a much larger and less well-defined environment can be avoided, whilst the advantages of use of a PKI can be retained. The PAN is assumed to contain at least one device acting as a 'Personal Certification Authority (Personal CA)', which is responsible for generating public key certificates for all devices within the PAN.

Sections 3.2 and 3.3 contain a discussion of requirements and issues. Section 3.4 provides a detailed discussion of the Personal CA and the corresponding issues. In Section 3.5, a protocol for device initialisation is introduced and analysed. Proof of possession (PoP) as a means of assuring the certifier of the possession of the private key related to the public key to be certified is discussed in Section 3.6, and Section 3.7 lists requirements and ideas about revocation in PAN environments. Finally, conclusions and issues for further research are provided in Section 3.8. Note that some of the issues discussed in this chapter have previously been discussed in Reference 7.

3.2   Issues in the personal PKI

We start by listing the various aspects of public key management for which solutions need to be found. More detailed discussions of the issues will be provided below.

- *Certificate and key pair update.* The public key certificates issued by the Personal CA will (almost certainly) have a specified expiry date. Once this date is reached the mobile device will need to be equipped with a new certificate. This certificate may be issued for the same key pair or for a new key pair.
- *Key status management.* At any time a mobile device's private key (or the mobile device itself) may be compromised or stolen. In such an event, all entities within the PAN will need to be informed that the public key certificate(s) assigned to this device should be revoked (i.e. no longer considered valid). In a similar way, the Personal CA may itself be compromised or stolen, in which case the Personal CA root key needs to be revoked. Information on which keys have been revoked will need to be distributed to mobile devices in a timely and efficient way.
- *Trust management.* The relationship between the mobile device and the personal CA will need to be managed, including CA (root) key update and the possible replacement of personal CA devices, especially in the event of lost or stolen personal CA devices.

3.2.1   Certificate and key pair update

If the mobile device merely wishes to obtain a new certificate for an existing public key then, because of the scale of the personal PKI, a simple solution is possible. Given that the total number of personal devices will be small, it is likely to be possible for the personal CA to securely retain a copy of all public keys for which it has generated certificates. It could even routinely check the certificates to see if any of them have expired. Once the need for a new certificate has been determined, the personal CA device simply asks the user if the certificate for an existing key pair should be renewed. Once the user has agreed, a new certificate can be generated and passed to the device concerned across the wireless interface at the next opportunity.

Even if storing all public keys at the personal CA is not feasible, in certain cases it may be possible to use a relatively simple certificate renewal process. The mobile device requiring a new certificate could pass the expired certificate to the personal CA, which would then pass the relevant information, that is, details of the device and the public key, to the user for a decision regarding whether or not the certificate should be renewed. If the user agrees a new certificate can be generated.

If a new key pair is to be assigned to the mobile device, then the renewal process becomes more difficult. In some cases it may be possible to use the old key pair to establish a secure exchange between the personal CA and the mobile device – however, if the key pair is still trusted and the parameters of the keys are still considered sufficient to secure this process, then it is not clear why it would need to be changed. Indeed, the default for many inexpensive mobile devices may simply be to use the same key pair indefinitely.

However, if a new key pair is definitely required, and if the old key pair cannot be used to secure the necessary interactions between the personal CA and the mobile device, then a new imprinting process will probably be necessary. However, given that this will involve relatively few user keystrokes, and given also that this will probably be a rare event, this should not present a huge practical problem for the user.

3.2.2   Key status management

We consider two different ways in which certificate status information can be disseminated to mobile devices. The choice between the two approaches depends on the online availability of the personal CA.

We will call the first approach *online status dissemination*. This is designed for use in the case where the personal CA is available online to every mobile device either permanently or at least at frequent intervals. In the case where the personal CA is permanently online, then an online status query protocol could be used, for example, a protocol along the lines of the Online Certificate Status Protocol (OCSP), as standardised in ISO/IEC 15945 [2]. However, because of the small scale and relatively closed nature of the personal PKI it may be possible to use a simplified version of OCSP.

In the case where the personal CA is not always online, but is nevertheless online at frequent regular intervals, the use of routinely distributed Certificate Revocation Lists (CRLs) – see, for example, X.509 – would appear to be appropriate. In this approach, the personal CA generates new CRLs at regular intervals and distributes them automatically to all mobile devices. Whilst the personal CA is not online permanently, and neither are all mobile devices, this approach will be appropriate in cases where the personal CA is online sufficiently often that the chances of every mobile device having the latest CRL is very high.

Another approach is *ad hoc status dissemination*. This is designed for use when the personal CA may only be online intermittently or rarely. In such a case, a mobile device may not be online at the same time as the personal CA very often, in which case directly distributed CRLs no longer appear appropriate. Thus, an alternative means for distributing CRLs appears to be necessary.

As in the previous case, we assume that the personal CA generates CRLs at regular intervals. We now suppose that the personal CA is online sufficiently often that it can distribute the latest CRL to at least one mobile device (if not then there is clearly no way of distributing timely status information). Subsequent distribution of CRLs is then assumed to occur in an ad hoc fashion between mobile devices. A more detailed discussion of these issues can be found in Section 3.7.

3.2.3   Trust management

We first consider the routine updating of root keys, that is, when an existing personal CA wishes to update its key pair. If the old root public key has not been revoked, this could be achieved by distributing a certificate for the new root public key signed using the old CA private key. Whilst this approach has dangers, it may be sufficiently secure for use in a PAN environment. The only alternative would appear to be to engage in a new imprinting process with all mobile devices, which could be a rather onerous process for the user.

The case of a compromised or stolen personal CA is rather more difficult. In such a case, there is a need to inform all mobile devices of this in a timely way. Of course, once the root key has been revoked, secure communications between devices will become impossible unless another root key (and a certificate signed using this key) is available. There would appear to be two main approaches to dealing with this issue.

The first approach is to use multiple personal CAs. In this case, every device will have multiple root keys and multiple certificates for their public key(s). If two or more Personal CAs are available at the time a mobile device is imprinted, then it should be possible to devise a special version of the imprinting protocol given in Section 3.5.1.1 to enable simultaneous registration and certificate generation. When one CA root public key is to be revoked, then the mobile devices can be informed by the remaining personal CAs, using the same mechanism as is used to disseminate revocation information for other mobile devices.

The second approach is to re-imprint every device with a replacement personal CA as soon as possible after the loss of the old personal CA. Such a process can be designed to simultaneously revoke the old CA and register with the new CA. An appropriately modified version of the imprinting protocol described in Section 3.5.1.1 will need to be used.

3.3   Personal PKI requirements

The underlying requirement is for two devices, which do not share any pre-existing secret keys or root certificates, to be able to securely exchange public keys that each device can verify. In this section, we identify the requirements that arise when a 'conventional' PKI solution is followed, albeit adapted to a PAN environment. In such a case, one of the devices within the PAN is defined as the 'personal CA' and is responsible for issuing public key certificates to other devices.

The following functional requirements, therefore, result (many are taken from Annex 2 of Reference 1):

a. the personal CA key pair can be securely generated within the device, or securely generated and transferred to the device at manufacture, and (in both cases) the private key is securely stored when in the device;

b. the root public key of the personal CA can be securely transferred to those devices that will have to verify certificates issued by the personal CA;

c. the personal CA can generate public key certificates for mobile devices (and in such a way that the security of the personal CA private key is not endangered);

d. mobile devices can verify certificates issued by the personal CA, and can check certificate validity and revocation status where appropriate.

The general security requirements applying to methods used in the personal PKI are:

e. no third party passive interceptor of communications can learn any secret information;

f. no third party active interceptor of communications can manipulate the exchanges between the mobile device and the personal CA so that a public key certificate is created for the incorrect device or that contains incorrect data (e.g. a public key other than that created by the mobile device);

g. for securing the transfer of the personal CA root certificate from the personal CA device to another mobile device, the interaction between a mobile device and a personal CA shall use at least a 'weak' shared secret, for example, a shared password or PIN, and the method of this use should be capable of resisting 'brute force' attacks on the shared secret; that is, one of the secure passkey protected mechanisms listed in Annex 2 of Reference 1, or a method of equivalent strength, should be used.

Additional and optional functional requirements are:

h. the security-critical personal CA functionality (including key generation and storage functions) should preferably be removable, personal and transferable;

i. the security-critical personal CA functionality can be directly verified and readily enabled/disabled from a single gateway and/or master user.

3.4   Personal CAs

3.4.1   Operation of a personal CA

In this section, we describe the operational processes of a personal CA.

3.4.1.1   CA initialisation

Before use, the personal CA must be initialised. This involves generating a signature key pair for the personal CA. The personal CA will, therefore, need to incorporate means for generating sufficient random material to enable it to securely generate a signature key pair.

The requirements for the personal CA functionality listed in Section 3.3 point towards the use of a smartcard or other portable tamper-resistant devices. Particular advantages could be obtained by combining this device with a device already used for global network access, for example, a GSM/UMTS SIM/USIM device.

3.4.1.2   Device initialisation

This will require a mobile device to perform the following steps – not necessarily in the order specified (note that some of these steps may be combined).

- The mobile device will generate any necessary key pairs (signature keys, encryption keys, etc.).
- At some point in this process the mobile device must import authentication material from its owner. As discussed below, for a variety of reasons this should require the minimum number of keystrokes by the user, that is, it should be a 'weak' passkey.
- The mobile device will be informed of which other device is the personal CA, or will have to 'discover' this device across the PAN.
- The personal CA root public key will be passed to the mobile device. This must be done in such a way that the mobile device can verify the integrity and origin of the CA public key.
- The mobile device will provide its public key(s) to the personal CA. This must be done in such a way that the personal CA can verify the integrity and origin of the public key(s) before it generates any public key certificates.
- The personal CA will generate a public key certificate for the mobile device.
- The newly created public key certificate will be passed to the mobile device. (The mobile device can verify the certificate using the CA root public key.)

3.4.1.3   Candidate mechanisms for password-based initialisation

There exists a considerable literature on protocols designed to enable two entities who share a password (a 'weak key') to use it to authenticate one another and (possibly) establish a shared secret key. A number of protocols of this type are known that are resistant to off-line searching attacks for the weak key, even if the attacker participates in the authentication protocol. A short discussion of such schemes can be found in Reference 1.

What is required here is slightly different, in that we wish to have a means for two entities to exchange public keys in an authenticated way, based on a weak (short) shared secret. Of course, one approach would be to first establish a shared secret key (as above) and then use this to establish an authenticated channel. However, other possibilities, if they exist, would also be of interest. In fact, the use of passwords for the PKI registration process is an issue of much more general application than for a PAN.

A possible candidate mechanism for password-based initialisation is discussed in Section 3.5.1. Further enquiries are provided in Chapter 9.

3.4.1.4   Public key status management

Once a mobile device has performed the exchange of public keys with the personal CA, the issue remains of managing the status of public keys, and disseminating public key status information. Specifically, if a public key is compromised, or suspicion of a possible compromise arises, how is this information disseminated to parties within the PAN? Solutions devised for conventional PKI scenarios, for example, OCSP, may not be appropriate within the PAN environment. Therefore, a discussion on revocation in PANs is provided (see Section 3.7).

3.4.2   Multiple personal CAs

Networks that consist entirely of mobile devices are necessarily of a more ad hoc nature than fixed networks. Mobile devices that perform certain tasks in the network may simply not be present at all times. For example, an extreme case is presented by the fact that mobile phones are prone to theft. In the context of the personal PKI, the device whose absence most dramatically affects the operation of the system is the one acting as a personal CA. For this reason, we consider the possibility of having several devices within one PAN that can act as CAs. This redundancy makes the system more robust, since there is no single device whose absence would make secure communication within the PAN impossible.

In every PKI an implicit (but fundamental) assumption is that the CA is secure. In traditional PKIs, this is a reasonable assumption, since a lot of effort is usually expended on keeping the CA physically secure. The same, however, cannot be guaranteed for the personal CA in a PAN. Mobile devices are prone to theft, and thus their security cannot be guaranteed. Therefore, we need to make the reason why a personal CA is not present in the PAN absolutely explicit. This case is typically characterised by one of the following two conditions.

1. The device is compromised or suspected of compromise, for example, as would be the case if it has been stolen (it may or may not be absent). In this case, the user positively knows that the device cannot be trusted any more, and needs to transfer the CA functionality to another device.
2. The device is not present, but its security is not compromised. This is a more common situation, where, for example, the device is simply switched off.

In this section, we describe (at a very high level) a solution for both situations. Throughout the section we make the following two assumptions.

1. There are two or more devices capable of acting as CAs. One is nominated as the Primary CA, and the others are Secondary CAs.
2. All CAs are known to every device in the PAN. More specifically, every device that enters the PAN is initialised with the primary CA, but is also given a trusted copy of every secondary CA's public verification key.

    Hence, in a PAN with multiple CAs, we suppose that the secondary CA(s) is (are) also known to every device in the PAN. Hence, if the secondary CA has to take over primary CA responsibilities, then every other device in the PAN will recognise it as a valid CA. At any time, all the CAs are kept synchronised. This can be achieved as follows: whenever the primary CA performs an operation, for example, issues a new certificate, it informs the secondary CA, which keeps a state practically indistinguishable from that of the primary CA (e.g. the secondary CA has a list with all certificates issued by the primary CA). We can also suppose that the primary CA equips every newly imprinted device with a copy of the public key of all secondary CAs, at the same time as it transfers its own public key.

Of course, some additional organisational aspects have to be taken into account. A policy for the PAN-PKI structure has to be specified. This needs to cover issues such as what happens when a CA is compromised, and who defines which component is the primary CA and which components are secondary CAs. Furthermore, an imprinting method for the CAs themselves has to be negotiated.

We now describe the actions that have to be taken when one of the above situations (point 1 or 2) arise.

3.4.2.1 Primary CA compromised

In situation 1, that is, when the primary CA is compromised, rather extreme measures have to be taken. This is because the corrupted CA may corrupt the secondary CAs or even take over the PAN, if the secondary CAs are not notified immediately. Clearly, little can be done to secure the inter-PAN communications after the primary CA is corrupted and before the PAN is notified. However, one must make sure that once the PAN is notified, secure communications can resume. To achieve that, the whole system must first be put into the state that existed prior to any initialisation. Thus, either a new primary CA has to be set up and used to imprint every device in the PAN (just as happened originally), or a list of valid CAs has to be entered into each device manually (or in any other secure way, e.g. as was used to originally initialise the devices). In the latter case, the list will obviously not contain the corrupted CA, and a new CA is nominated as primary, and the devices are initialised one by one with the new primary CA.

3.4.2.2 Primary CA switched off

In situation 2, that is, when the primary CA is secure but not present, the transition is smoother. Of course, one could treat this situation in the same way as the previous one, but this would incur unnecessary re-initialisations. After all, the absence of a CA may not be noticed at all by the devices in the PAN (i.e. if no device in the PAN needs the CA during its absence). This should be taken into account in the proposed solution; actions (and thus computational and communications overheads) should be kept to a minimum, and taken only when necessary. Following this principle, no action will be taken unless the secondary CA is contacted.

3.4.2.3 Synchronisation issues

Finally, we consider issues of synchronisation between the CAs of the system. As was mentioned earlier, all secondary CAs are kept updated by the primary CA while they are in the PAN. An issue that was left open arises in the situation where a (secondary) CA re-enters the PAN after a period of absence. This CA has not been updated for the duration of its absence, and, therefore, keeping it updated from this point on is not enough. Thus, at the time it re-enters the PAN, the secondary CA contacts the primary CA, and receives a signed list of the public keys of all devices, possibly including those that may not currently be in the PAN, but whose public keys are valid, and a CRL. From this point on, the secondary CA is kept updated as discussed earlier.

3.5 Device initialisation

3.5.1 *A protocol for device initialisation*

The security requirements for the device initialisation process have been listed in Section 3.3. A protocol has been proposed by Gehrmann and Nyberg to meet the identified requirements – note also that this protocol has been previously described in annex 2 to Reference 1 (see also Chapter 9). We sketch this protocol below. The results in Reference 1 and Chapter 9 provide in full detail examples of protocols that can be used to securely transfer security parameters (e.g. a root certificate) from one device to another and/or ensure that both devices possess the same particular security parameter. These protocols can be used for two devices to exchange P_{CA} and P_M, as described below.

Before giving this protocol observe that, in order to operate successfully, the mobile device and CA must meet certain minimum requirements.

- The personal CA must be equipped with a display and a simple input device for giving it commands.
- The mobile device must possess a moderately sophisticated user interface – that is, it must possess both the means for a user to input a sequence of digits (e.g. a numeric keypad or at least two buttons to insert a sequence of zeros and ones), and a simple output device, for example, an audio output, to indicate success or failure of the initialisation process.

The question of how to perform the initialisation process for mobile devices that do not possess a numeric keypad (or similar) is discussed in Section 3.5.2.

Finally, note that we also assume that the mobile device and the personal CA can communicate via an unsecured wireless interface.

3.5.1.1 Protocol specification

The protocol operates as follows.

1. The personal CA must be reliably informed of the identifier for the mobile device. This could, for example, be achieved by the user typing the identifier for the mobile device into the keyboard of the personal CA. However, it could also be achieved as part of the protocol itself (see below).
2. The personal CA sends its public key P_{CA} to the mobile device, and the mobile device sends its public key P_M to the personal CA. This transfer is assumed to take place via the wireless interface. Along with P_M, the mobile device can send (again via the wireless interface) any other information it wishes to have included in the public key certificate that the personal CA will generate. This could, for example, include the identifier for the mobile device.
3. The personal CA now generates a random key K, where K is suitable for use with a MAC function shared by the personal CA and the mobile device. Using this key K, the personal CA computes a MAC as a function of P_{CA}, P_M and any other data supplied by the mobile device. The MAC and the key K are then output by the personal CA (e.g. via a display attached to the personal CA).

4. The user now types the MAC and key K into the mobile device, which uses the key K to recompute the MAC value (using its stored versions of the public keys and associated data). If the two values match then the mobile device gives a success signal to the user. Otherwise it gives a failure signal.

5. If (and only if) the mobile device emits a success indication, the user instructs the personal CA to generate an appropriate public key certificate. This certificate generation must only take place after the mobile device has given the required positive indication. This certificate can then be sent (unprotected) to the mobile device via the wireless interface.

6. The mobile device now performs two checks before accepting the certificate. First, the mobile device checks the signature using the personal CA's public key (P_{CA}). Second, the mobile device verifies that the data fields within the certificate (including the public key P_M and the identifier for the mobile device) are all as expected. The protocol is now complete.

3.5.1.2   Implementation considerations

Apart from meeting the security objectives of the initialisation process, a further primary objective for the design process is to minimise the length of the data strings that the user has to type into the mobile device. This is important for several reasons.

- First, the user will wish the initialisation process to be as quick and simple as possible, arguing in favour of the minimum number of required keystrokes. This is accentuated by the fact that the keypad on the mobile device may be rather small and awkward to use for large strings of data (notwithstanding the ability of many users of existing mobile devices to send text messages using small numeric-only keypads).
- Second, the initialisation process should have a high probability of successful completion. This will clearly not be the case if the user is required to enter a large number of digits, especially using a small keypad and/or with a small or non-existent display to give feedback.
- Third, if typing in long data strings is necessitated by the scheme, then it might be just as simple to type in the respective public keys, thus avoiding the threats that arise from use of the wireless interface.

In the protocol specified in Section 3.5.1.1, this minimisation of data entry can be achieved by using a very short key K and a very short MAC. For example, if the key and MAC both contain four decimal digits, then the probability that an attacker can successfully manipulate any of the information protected by the MAC is very small. (The precise effects of particular parameter choices on the security level of the protocol are discussed in more detail in Section 3.5.1.4 and Chapter 9.)

3.5.1.3   Proof of possession requirements

In some circumstances, before generating a certificate, it is necessary for a CA to ensure that the requester of a public key certificate knows the private key corresponding to the submitted public key. To provide this service, the mobile device could supply a 'proof of possession' of the private key in step (2) of the protocol specified in Section 3.5.1.1.

The nature of this PoP will vary depending on the 'type' of the mobile device's public/private key pair. For example, if it is a signature key pair, then the private key can be used to create a 'self-signed certificate', that is, a signature generated using the mobile device's private key on a string containing the mobile device's public key and the mobile device's identifier.

A detailed discussion of PoP in PAN scenarios is given in Section 3.6.

3.5.1.4 Analysis of protocol

The purpose of the protocol described in Section 3.5.1.1 is to transfer the public keys and other data needed for production of the certificate. All data to be transferred are assumed to be public. Therefore, the security goal is to protect the integrity of the data, not the confidentiality. The necessary integrity protection is performed using the MAC-based checking procedure in steps 3 and 4 of the protocol.

The security threat against the protocol is an active adversary who by any possible means tries to modify the data exchanged between the CA and the mobile device in step 2. If such a modification, insertion of new data or deletion of data takes place on the wireless communication between the devices, then the data sent by one party will be different from the data received by the other party.

The adversary is successful if the integrity protection method fails to detect modification of data. In what follows the probability of failure is determined.

For the security analysis of the protocol it is essential to observe that the communication channel used for the checking procedure in steps 3 and 4 is completely independent of the wireless communication channel used for other exchanges of data in the protocol.

Also, different instances of the protocol are independent. This is due to the fact that for each protocol instance the key K is randomly generated. The key is generated independently for each protocol instance and for each MAC computation. This means, in particular, that even if the data between two protocol instances are strongly related, the respective MAC values computed using different keys are independent. To achieve this randomisation property of the MAC, the length of the key should be larger than or equal to the length of the MAC value.

Let m be the bit length of the MAC and k the bit length of the key. Then, the adversary is successful either if he guesses the key K correctly or if the guess for the key is not correct, but the MAC values for the different data happen to be the same. Hence, the probability of success is

$$\frac{1}{2^k} + \left(1 - \frac{1}{2^k}\right) \times \frac{1}{2^m} = \frac{2^m + 2^k - 1}{2^{m+k}}.$$

For a fixed total length of the bit string to be entered into the mobile device, this probability is minimised if the lengths of the MAC and the key K are equal, that is, if $m = k$, in which case the success probability for an adversary is approximately equal to 2^{1-k}.

3.5.2   *Initialisation methods for limited devices*

In Section 3.5.1, it was demonstrated how device initialisation can be achieved provided that the two communicating devices have sufficient input/output capabilities. In particular, it was assumed that they have numerical keypads and displays. The purpose of this section is to study the same problem in the case where one of the two devices has very limited input/output capabilities. For the rest of the section we assume that one of the devices (the one acting as the personal CA) has both a numerical keypad and a display.

3.5.2.1   Case 1: no numerical keypad

Here, we assume that the very limited device does not have a numerical keypad, but does possess a display. In this case, essentially the same protocol that was given in Section 3.5.1 is used. In the description of the protocol, we adopt the following convention: device A is the very limited device, and device B is the device with both display and keypad, that is, the CA. The protocol is as follows.

1. Device A sends to device B its value X_A.
2. Device B sends to device A its value X_B.
3. Device A generates a temporary PIN K, and displays it.
4. Device A sends $\mathrm{MAC}_K(X_A, Y_B)$ to device B, where Y_B is the value received by A.
5. The user enters K into device B.
6. Device B uses K to compute $\mathrm{MAC}_K(Y_A, X_B)$, where Y_A is the value received by B.
7. If the received MAC matches the computed MAC then device B accepts, and notifies the user. If not, device B rejects and notifies the user.

The correctness of the protocol rests on the observation that if the values X_A and X_B are not tampered with then the protocol terminates successfully. The protocol is also secure against online attacks because the attacker would have to intercept the MAC and substitute it with a value M, such that $M = \mathrm{MAC}_K(Y_A, X_B)$, if device B is to accept (and the attack to be successful). However, finding such a value M can be done with a very small probability, since the key K is not exposed to the attacker. Of course, the key K can be recovered by an offline search (after the key exchange has been completed successfully), but this is not a problem, since a new value K is used for each execution of the protocol.

3.5.2.2   Case 2: no numerical keypad/display

Now, we assume that the limited device has neither a numerical keypad nor a display. The problem now becomes considerably harder, as device A can only communicate over an unauthenticated channel. Note that in the protocol of Section 3.5.2.1, the assumption that A has a display provided an authenticated channel (namely the user), which could be used for very limited data (namely a short PIN). We see no way to achieve our goal unless we assume that this 'user channel' is available. Our assumption for this section is that the very limited device A comes with a pre-installed

PIN, which is known to the user. Then, the protocol of Section 3.5.2.1 can be used safely, but only once! This is because an offline attack will reveal the password to the attacker, who can then use it in subsequent executions of the protocol. One solution to this problem is the following: execute the protocol for the first time, to exchange the authenticated data. These data essentially give public key capabilities to the device *A*. The first thing to do then for device *A* is to send a *new* encrypted PIN to device *B* (the personal CA). This is the new PIN to be used if the private key of device *A* is compromised, and there is no other way to exchange authenticated data with device *B*.

Note that the new PIN will need to be displayed to the user by device *B*, who will need to write it down and store it securely. The new PIN should not be stored by the CA device, since this would potentially make the PIN available to anyone who steals or compromises the CA, preventing the secure re-initialisation of device *A*.

3.6 Proof of possession

Proof of possession (PoP) is required to demonstrate knowledge of the private key corresponding to the public key sent in a request to a certifying party. This concept has been used in conventional PKIs for a long time. PoP in the personal PKI may be performed in a similar way. Nevertheless, we need to analyse whether the assumptions and requirements for the personal PKI will lead to a different view of PoP. Furthermore, the scenarios where PoP is relevant in a PAN have to be identified.

The idea of 'proof of possession' of a private key as part of the public key certification process now appears to be well established. That is, to avoid certain 'source substitution' attacks on cryptographic protocols, it is generally accepted that it is good practice for a CA to ensure that the submitter of a public key knows the corresponding private key. This idea is now incorporated into PKI standards – see, for example, ISO/IEC 15945 [2].

Generally, one can think of two different scenarios for source substitution attacks:

- an attacker may request a certificate for a public key of another person, at the same time spoofing the other person's identity;
- an attacker may request a certificate for a public key of another person without spoofing the other person's identity, that is, using another identity.

3.6.1 Motivation for establishing proof of possession

One example of why such a PoP might be useful is provided by the following description of a source substitution attack on the MTI/A0 key establishment protocol,[2] taken from Note 12.54 (pp. 518, 519) of the *Handbook of Applied Cryptography* [3]. To put this attack into context, we also provide the description of the MTI/A0 protocol provided in Note 12.53 of the Handbook.

We first give the protocol description.

Protocol MTI/A0 key agreement

SUMMARY: two-pass Diffie–Hellman key agreement secure against passive attacks.

RESULT: shared secret K known to both parties A and B.

1. *One-time setup.* Select and publish (in a manner guaranteeing authenticity) an appropriate system prime p and generator α of Z_p^*, $2 \leq \alpha \leq p - 2$. A selects as a long-term private key a random integer a, $1 \leq a \leq p - 2$, and computes a long-term public key $z_A = \alpha^a \bmod p$. (B has analogous keys b, z_B.) A and B have access to authenticated copies of each other's long-term public key.

2. *Protocol messages*:

$$A \rightarrow B : \alpha^x \bmod p \tag{3.1}$$

$$A \leftarrow B : \alpha^y \bmod p \tag{3.2}$$

3. *Protocol actions.* Perform the following steps each time a shared key is required:
   (a) A chooses a random secret x, $1 \leq x \leq p - 2$, and sends B message (1).
   (b) B chooses a random secret y, $1 \leq y \leq p - 2$, and sends A message (2).
   (c) A computes the key $k = (\alpha^y)^a (z_B)^x \bmod p$.
   (d) B computes the key $k = (\alpha^x)^b (z_A)^y \bmod p$. (Both parties now share the key $k = \alpha^{bx+ay} \bmod p$.)

The attack is then as follows.

Source-substitution attack on MTI/A0

As a general rule in all public key protocols, prior to accepting the authenticated public key of a party A, a party B should have assurance (either direct or through a trusted third party) that A actually knows the corresponding private key. Otherwise, an adversary C may claim A's public key as its own, allowing possible attacks, such as that on MTI/A0, as follows.

Assume that in a particular implementation, A sends to B its certified public key in a certificate appended to message (3.1). C registers A's public key as its own (legitimately proving its own identity to the certificate-creating party). When A sends B message (3.1), C replaces A's certificate with its own, effectively changing the source indication (but leaving the exponential α^x sent by A to B unchanged). C forwards B's response α^y to A. B concludes that subsequently received messages encrypted using the key $k = \alpha^{bx+ay}$ originated from C, whereas, in fact, it is only A who knows k and can originate such messages.

A more complicated attack achieves the same objective, this time with C's public key differing from A's public key z_A. C selects an integer e, computes $(z_A)^e = \alpha^{ae}$, and registers the public key α^{ae}. C then modifies α^y sent by B in message (3.2) to $(\alpha^y)^e$. A and B each compute the key $k = \alpha^{aey}\alpha^{xb}$, which A believes is shared with B (and is), while B believes it is shared with C.

In both variations, C is not actually able to compute k itself, but rather causes B to have false beliefs. Such attacks may be prevented by modifying the protocol such that the exponentials are authenticated, and binding key confirmation evidence to an authenticated source indication, for example, through a digital signature.

The Handbook [3, p. 537] also points out that active attacks related to the above attack are considered by Diffie *et al.* [4], and Menezes *et al.* [5].

Although the above attack only applies to key pairs used for key establishment, other attacks can be constructed for protocols based on signature and/or encryption/decryption key pairs. One very naïve attack applying to signature key pairs is as follows.

> Suppose A wishes to send a secret message to B, and wishes B to make an appropriate reply. In order to ensure that the message is not available to anyone other than B, A encrypts the message using B's public encryption key. In addition, in order that B can verify the origin of the message, A signs it using the private signature key of A.
>
> Meanwhile, malicious eavesdropper C has, by some means, arranged for A's public signature verification key to be certified as belonging to C. C now intercepts the signed encrypted message and prevents it from reaching B. C now resends the message to B, claiming that it originates from C. On receipt of the message, B verifies the signature using C's public key, and verifies that it does indeed come from C. B now replies to C (instead of to A), and in doing so may reveal the contents of the secret message.

Finally, note that it is considered good practice to design cryptographic protocols that are resistant to source substitution attacks – see, for example, Reference 6. Nevertheless, this does not mean that, for the moment at least, it is safe to omit the PoP step, since protocols not protecting against such attacks may still be in use.

3.6.2   *Assumptions and requirement for personal PKIs*

We start by considering the issue of key generation. There are various different ways in which a key could be generated. When proof of possession is being considered, it is important to take into account the place and the time that a key is generated. There are three main cases to consider.

- The key pair is generated by the user's device. This situation is essentially the same as in fixed network scenarios. The certifying party has to use a PoP algorithm to obtain assurance that the requesting party is using a legitimate public key in the certificate request.
- The asymmetric key pair is generated by the manufacturing party before the device is delivered to the customer. In this situation, the need for PoP depends on whether the user is able to read out the public and/or private key from the device. Reading out the private key should not be possible for the user or an attacker. If the key to be certified is sent in a secured way and can be linked to the requesting device, PoP might not be necessary.

- The asymmetric key pair is generated by the certifying party when the certificate is requested. PoP is not necessary in this context as both parts of the key pair are generated by the certifying party. In this case, the establishment of an authentic and confidential channel for the transport of the private key has to be supported.

The second major issue concerns the type of the public key to be certified, typically one of Encryption, Signature verification and/or Key establishment (e.g. as used in an authenticated Diffie–Hellman key agreement protocol).

If a public key to be certified is to be used for a particular purpose, then there may be restrictions on the way PoP is performed. For example, if a private key is used for signing, then the request for a certificate for the corresponding public key may be signed with the private key, whereas a private key only permitted to be used to perform decryption operations may not be used to sign such a request. Use of the wrong kind of PoP technique may result in a breach of security.

In some cases, the use of the private key to sign the message may merely break key separation rules; in other cases it may simply not be possible, for example, if there is no known digital signature algorithm that employs key pairs of the appropriate form. The typical case will probably be somewhere between these two extremes, in that, although a public key may be usable with a signature scheme, it may be usable with many such schemes, and it may not be simple to choose one. For example, in most public key cryptosystems based on discrete logarithms, the public key is equal to a base value raised to the power of the private key – in such a case, there will be a very large number of signature schemes for which the key pair would be a valid key pair. The problem would then be coming to an agreement between signer and verifier about precisely which signature scheme should be used.

For any discrete logarithm based public key cryptosystems (including elliptic curve cryptosystems) the private key can be used to create an ElGamal signature on the certificate request, even if it is to be used subsequently as an encryption or key agreement key. However, this might be a problem, since it breaks the usual key separation requirements. One way of avoiding any problems might be as follows. Prior to computing the signature, generate a random value x, and if a is the client's private key, then generate the signature using $a + x$ *as the input*, and send x to the CA along with the signature and the public key g^a. The public key to verify the signature will simply be $g^a g^x$.

The various PoP techniques, as discussed in Section 3.6.4, must, therefore, be mapped to the kind of key that they can be used for.

The third major issue is the nature of the algorithm to be used with the public keys. It needs to be investigated whether the nature of the algorithm specified to be used with a certain key makes a difference as to how PoP should be achieved. One possible criterion might be whether or not a signing algorithm always produces the same signature for a certain input, or whether the signature is different every time, as with RSA and El-Gamal based signatures, respectively.

The fourth and final issue concerns what information is passed to the verifier. The party proving the possession of a private key sends some kind of information to the requesting party, that is, the certifying party. The sensitivity of this information can be different depending on the PoP mechanism in use. Whereas in some cases the information may be non-critical, there may be mechanisms where the proving party is required to perform an action that enables an attacker to decrypt a certain message or, in the worst case, to compromise secret information. The proposed PoP mechanisms must, therefore, be further investigated with respect to zero-knowledge properties.

3.6.3   Analysis of the necessity of proof of possession in different scenarios

Certain scenarios may require PoP mechanisms, whereas in other scenarios PoP is not necessary. Different factors may influence the necessity of PoP.

As already mentioned, it is essential to consider where a key is generated. When key-generation is performed by the certifying party, and the private key is sent to the client together with the certificate, a PoP mechanism is certainly not necessary. If a key is generated by the manufacturer and pre-installed on a device before shipping, PoP may be necessary if an attacker could get hold of the public key before the device reaches its owner. PoP is not necessary if the authenticity of the key sent to be certified is secured with another mechanism. One example of such a mechanism could be as follows.

A manufacturer of smartcards pre-installs cryptographic keys on his cards. In addition, a non-personal certificate is generated and included on the card simply to prove that the public key sent for certification was generated by the manufacturing party. During the personal PKI certification request, this non-personalised certificate is sent with the request. The certifying party then is at least sure that the owner of the card has sent a request, and will get suspicious if he does not receive a proper certificate from the certifying party. That means that the only risk left is a man-in-the-middle attack, which is likely to be discovered very quickly by the owner of the card. Furthermore, the issuing party can be sure that the public key to be certified has cryptographically good properties (as it was generated by the issuing party).

A variant of the mechanism described above could be as follows: the manufacturer stores a key pair on the card and certifies the public key before the device is delivered to the customer, as described above. But, in contrast to the last approach, this key pair is not intended to be used by the customer for purposes other than proving the possession of a key or the possession of the device.

Scenarios in which the key is generated by the requesting party, either on a hardware device like a smartcard, or with software mechanisms, will be the most interesting scenarios for PoP considerations. Nevertheless, even in this case there may be scenarios where PoP is unnecessary. An example of the latter could be a PAN scenario where people are in a local environment and can trust the transmission of data or the authenticity of the sending party with out-of-band mechanisms.

3.6.4   Proof of possession mechanisms

We now consider a number of different mechanisms for establishing PoP. All these mechanisms are in some way specialised in that they only apply to certain types of key pairs. It would appear rather difficult, if not impossible, to devise general purpose mechanisms for PoP, applicable for all key pairs.

1. *Signature of the request or a part of the request.* In this approach, a user generates a key pair or uses a key pair that is already on his device or token. Before sending the request for certification of the public key, the request itself, or a certain part of the request, is signed by the corresponding private key, and the signature is added to the request. If only a part of the request is signed, a dedicated solution might have to be developed, whereas signing the whole request may be done with standard signature mechanisms and standard applications.

2. *Signature of a certain value derived from the request.* This solution works like the one described in point 1. The difference here is that the value to be signed is now not a direct part of the data of the request.

3. *Signature of a value that is independent of the request.* This solution works like the one described in point 1. The difference here is that the signed value is independent of the request data.

4. *Prompt the user to decrypt a specified challenge.* This approach has to be used with care, as the user is giving away information by decrypting a value selected by the CA. This means that, in standard cryptographic terminology, the user is acting as an oracle. As the PoP process is only performed once, that is, during the certification process, it may nevertheless be a useful method. This is especially likely to be the case if the decrypted message has to be in a pre-agreed format. In the case where an attacker wants to use the user to decrypt an arbitrary challenge, the decrypted value will, with very high probability, not match the format, and the answer could be discarded by the party requiring proof of possession of the private key.

5. *Prompt the user to decrypt a certain value and send back a value derived from the result.* To overcome the dangers as described in point 4, the decrypted value could be input to a one-way function before returning it to the CA. So, if an attacker wanted to use the requesting party as an oracle, he would not get the original message but only the output from the one-way function, and this will almost certainly not be useful to the attacker.

6. *Prompt the user to decrypt a certain value and to prove knowledge of this value with a zero-knowledge protocol.* To avoid sending any information related to the private key but yet prove possession of the private key to the requesting party, a zero-knowledge protocol may be used. Since no information about the secret (i.e. the decrypted value) is given away to the requesting party, a potential attacker gets no information at all. Only the legitimate requester can use the information to verify that the proving party is in possession of the decrypted value. The use of a zero-knowledge proof may, unfortunately, result in a more complex protocol, especially with respect to the number of steps to be performed.

7. *Issue the certificate in an encrypted format, so that the requester is only able to get hold of the certificate if he is the owner of the private key.* This method may only be used in scenarios where the certificate is not published automatically. If automatic publication of the certificate takes place, the mechanism is of no use, as an attacker will be able to retrieve the certificate from the directory. If the

certificate is only distributed by pushing it to the requester, and is not available to pull from some directory, then this method is potentially very efficient.

3.6.5   *Proof of possession in standards*

Proof of possession is also an issue addressed in standards. The IETF has produced two RFCs that deal with proof of possession.

- *IETF RFC 2511: Internet X.509 CRMF (Certificate Request Message Format) [8].* RFC 2511 discusses the PoP issue briefly and proposes some of the methods mentioned above. As discussed here, RFC 2511 distinguishes between the uses of a key, and signing and decrypting are essentially the proposed mechanisms. As a third alternative, the computation of a MAC on the certificate request with a key derived from a secret, shared between the CA/RA and the requesting party, is proposed. This approach may not be of use in our scenario, as this assumption will probably not hold in PAN scenarios. A message format for PoP is given as follows. The general structure lists the type of possible PoP-mechanisms.

```
ProofOfPossession ::= CHOICE {
   raVerified              [0] NULL,
   signature               [1] POPOSigningKey,
   keyEncipherment         [2] POPOPrivKey,
   keyAgreement            [3] POPOPrivKey }
```

The format of the different mechanisms is as follows (some parts have been omitted; for a full description see RFC 2511).

```
POPOSigningKey ::= SEQUENCE {
   poposkInput             [0] POPOSigningKeyInput
                               OPTIONAL,
   algorithmIdentifier     AlgorithmIdentifier,
   signature               BIT STRING }

POPOSigningKeyInput ::= SEQUENCE {
   authInfo                CHOICE {
      sender                  [0] GeneralName,
      publicKeyMAC            PKMACValue },
   publicKey               SubjectPublicKeyInfo }
POPOPrivKey ::= CHOICE {
   thisMessage             [0] BIT STRING,
   subsequentMessage       [1] SubsequentMessage,
   dhMAC                   [2] BIT STRING }

SubsequentMessage ::= INTEGER {
encrCert (0),
challengeResp (1) }
```

- *IETF RFC 2875: Diffie–Hellman proof of possession algorithms [9].* This RFC provides two methods for generating an integrity check value from a Diffie–Hellman key pair. The two different approaches differ depending on whether or not they use information concerning the receiver. The first solution produces a PoP value that can only be verified by the intended recipient, whereas in the second solution a PoP value is generated that everyone can verify.

- *ISO/IEC 15945: information technology – security techniques – specification of TTP services to support the application of digital signatures [2].* As this standard only applies to signature keys, PoP is viewed purely from this perspective. The mechanisms and the syntax used for PoP of the signature keys are similar to the ones from RFC2511, as described above. Of course, the syntax in this standard is reduced to the relevant parts, that is, the parts describing the use of PoP for signature keys (signature [1] POPOSigningKey).

3.6.6   Efficiency of PoP mechanisms

In Table 3.1, the main properties of the mechanisms described in Section 3.6.4 are summarised.

Table 3.1   Properties of proof of possession mechanisms.

Mechanisms	Steps to perform	Computational complexity
1. Signature of the request or a certain value	Client: sign CA: verify	Two PK operations
2. Signature of value derived from request	Client: sign CA: verify	Two PK operations
3. Signature of an independent value	Client: sign CA: verify	Two PK operations
4. Prompt the user to decrypt a challenge	Client: Send public key CA: send challenge Client: decrypt and return CA: verify	Two PK operations
5. Prompt user to hash a decrypted value X	Client: Send public key CA: send challenge Client: decrypt X & hash CA: verify	Two PK operations, hash computation is negligible
6. Zero-Knowledge proof	Depending on the concrete implementation	Depending on the concrete implementation, but probably higher than the other mechanisms
7. Encrypted issued certificate	CA: encrypt Client: decrypt	Two PK operations

3.6.7 Suitability of the methods for mobile environments

The most convenient mechanism to prove possession of a private signing key is probably to sign the complete request for certification. It is possible that the signature is only on the request format, for example, on a PKCS#10 structure, or over a whole message containing the request. No particular dangers have been identified for this mechanism. To sign only parts of the request or an independent value is only slightly more efficient, as the hash-function before signing might be performed faster, but the signature process is certainly less standardised than signing the whole message.

In the case of PoP for an encryption key, the selection of mechanism will depend on a careful analysis of the precise implementation environment. The most critical factor in this context is the leakage of information from the proving party. This potential danger can be overcome by using zero-knowledge protocols, but these protocols are generally more complex and time consuming.

The efficiency of the proposed methods can be summarised as follows. Just signing the request or parts of it requires two steps and two PK operations to be performed, the signature by the requester and the verification of the signature by the certifying party. The number of steps grows to four where the decryption of a challenge is part of the PoP process, although the number of PK operations to be performed does not change and is, therefore, two.

3.7 Revocation in personal PKIs

3.7.1 Assumptions and requirements for personal PKIs

- *Structure of the PAN.* A PAN is usually a relatively small structure so it may be possible to implement the management of revocation information in a different way to conventional methods, where a large number of entities are involved.
- *Availability of all PKI users to the CA.* In a PAN, the availability of the users might be greater than in conventional PKI environments, as the number of users is likely to be much smaller. This means all users can be reached at a certain time, or the CA can keep track of which users have not received revocation information, so that the CA can pass on the information at the next login.
- *Structure of the used certificates.* The structure of the used certificates might be adapted to the particular requirements of a PAN.

3.7.2 PAN-specific versus general revocation mechanisms

In conventional PKIs, 'pull mechanisms', such as the provision of revocation lists or online status mechanisms, are typically used to provide access to revocation information. This is the case for the following reasons.

- Often a large number of clients take part in the PKI. Thus, 'push' services would lead to a bandwidth problem and associated management problems for the CA.

A broadcast service could be implemented, but then clients not logged in during the broadcast would not receive the most recent revocation information.

- Large PKIs are usually long-term PKIs, where revocation information has to be provided covering a lengthy time period.
- Not all clients are always online, and so some will not receive all revocation information that is distributed.
- Not all clients are interested in revocation information at all times. This would make a broadcast potentially inefficient.
- The revocation information gets so large that not all clients are able to cache the information. Thus, it must be possible to load the information when required.

In contrast to conventional PKIs, push services can probably be implemented in personal PKIs, as the following assumptions hold.

- Personal PKIs may be short-term PKIs. Therefore, revocation information may only be relevant for a short time. The issuing of 'renewed' certificates on a more frequent basis might not be a problem due to the small number of certificates in the structure. In consequence, one might introduce short-term certificates that are only valid for a short period of time and have to be renewed frequently, for example, every day, as is already done in WAP in the SSL-server-certificate context. This could result in the removal of any need for revocation mechanisms.
- Only a restricted number of parties take part in the personal PKI. Therefore, more time or resource consuming mechanisms (such as the issuing of short-term certificates as described above), which do not scale in large PKIs, might be feasible for implementation in the PAN scenario.
- When components make use of Personal PKI they are online. Therefore, revocation information can be pushed to them when logging in to the personal PKI. The limited number of PKI users in a PAN makes it possible for the CA to keep track of the users and the revocation information that they have already received.

3.7.2.1   Mechanisms adapted from conventional PKIs

Revocation mechanisms used in traditional PKIs can be adapted to also perform revocation management in Personal PKIs. The two general approaches, that is, using CRLs or requesting status information online, can be implemented. A short summary of the discussion of relevant issues follows.

CRLs

A major problem with CRLs is that, in the mobile domain, they cannot be used to provide up-to-date certificate revocation information because their size means that mobile bandwidth considerations prevent very frequent updates of CRLs, and infrequent CRL updates considerably reduces the effectiveness of CRL use. (Of course, this does not necessarily rule out the use of delta-CRLs, but they carry their own significant management overhead.)

Therefore, we focus on the online status protocols OCSP and XKMS.

OCSP. OCSP, as an online revocation method, uses signed messages from the OCSP responder to the client (in our case a mobile handset) to convey revocation information. The purpose of OCSP is to provide revocation status and nothing else. OSCP has been developed by the IETF PKIX group. Several vendors, such as Baltimore, alicert, VeriSign and Entrust, provide OCSP client and server implementations. OCSP provides server authenticity, as in OCSP it is mandatory for all responses to be signed. OCSP also offers optional client authenticity, in that the client may sign OCSP requests. This could be used if the OCSP responder only wishes to give responses to authorised requesters. OCSP offers protection against replay attacks by including a nonce within every message sent. The requester includes a randomly chosen nonce in his response, and the responder extracts this nonce and places it in the response. The requester can then check if the packet has been replayed by verifying that the nonce in the response is that sent in the request. The inclusion of the nonce is an optional feature.

XKMS. XKMS, like OCSP, provides an online certificate revocation checking method. XKMS, however, offers more than just certificate revocation; it can also check the certificate validity and process a certificate chain. It also allows for key registration. Compared to OCSP, XKMS is a fairly new specification. It has been published within the World Wide Consortium (W3C) as a 'technical note', which means it is not a standard as yet. A client supporting XKMS will have to support the verification of XML digital signatures and will have to support XML. All XKMS responses are signed with XML digital signatures. The revocation status of the public key corresponding to the XKMS signed responses is ambiguous, as the specification does not define a way of validating the corresponding public key certificate, it is simply assumed to be trusted. XKMS protects against replay attacks by using a transaction ID in each request. The transaction ID is comparable to the nonce issued within OCSP. This is not a mandatory feature within XKMS. The transaction ID should be unique within a client with regard to a particular certificate. The use of the term 'transaction ID' suggests that the client must use the transaction ID as a sequence number but, in practice, the client could just generate a nonce in each case.

If the two online status protocols are compared, the following conclusions can be drawn. Signed OCSP responses and requests are nearly four times shorter than XKMS messages. The size differences between XKMS and OCSP are purely based on the encoding and format of the two schemes, and do not depend on any differences in security functionality offered. It is clear, therefore, that, on memory and bandwidth grounds, OCSP requests and responses are preferable in the wireless world, as they use less bandwidth (typically all responses will be signed to authenticate the responders). The encoding method for OCSP messages is preferred to that of XKMS as the mobile world has already limited ASN.1 encoding support, whereas support for XML within the mobile world is still very scarce. This fact would enable OCSP implementation (particularly on the client side) to be developed more quickly than XKMS implementations. However, XKMS provides more services than OCSP. OCSP only provides a certificate revocation service, where XKMS provides a complete certificate revocation, validation, key registration solution that will look more desirable as technology improves. Overall, OCSP seems to be the preferable solution for certificate revocation in the short-term future because of the significantly smaller size of its messages, and the fact that it can be implemented more easily on the client device.

3.7.2.2   PAN-specific mechanisms

Because of the particular characteristics of a personal PKI, new mechanisms to manage revocation, not appropriate for traditional PKIs, can be used. Generally, the new situation is that the PKI is a relatively small structure. Therefore, it may be possible to implement some kind of push mechanism. Thus, it means that a member of the PKI gets informed automatically about recent revocation incidents. Several possible models for this can be devised.

- CA-based distribution models.
  1. *Automatic distribution of newly generated CRLs.* This requires the CRLs not to be too big. It is not the most elegant approach, as the CRL concept was intended to be a pull concept. The major advantage could be that the CRL is more up-to-date as the lists can be pushed when a new entry has been added. Of course, this mechanism is only efficient when there are not too many revocations.
  2. *Automatic distribution of new revocation incidents.* This solution requires the introduction of a new protocol/application that is able to store single revocation incidents or put them together and store them authentically on the client device. In comparison to the distribution of complete CRLs it is more efficient to distribute only recent incidents, but as past experience indicates, the introduction of new functionality on the client is always problematic.
  3. *Automatic distribution of CILs (Current Identity Lists).* Instead of 'black-lists' we propose the distribution of 'white-lists'. This has the advantage that a user can be sure that a certificate is not revoked *and* that this certificate was issued by the CA concerned. The use of white-lists only makes sense when the number of participants is not too big. The time at which white-lists are published must be specified. Of course, the current list has to be sent to new members and members re-entering the PAN. A question in this context is whether clients already logged in get the whole list again when a new member enters the PAN, or whether there may be another mechanism just announcing the new member, as for the proposed mechanisms to distribute revocation incidents only (see above).
- Ad hoc distribution of CRLs (as already discussed in Section 3.2.2).
  4. The idea behind this approach is that the distribution of CRLs is done between the clients, so that the necessity to contact a directory is removed. Determining the currency of CRLs can be handled by introducing serial numbers or time stamps, and clients always update to the most recent version. That is, whenever mobile devices communicate, they exchange the serial number (or time

stamp) of the CRL they possess. If one device has a higher serial number than the other then it passes the latest CRL to the other device. Thus, the latest CRL should disseminate across the PAN very rapidly, without requiring any active support from the personal CA. Such an approach may even be appropriate in other networks, although that is outside the scope of this discussion.

3.7.2.3   Issues with local caching in clients

An important question for revocation in PAN scenarios is whether very limited devices can cope with certain revocation information at all. In the case of very limited devices, storage space may be an issue. Therefore, the caching of CRLs may not be possible on every device. Usually, revocation lists will not get very large in PAN scenarios, as there are only a small number of components taking part in the local PKI; nevertheless, there may be scenarios where a list can get longer. One example of this latter case may arise in a PAN having a lot of guest members. The question here is whether the personal CA has the capability to issue a certificate to those parties having only a short validity period, thereby avoiding the necessity to revoke many certificates.

3.7.2.4   Support of multiple mechanisms

As discussed in the previous section, mechanisms exist that are attractive for a certain class of devices, whereas they may be less attractive or even infeasible for other classes. Thus, there may be a need to integrate at least two revocation mechanisms in one PAN. New issues may arise from the introduction of different mechanisms:

- it could be necessary to implement additional logic on the personal CA device;
- if push mechanisms are used, devices in the PAN must be able to cope with them, that is, evaluate the pushed information or discard it if an evaluation is not possible.

3.7.3   Suitability of the methods for mobile environments

PANs offer many possibilities for implementing revocation mechanisms differing from the ones familiar in fixed network scenarios. Currently, these novel mechanisms are only proposals, and have not yet been implemented. It will be difficult to implement mechanisms that relate to the client software, as this functionality must first be standardised. Nevertheless, as PAN scenarios get more and more attention in the mobile world, and given that PANs will be an essential part of tomorrow's mobile infrastructure, it is necessary to consider these concepts and to develop solutions that may be more efficient than the ones available today.

3.8   Summary and conclusions

After defining the requirements for PKI in a PAN, we have looked at concrete issues such as imprinting devices, management of certification authorities and revocation mechanisms. The discussions in this chapter have shown that whilst many mechanisms and protocols from the fixed network environment may be used or adapted for the PAN environment, new mechanisms, not feasible in conventional fixed network PKI scenarios, may be advantageous in PANs. The latter category of mechanism is probably best represented by the new imprinting protocols where the user has to act as a trusted channel and by the various scenarios proposed for revocation checking based on push-mechanisms.

Acknowledgements

The authors would like to thank all their colleagues in the IST SHAMAN project who have contributed to the development of the results described in this paper, especially Jozef Dankers, Theo Garefalakis, Heiko Knospe, Scarlet Schwiderski-Grosche and Tim Wright. Further details on the results presented in this paper may be found in the SHAMAN project deliverables, available at http://www.ist-shaman.org.

Notes

1 The work described in this chapter has been supported by the European Commission through the IST Programme under Contract IST-2000-25350. The information in this chapter is provided as is, and no guarantee or warranty is given or implied that the information is fit for any particular purpose. The user thereof uses the information at his/her sole risk and liability.
2 This particular attack applies to key establishment key pairs.

References

1 SHAMAN Deliverable D13: 'Final technical report – results, specifications and conclusions (Available at http://www.ist-shaman.org)
2 ISO/IEC 15945: 'Information technology – Security techniques – Specification of TTP services to support the application of digital signatures'. ISO/IEC, 2002
3 MENEZES, A. J., VAN OORSCHOT, P. C. and VANSTONE, S. A.: 'Handbook of applied cryptography' (CRC Press, Boca Raton, FL, 1997)
4 DIFFIE, W., VAN OORSCHOT, P. C. and WIENER, M. J.: 'Authentication and authenticated key exchanges', *Designs, Codes and Cryptography*, 1992, **2**, pp. 107–25
5 MENEZES, A. J., QU, M. and VANSTONE, S. A.: 'Some new key agreement protocols providing implicit authentication'. Workshop Record of the 2nd Workshop on *Selected Areas in Cryptography* (SAC '95), Ottawa, Canada, May 1995
6 HORN, G., MARTIN, K. M. and MITCHELL, C. J.: 'Authentication protocols for mobile network environment value-added services'. *IEEE Transactions on Vehicular Technology*, 2002, **51**, pp. 383–92

7   GEHRMANN, C., NYBERG, K. and MITCHELL, C. J.: 'The personal CA – PKI for a personal area network'. Proceedings – *IST Mobile & Wireless Communications Summit 2002*, Thessaloniki, Greece, June 2002, pp. 31–5

8   MYERS, M., ADAMS, C., SOLO, D. and KEMP, D.: 'RFC 2511: Internet X. 509 certificate request message format'. IETF, March 1999

9   PRAFULLCHANDRA, H. and SCHAAD, J.: 'RFC 2875: Diffie – Hellman Proof-of-Possession algorithms'. IETF, July 2000

Chapter 4

The smartcard as a mobile security device

Dirk Scheuermann

Modern society would like to replace paper material by electronic data carriers and mechanical processes by electronic processes. Smartcards offer one means to this end in the form of a personal mobile security device. Personal data can be stored in a mobile personal environment instead of a central database, and processor smartcards additionally provide a sort of pocket PC that can perform security functions with higher security than an ordinary PC. Interoperability of different smartcards with different smartcard readers and data terminals is very important; therefore, there exist certain standards for the structure of data objects on the card and for coding the commands sent to the card. Biometric user-authentication is becoming increasingly important for smartcards as an alternative to the previously used PIN or password authentication. Additional convenience for the user can be provided by contactless cards.

4.1 Introduction

Storing and filling out paper forms is troublesome work. It is not surprising, therefore, that there is much interest nowadays in replacing paper by devices that can store and process data electronically. One such device is the smartcard, whose portability and size make it particularly suitable for small personal data records. Considerable work related to filling out paper forms can be saved simply by presenting a smartcard; in addition, storing data on a smartcard provides more privacy than storing them in a central database.

PCs are often not secure enough for specific applications. Intelligent processor smartcards, however, provide the possibility to build up security infrastructures and to execute security functions in a personal environment on a mobile device. They represent a sort of mobile pocket PC with high-security features delivering a good contribution to the desired security for mobility.

This chapter gives an overview of the smartcard technologies and some of their applications.

4.2   Storage cards and processor cards

For smartcards, different general categories have to be distinguished: the card may be a data storage device only or contain a processor to execute some functions. And, among the processor cards, there are certain sorts of cards to enable the user to implement individual programs.

4.2.1   Storage cards

Storage cards are the simplest form of smartcard. They are only used to store data that can be read out at any time. They do not contain any processor to execute functions and do not have any security concepts to protect the data from being read out.

Storage cards provide the convenience of being smaller and easier to handle than paper documents; and data can also be directly read and processed electronically by a PC.

The only security feature is the fact that once manufactured it is not possible to write on the card, that is, to change the data on it. However, this represents a disadvantage to processor cards (described in Section 4.2.2): once the validity of the data on the card expires and the data need to be changed, the card becomes useless and needs to be thrown away and replaced by a new card, just in the same way as paper documents need to be renewed.

4.2.2   Processor cards

Processor cards are intelligent smartcards that not only provide a device to store data electronically, but also represent a small mobile PC, powered and handled by an appropriate data terminal, to process the data. There exist certain security features to protect the access to the data and to certain functions. With the aid of these features, a smartcard can be personalised, that is, bound to a certain person. Similar to an ordinary PC, a processor smartcard has the following fundamental components:

- the random access memory (RAM),
- a read only memory part (ROM) containing the operating system,
- non-volatile memory (EEPROM),
- a processor (CPU).

Currently available smartcards have the following amount of memory and computing power:

- RAM: up to 5 kB,
- ROM: up to 246 kB,
- EEPROM: up to 128 kB,
- 32-bit CPU.

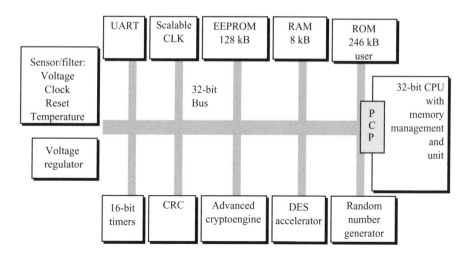

*Figure 4.1   Block diagram of a smartcard chip with 32-bit CPU. PCP, Peripheral
Control Processor; DES, Data Encryption Standard; UART, Universal
Asynchronous Receiver Transmitter*

Figure 4.1 shows the block diagram of such a smartcard.
The following general security functions are provided:

- application of cryptographic functions (encipherment, decipherment, electronic signatures);
- user authentication (PIN or password);
- device authentication with the aid of cryptographic protocols.

The storage capacity and computing power of these mobile devices is still limited, but, in exchange, they provide higher security than ordinary PCs.

4.2.3   Cards with downloadable program code

Among the processor smartcards, a certain type plays an important role for users who want more flexibility to design their own security applications: the smartcards with downloadable program code. If considered as a mobile PC, these cards follow the strategy of PCs with RISC architecture: only a few basic functions have to be provided by the hardware or the operating system, whereas the variety of possible functions lies in the software to be installed in the non-volatile memory by the user.

One important type of these cards is provided by Java cards, which allow implementing Java byte code in EEPROM. Standardised smartcard commands (as defined in ISO/IEC 7816, see Section 4.3.1) are not handled directly in ROM any more, but by a Java applet in EEPROM. By installing individual applets on the card, standardised commands may be executed in different ways on the same card, depending on the desired application context.

4.3   Standardised data objects and commands

4.3.1   Standards by ISO/IEC 7816

Important features of smartcards that need to be standardised are the data objects present on the card and the structure of the commands sent to the card. Besides the specification of general physical characteristics of smartcards, the standardisation of these features is one important task of the series ISO/IEC 7816. General transmission protocols for exchanging data between the card and the data terminal are specified in part 3 (ISO 7816–3). Part 4 (ISO 7816–4) [1] covers file structures on the card, the structure of commands and the specification of some basic commands. Part 8 (ISO 7816–8) [2] specifies special commands for cryptographic purposes. And, for the increasing importance of biometric user verification with smartcards, part 11 (ISO 7816–11) [3] is under development to have a closer look on command sequences and data objects for this special purpose; more details are presented in Section 4.5. We next focus on two important general features, namely the system of files on the card (Section 4.3.2) and the structure of the smartcard commands (Section 4.3.3).

4.3.2   File system

In a smartcard, there exists a file system similar to that on a PC. The following categories of files exist: one single master file, dedicated files and elementary files. The mandatory master file, which is implicitly selected after the activation of the card, is the root file on top of all other files; it is comparable to the root file on a PC system. Dedicated files represent directories, and elementary files correspond to individual files containing data. In contrast to PCs, the operating system of smartcards already provides different structures for files that are useful for certain applications. The following structures exist (see Figure 4.2).

Transparent files are amorphous files similar to files on ordinary PCs; they have no inner structure, and each byte is individually addressable for being read or written. For some applications, for example, storing tables, it is useful to have structured files. For this purpose, on smartcards there exist file structures *linear fixed* and *linear variable*. These files are separated into records; each byte is addressed by a record ID and an offset number giving the exact location inside the record. In the case of linear fixed, all records must have the same length (which is one specified file parameter), whereas for linear variable, different record sizes are allowed giving more flexibility. One additional file structure for smartcards, also divided into records, is provided by the structure *cyclic*, which represents a sort of drum storage: if the data of a writing command exceed the space remaining up to the end of the file, its first record is

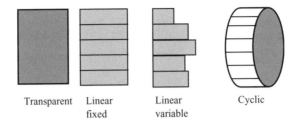

Transparent   Linear   Linear   Cyclic
fixed   variable

Figure 4.2   *Data structures in smartcards*

Header				Body		
CLA	INS	$P1$	$P2$	L_c	Data	L_e

Figure 4.3   *Structure of smartcard commands*

overwritten again. Cyclic files are useful, for example, for storing a certain number of past transactions performed with the card.

Structure, size and security features are specified during the creation of the file and are unchangeable afterwards. With the security features, the access rights to the files are specified: read or write access or the possibility to delete the file may be protected by knowledge-based or biometric user authentication as well as by cryptographic device authentication, or may be denied completely. With these unchangeable features, a much higher security is provided than on a PC since accidental modification of data or reading out critical data can be avoided. For example, cryptographic keys, PINs or passwords are stored in files with no read or write access; they can only be accessed by special security commands (see Section 4.3.3) using or modifying the data without reading them out of the card.

4.3.3   Smartcard commands

According to ISO/IEC 7816–4 [1], a command sent to the smartcard consists of a so-called application protocol data unit (APDU). Such an APDU consists of two major parts: a mandatory, 4 byte command header that must be present for all commands, and a command body that can be missing for some commands. The individual fields of the APDU are as follows (see Figure 4.3).

A class byte (CLA) at the beginning of the header contains some general informations about the command including regulations to interpret the subsequent fields. The instruction byte (INS) specifies the fundamental function of the command. Besides the inter-industry commands defined in References 1 and 2, a card manufacturer or programmer may also define individual commands, the so-called private use commands; the information, to which sort of commands the instruction code belongs to is indicated in the class byte. Two further bytes in the header, $P1$ and $P2$, specify further parameters for the command execution. For commands with input

Optional part	Mandatory part	
Data	SW1	SW2

Figure 4.4   Structure of responses to smarcard commands

data, the command body contains a data field preceded by an L_c byte defining the length of the data. For commands delivering back output data, an L_e byte indicates the expected length of these data.

Compared with functions executed on a PC, the instruction byte represents the function name, whereas $P1$, $P2$ and the optional command body represent the input parameter list of the function.

To always give feedback to the user, the smartcard delivers a response string for each command whose structure is also standardised. This response also contains a mandatory part and an optional body (see Figure 4.4). The mandatory status bytes SW1 and SW2 indicate the correct or incorrect execution of the command with precise error specification; they are comparable to predefined return values of functions indicating certain errors. Additional response data, if indicated by the L_e byte of the command, are contained in the optional response body.

4.3.4   Secure messaging

For certain security applications, data transmitted to or obtained from the card are strictly confidential, and additionally, the integrity or authentic origin of the data must be ensured. In Reference 1, a solution for this problem is also provided; the general concept is called *secure messaging*. With the aid of this concept, data may be cryptographically protected: the command data may be enciphered to ensure confidentiality and a cryptographic checksum can be calculated and appended to ensure integrity. The use of secure messaging and its exact mode is indicated within the CLA byte of the command.

4.4   Sample applications

4.4.1   Existing applications for storage cards

People are already accustomed to some applications where the smartcard is just used to store data. Two common examples are telephone cards and health insurance cards. Telephone cards are convenient as there is no need for small coins, and insurance cards save time as these is no need to fill out paper forms. However, the use of the data stored on it has no specific protection, that is, the data can be read out and used by everyone finding the card. And, it is known that these cards are not reloadable or rewritable: if the telephone card is empty, it has to be thrown away, and a new one has to be bought. Moreover, after the end of the validation period of an insurance card or after any change of the patient data, for example, name or address, the patient has to apply for a new card with his/her insurance company.

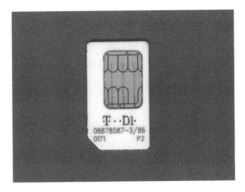

Figure 4.5   Example of a German SIM

Nevertheless, the current data terminals licensed to be used for reading patient data cards make them compatible with the new concept of intelligent insurance cards with increased functions (see Section 4.4.5) by translating the commands for processor cards into commands for storage cards. Modification of the data on the card is only possible with the aid of special terminals owned by the insurance companies but not with terminals licensed for medical doctor's surgeries.

4.4.2   Identification cards for mobile phones

The most popular application for processor cards currently is with mobile phones. The central security processor of a mobile phone is provided by a GSM subscriber identification module (SIM). Unique identification of such a SIM on a worldwide basis is provided by an 8 byte 'International Mobile Subscriber Identity'. Figure 4.5 shows the example of a German SIM.

In this context, the smartcard is used together with a series of specifications for wireless communication network applications, namely the 'Wireless Application Protocol' (WAP). This is done by inserting, at manufacture, the WAP Identification Module (WIM) application [4] onto the same smartcards as the SIM application.

4.4.3   Signature cards

An important security application for smartcards is the creation of electronic signatures (some more specific applications are mentioned in Sections 4.4.4 and 4.4.5). With the signature function implemented, smartcards provide a personal mobile device for performing signatures. If the smartcard meets rigorous security requirements it may be classed as a so-called 'Secure Signature Creation Device' as defined in Reference 5. Signature cards must be personalised and contain a certificate of the owner.

The German card manufacturer Giesecke & Devrient developed a signature card based on the smartcard operating system STARCOS SPK 2.3, which has been certified and evaluated as meeting SOF high on level E4 according to [ITSEC1.2] [10].

Besides smartcards as universal signature creation devices (to be used for signing arbitrary documents), it is also possible to integrate the signature function into more smartcard applications (see Sections 4.4.5 and 4.4.6).

4.4.4   Electronic identity cards

Smartcards may be used in the place of paper documents. They may be used for automatic access control, thus avoiding the need to employ people to watch entrances. An office identity card is a typical application.

Access control could be provided by simple data storage cards. However, processor cards are more useful as office identity cards as they may provide different individual access rights or authorisations to perform certain security functions.

The German consortium TeleTrusT established a specification of a German office identity card [6]. Certain cryptographic security functions including electronic signatures are provided as basic services for the office identity card.

Some European countries are planning to introduce a general electronic identiy card for everyone [7]. Finland and Sweden were the first countries to introduce it.

4.4.5   Health applications

In the future, smartcards will play an important role in health applications. Basically, the following two cards will be involved: a patient data (health insurance) card and a health professional card (to be used by a pharmacist or a physician).

A main reason for using smartcards for health applications is the enlargement of the functions of the patient data card: the previous data storage card (see Section 4.4.1) is to be replaced by an intelligent processor card with security features to regulate the access to the card. Access may be granted, for example, to the patient after knowledge based or biometric user authentication or to a physician or pharmacist after cryptographic device authentication with the aid of the health professional card. Besides the authentication to get read or write access to the patient data card, the health professional card will also be equipped with an electronic signature function, for example, for signing electronic prescriptions. In the meantime, a precise specification of a health professional card – together with an additional security module card for protection of authentication data – has been created [8].

4.5   Smartcards and biometrics

Traditionally, user verification is performed with the help of PINs or passwords. Presently, biometric methods, where a user is recognised by personal features, are gaining considerable importance. Biometric features provide the advantage of not being lost, forgotten or unintentionally given to someone else.

Biometric verification and identification methods are also gaining importance in connection with smartcards; they provide a way to not only relate but also bind these mobile security devices to a certain person.

With respect to the processing of the biometric data, the smartcard may fulfil two general functions: it may be simply a device for storing the data or it may process the data, that is, perform the biometric verification. In which way the smartcard is used strongly depends on the application scenario: the biometric verification or identification process has to be performed by the device holding the protected security application or the protected data since this represents the verifying authority.

4.5.1 Smartcards as data storing device

If the security application to be protected by biometric verification or identification is located outside the smartcard, the card itself is used as a data storing device only. Typical examples are access control systems or cash dispensers: not the smartcard, but the access control system and the banking system have to check the identity of the user.

In these scenarios, the smartcard provides a mobile personal storage device avoiding the necessity to store the biometric data in a central database. This provides advantages with respect to data privacy. Users will have better trust in biometrics if they are not afraid that their personal data stored in a central database could be misused.

4.5.2 Smartcards as verification device

If the security application or valuable and critical data are contained in the smartcard, it is important that the smartcard itself is able to recognise its legal user. For this purpose, the biometric verification process has to be implemented into the smartcard, too (on-card matching). A typical example is the signature card described in Section 4.4.3.

With this functionality, the smartcard becomes a mobile security device that is more strongly bound to its owner than previously used PIN or password verification.

Basic smartcard commands for biometric user verification – based on commands previously used for PINs or passwords – are already defined in Reference 1. Due to the importance of this new technology, a new standard considering the necessary command sequences and data objects is now under development [3].

On-card matching provides a broad market for smartcards with downloadable program code: standardisation activities in biometrics are currently restricted to data objects and general command sequences, whereas there are no standards for biometric algorithms. Therefore, biometric algorithms in most cases are not provided directly by card manufacturers within the operating system, that is, inside an expensive to be developed ROM mask. Instead, individual algorithms are implemented in smartcards by software manufacturers or application providers. One of the first known developments of a ROM mask with biometric matching algorithms is provided by the integration of the SmartMatch algorithm of IKENDI into the Apollo Operating System of SC^2 Ltd [9].

It should be noted that some biometric data – for example, fingerprints – are no secrets, but publicly available. Even though the biometric data will be treated in the same way as PINs or passwords (i.e. stored in a file with no read or write access), the security of the biometric verification system must not depend on keeping the biometric reference data safely inside the card. For avoiding replay or data acquisition attacks, for example, obtaining a fingerprint from a glass and sending the digitised data to the smartcard, secure messaging can be used as described in Section 4.3.4.

4.5.3   Implementation of further steps

Besides the pure comparison (matching) of the biometric data, the complete biometric verification process contains two further preliminary steps: first, the data need to be captured by a biometric sensor. Afterwards, specific data are extracted out of the raw data. For the future, it is intended to also implement the feature extraction process and the biometric sensor into the smartcard. This makes the mobile security device more secure and independent by avoiding the transportation of biometric data between the card and the data terminal.

These steps have not yet been realised, but can be expected in the future due to the continuous development of smartcard technology. Four possible roles of the smartcard are shown in Figure 4.6:

- data storage only (1);
- biometric verification (feature matching and decision processing) on card, sensor and feature extraction outside it (2);
- feature extraction and feature matching with decision processing on card, sensor outside it (3);
- complete system on card (4).

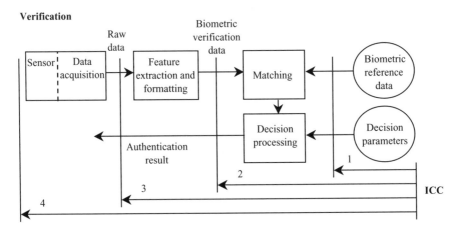

Figure 4.6   Possible roles of smartcards in biometric verification systems

4.6   Contactless cards

The standard series ISO/IEC 7816, which also involves biometric verification on smartcards, covers contact cards only. For some applications, where certain actions are to be performed very quickly, the convenience for the user can be increased by using contactless cards.

4.6.1   Standards for contactless cards

The maximal distance between the card and the data terminal varies depending on the application. There already exist international standards for specifying the following three fundamental types of contactless cards:

- close coupled cards: distance up to 1 cm (defined in ISO/IEC 10536);
- proximity cards: distance up to 10 cm (defined in ISO/IEC 14443);
- vicinity cards: distance up to 1 m (defined in ISO/IEC 15693).

The market for close coupled cards is small as some effort is required to position the card correctly and hence they do not offer much improvement in convenience. However, proximity cards or vicinity cards are not suitable for high-security applications where critical data are transmitted. A typical application for close coupled cards is provided by access control systems; the other contactless cards with higher distances may be used for some payment processed, for example, fare payment when boarding a bus or payment for use of toll roads. Regarding payment for public transportation, some cities (e.g. Hong Kong, Singapore and San Francisco) have already introduced payment systems based on contactless smartcdards; some field trials were also performed in Germany.

4.6.2   Dual interface cards

For some critical applications, it is still desirable to use contact cards. Therefore, it is planned to design multi-application smartcards as dual interface cards with interfaces for both contact and contactless communications. In a first step, the card will contain two different chips. The next step will be a single dual interface chip that is able to handle both modes of communication. The office identity card (see Section 4.4.4) represents one possible example: the embedded signature function designated in Reference 6 should be handled over a contact interface while the access control function is better suited for a contactless interface.

4.7   Summary

The smartcard is a well-suited personal mobile security device for improving the security of mobile applications. Besides the convenience of avoiding handling paper documents, intelligent processor cards also provide a security infrastructure at a higher level than a PC. The option to download own programs to the card provides even more flexibility in designing multi-application smartcards with individual security applications. The most common smartcard application currently in use is that provided by a mobile application, namely mobile phones with GSM cards.

As outlined in Section 4.3, a couple of standards already exist that provide interoperability between different smartcard operating systems and data terminals as well as standardised security level. The security of smartcards is further enhanced by the use of biometric user verification, which offers a mobile security device not only related to, but strongly bound to a certain person.

The computing power and memory capacity of smartcards is still limited, but increasing continuously. Future developments will concentrate on placing complete systems, for example, a biometric verification system with all steps, on the card to avoid the exchange of critical data with the data terminal.

The convenience of using smartcards will increase with the further development of contactless cards. The future belongs to dual interface cards being able to handle different applications, one that is better suited for contact cards, another that is better suited for contactless cards.

Smartcards are further developing as a universal mobile security device for a continuously increasing number of security applications.

References

1   ISO/IEC 7816–4: 'Information technology – Identification cards – Integrated circuit(s) cards with contacts – Part 4: Interindustry commands for interchange'. Revised Version: ISO/IEC FCD 7816–4: 01/2003, 1995

2   ISO/IEC 7816–8: 'Information technology – Identification cards – Integrated circuit(s) cards with contacts – Part 8: Security related interindustry commands'. Revised Version: ISO/IEC FCD 7816–8: 01/2003. Information technology – Identification cards – Integrated circuit(s) cards with contacts – Part 8: Interindustry commands for a crypto-graphic toolbox , 1999

3   ISO/IEC FCD 7816–11: 07/2002. 'Information technology – Identification cards – Integrated circuit(s) cards with contacts – Part 11: Personal verification through biometric methods'

4   Wireless Identity Module Specification, WAP-260-WIM-20010712-a, July 2001

5   EU Directive 1999/93/EC of the European Parliament and the council of 13 December 1999 on a Community framework for electronic signatures

6   German Office Identity Card, V.1.0, 6.7.2000

7   The Electronic Identification of Citizens and Organisations in the European Union: State of Affairs. Report of 37th Meeting of the Directors General of the Public Service of the Member States of the European Union, Brugges, 26–27 November 2001

8   German Health Professional Card and Security Module Card Specification – Pharmacist & Physician, V.2.0, July, 2003

9   IKENDI®Smart Match – Fingerprint Match on Card for Infineon Smart Card and Security ICs. SECURE – The Silicon Trust Report, 02/2002

10 ITSEC 1.2: Information Technology Security Evaluation Criteria, V.1.2, June, 1991

Chapter 5

Secure mobile tokens – the future

Hubert Ertl

Secure mobile tokens are widely used in consumer products. The most popular one is the Subscriber Identity Module (SIM) security module used in GSM mobile phones. In the future, IUCC security modules would be used for 3G mobile phones. The use of secure mobile tokens in commercial environments besides mobile voice call and SMS-based services is still an emerging market. Existing security module technology will follow the technology roadmap in the near future and expand functionality and performance while still under heavy cost pressure in the consumer market. New technologies combining existing services or integrating standard Internet protocols would show up and allow for new devices and scenarios based on advance secure mobile tokens like secure multimedia card (SMMC) or secured FLASH disks. The success story of the SIM technology in GSM mobile networks is currently followed by UICC and USIM technology in 3G networks and plans exist to make use of secure mobile tokens for securing WLAN access and usage as well as to make use of it for general heterogeneous network access, regardless of it being DSL, fixed line, GSM, WLAN or Universal Mobile Telecommunications Systems (UMTS).

5.1   Introduction

Secure mobile tokens are in widespread use in today's communication devices. GSM mobile phones are equipped with SIM cards for both pre-paid and post-paid subscribers, whereas 3G mobiles use USIM cards. These cards are issued by the mobile network operators, to manage their customer identities, to protect network access, to manage services and to provide a reliable basis for billing. Besides these basic features new services and features are added. Some of these new services include providing information service, adding security functionality for banking, mostly based on 3DES schemes, signature schemes to be integrated in Public Key Infrastructures (PKI) or fully equipped crypto devices based on PKCS#15 like the OMA WIM.

This chapter gives an overview of how security tokens work in these environments and how they will evolve in the future. The text is mainly based on the results from European Research Project Shaman. Section 5.2 gives the definition of a security module. Section 5.3 describes the current use of security modules, while Section 5.4 describes security modules technology and gives a roadmap for future technology development. Section 5.5 outlines the basics of prevalent mobile tokens like GSM SIM, 3G USIM and OMA WIM and finally, Section 5.6 focuses on the emerging use of personal security tokens in the future.

5.2   Security modules

Secure mobile tokens are a flexible and secure mechanism to allow for secure and personalised mobile communication within mobile communication infrastructures. Besides providing the basis for individual phone numbers and accountability of phone time, recent advances in technology make use of security modules for secure transmission of data (e.g. OTA servers), storage of data (e.g. SMS) and digital signing procedures (e.g. Wireless Identity Module, WIM).

A tamper-resistant implementation of security modules is a precondition for easy administration of removable and exchangeable security modules, since they are faced with various unauthorised intrusions during their lifetime.

The hardware of secure mobile tokens is based on security modules containing single-chip micro-controller technology. Security modules will continuously increase its capabilities regarding processing power, storage sizes and interface bandwidth in the future, since it is directly linked to the reuse of existing memory and processor technology with approximately eight years in delay and directly scales with the investment in the modules.

A similar situation exists with software technology for security modules since the telecommunication community is now switching from proprietary operating system technology to open systems technology. Again, the same scenario that existed in general purpose systems earlier.

Thus, from basic technology development there is a clear picture of which devices will be available in the future. As a result, future devices will not be dependent on what is feasible from a technical point of view, but from requirements imposed regarding standards, functionality and cost. Thus, the essential question is on the functional requirements for security modules and how standardisation of devices will evolve in the future.

5.2.1   Definition of security module

A security module is a tamper-resistant device that is both physically and logically secure and has the ability to contain data and/or perform functions for certain security systems. A security module is capable of storing secret data and executing security functions in such a manner that no information about the secret data that could be used to break the security system is leaked out from the security module.

For example, smartcards are regarded as suitable devices to be used as a security module. It is not necessary that the module must perform alone all functions that use secret data. In fact, in some cases execution of security functions can be distributed between the security module and some other devices without revealing information about the secret data outside the security module.

The module can be a single smartcard, but does not have to be. It can also consist of multiple smartcards or a smartcard and a special server functionality it communicates with. By embedding the security module in larger casings they can look like a key ring or in comprising a USB connector it can look like a USB stick. Today, the most commonly used version is an ID-000 SIM plug-in or a credit card sized ID-1 smartcard.

5.3 Current use of security modules

The use of security modules for holding sensitive data is already widespread in a variety of applications supporting a number of services. Security modules have evolved and are now capable of doing more than just holding data securely. Apart from just holding data, many security modules are used for cryptographic computations, generating encryption keys, carrying out authentication, etc.

5.3.1 Support for services to the terminal

Security modules are defined to be tamper-resistant devices that can perform data processing and store confidential information. A terminal (PDA, mobile handset) is generally not as physically or logically secure as a dedicated security module when it comes to performing specific security operations. It is typically easier to extract data from a terminal than from a security module. However, on the other hand, a terminal can hold more data, is more powerful and is capable of supporting a larger range of security functionality.

It is, therefore, often necessary to split security functions between the security module and the terminal to ensure an adequately secure yet efficient implementation. In this case, the security module has to support some security functionality and the terminal may use the result of this computation. An example of such integration is the Wireless Identity Module (WIM), as specified by the Open Mobile Alliance. The WIM has support for processing and storing security data, is able to do digital signing and will soon support encryption where the signed (and encrypted) data are passed to the terminal; additional security mechanisms can also be applied (e.g. transport layer encryption using WTLS or TLS). For example, the WIM is used for signing data with a user's private key that is stored in the WIM; the terminal carries out the verification of digital signatures where the root certificates can be stored on the terminal or in the WIM.

Another example is the GSM encryption method. The terminal is required to use the session key generated by the SIM to encrypt the voice data. Figure 5.1 shows that the session (encryption key) is generated in the SIM by inputting the master

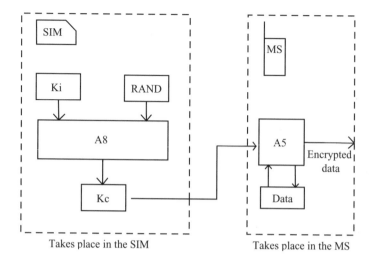

Takes place in the SIM   Takes place in the MS

Figure 5.1   Distribution between the mobile and the security module

encryption key, Ki, and a random that has been received from the mobile switching centre (MSC). The session key is then passed on to the mobile station (MS). The MS uses this session key to encrypt the voice data using the A5 encryption algorithm, which the MS must support. The session key is inputted into the A5 encryption generator along with the voice data, the outcome being scrambled data. The master encryption key, Ki, never leaves the SIM, but is used to generate a session key, Kc.

5.3.2   Support for services to the network

Providing services to networks is the most common use for security modules today. The wireless world, in particular GSM, relies on the security module (or SIM) for authentication to the network based on a long-term subscriber authentication key before a user is allowed to use any service from that network. The security module is required to encrypt the voice call with the symmetric key generated in the SIM. The SIM, therefore, is an integral part of the GSM architecture, providing security data storage and cryptographic processing.

The USIM is the 3G version of the SIM to work with the UMTS network. It supports 3G protocols and is backward compatible to support 2G authentication methods for access to 2G networks. Although the USIM is not regarded as a physical entity, it is seen as a logical application that resides on a UICC. The UICC contains one or more USIMs and possibly other applications (e.g. credit card functionality or WIM). By inserting the USIM card into a UMTS terminal the user is recognised by the UMTS network and can be addressed on this terminal via his personal telephone number (MSISDN).

In contrast to GSM, there will be a multitude of different types of terminals in UMTS, for example, multi-mode or multi-band handsets, notebook-like communicators or UMTS laptops with camera, speakers and microphone all equipped with a USIM card. There will be terminals too where more than one USIM card can be inserted. This means that some terminals (e.g. fax terminals) shall be used by several UMTS customers simultaneously. The USIM stores the identity of the subscriber (user), operator and service provider and (at least one) user service profile. This service profile defines the services that a customer is subscribed to, the time and the network where he can use them.

LAN services have also adopted the use of security modules for secure log on. The security module, like in the GSM world, authenticates the user to the network in order for the user to use the services offered by the network. The security module can also be used to sign e-mails and hold encryption keys to encrypt e-mails and other data.

5.3.3 Support for services to the end user

The use of security modules for the end user is a must when storing private and confidential data. Presently, security modules are used in various ways to hold different types of information for a user.

The key benefits in all scenarios are:

- to provide a device with proven security for basic authentication;
- to personalise this device for individual users or identities;
- to hand over the device to the customer or from person to person;
- to provide two-factor authentication (possession and knowledge).

Cards are processor cards based on standard CISC architectures know form other embedded devices, but RISC schemes emerging, even based on generic architectures like ARM. Crypto co-processors are still optional and external interfaces are emerging providing advanced throughput and bandwidth beside the basic serial line transfer capabilities.

When secure communication between parties is established, security algorithms and keys for these algorithms are needed. Since security of communication is based on the fact that only authorised entities know secret keys, it is clear that these secret keys must not be revealed to other entities. So there must be a secure way to store and use these secret keys, a solution for this problem is to use security modules that are tamper-resistant devices, which provide storing and using secret keys safely. Security modules must be resistant to physical attacks.

5.4 Security module technology

Security modules for mobile telecommunication are currently based on SIM cards containing standard low-cost and low-power microprocessors equipped with ROM, RAM, EEPROM storage and hard wired security algorithms. The security algorithms are based on symmetric key cryptography with individual permanent secret keys securely stored on both the SIM cards contained in the mobile device and the network operators' centralised databases.

While the operators' centralised data processing centres can easily be protected and secured by standard means, the secret keys stored in the mobile device need to be protected by a tamper-resistant and protected SIM card device against various attacks (e.g. spying out the key) and potential abuse (grabbing another person's identity).

SIM cards are tamper-resistant and trusted devices, produced by certified and trusted companies following agreed specifications and state-of-the-art in technology. SIM cards act as proxy for an individual user's identity subscribed for the service and can be easily transferred on to other devices by this single person. Besides this the SIM card is the property of the operator and all security features and services are set up, guaranteed and certified by the individual operators.

Hardware technology of smartcards as a low-cost, embedded consumer product can be estimated to follow the hardware technology of high-end standard off-the-shelf microprocessor technology with a delay of approximately eight years. Production technology is at least two generations behind current state-of-the-art in semiconductor production. First-generation high-end productions are used for memory components, second-generation fabs are in use for standard microprocessor products, while third- and fourth-generation fabs are used for telecom and embedded devices.

Despite the delay in semiconductor core technology of approximately eight years, these is a clear outlook of what can be expected from core technology in the future. Additionally, specialised and standardised components are introduced faster than the eight-year delay into smartcard technology. This effect can currently be seen with additional interfaces (e.g. USB), crypto-coprocessors (e.g. for RSA or PKI) or Java byte code interpreters.

5.4.1   Processing power

High-end devices for smartcard products are currently using up to 33 MHz CPU clock and will reach 100 MHz by 2005. Restrictions from the mobile equipment and by the telecommunications specifications limit the used clock rate. Adapting the CPU clock rate to the individual processing power needed during peak times and allowing slower low-power operation during off-peak times is available from today's chip technology. Additional restrictions arise from heat dissipation and power consumption of individual devices. Currently, 8-bit processor technology is being replaced by 16-bit RISC technology and the transition to 32-bit technology can be foreseen in the near future.

5.4.2   Storage

Today's standard devices are still using 64 kB ROM, 4 kB RAM and 32 kB EEPROM. While low-cost products are still using smaller amounts of memory there are high-end smartcard products available providing 128 kB of ROM, 6 kB RAM and 64 kB EEPROM. Any scale in memory size directly scales to the chip area needed for the device and, therefore, scales the price for the product. This especially holds for high-volume production periods like in 2000, but has a major impact on prices in lower production volume periods as well. The roadmaps of various chip suppliers show new products during next few years mainly increasing the size of RAM (up to 10 KB), while retaining the ROM and EEPROM sizes at today's high-end sizes (128 kB ROM, 64 KB EEPROM). As a new feature, the FLASH ROM technology is introduced with additional 196 kB of memory available in FLASH technology. Besides the additional memory space available the devices can now be programmed and used more individually. Updates, enhancements or replacement of operating systems and services can be completely done after initial chip production without the need of an additional ROM mask redesign and chip production cycle. But a complete update of operating systems during smartcard lifetime cannot be expected during the next few years for security reasons as well as for technical reasons.

Additionally, memory management and protection units (MMUs) are currently introduced to the consumer market allowing for hardware-based protection and mapping of individual memory areas to the applications and operating systems needs. Up to now, these protections are assured by a secured operating system implementing access restrictions by software means, which is technically feasible, but causes overheads in verification of the mechanisms. By using MMUs the operating system can now make use of individual hardware protection mechanisms to protect certain memory, code and interface areas from program access at operating system or applications program level.

In the long term, FERAM technology might replace the EEPROM technology and provided larger memory sizes with characteristics only known from today's RAM technology. This includes memory access cycles of $1\,\mu s$ compared to EEPROM access cycle times of 1–3 ms.

5.4.3 Interfaces and bandwidth

While there are several attempts to provide a faster interface for other uses of cards (e.g. USB interface for signature or computer logon cards), the standardisation of mobile equipment is restricted to the available serial transmission rates on the existing contact pads of GSM smartcards. The serial line of state-of-the-art smartcard processors can be programmed to provide up to 115.2 kbit/s and as an extension two existing but commonly unused pads on the external interface can be re-used for a second serial I/O-channel. Thus, the I/O bandwidth can be significantly increased for special usage without violating the standard interfaces definitions for the regular I/O line.

Of additional importance is the memory bandwidth if larger amount of data have to be stored into secured memory areas. Currently, the memory bandwidth is low compared to exiting DRAM technology and no significant improvement can be expected in the near future. Storing single byte or single page data currently takes some milliseconds with today's EEPROM technology. In fact, even the existing processing power of the most recent smartcard CPU cannot be fully used, since both the external serial interface and the internal memory interface are slow compared to today's demand from data streaming format.

For example, full data stream encryption is only possible for slower streams by today's technology. The external I/O bottleneck will be removed soon by USB-like interfaces, but GSM specifications do not allow its use in practice. The memory bandwidth will be significantly increased by FERAM technology in the longer run (five years), allowing smartcards to make use of their already existing processing power.

5.4.4   Architectural support for software

Architectural support for software is provided by various means. The MMUs mentioned earlier as well as basic segmentation concepts support the implementation of access checks in operating systems software. Another important area is the provision of architectural support of cryptographic algorithms by implementing cryptographic co-processors for DES and consequently for 3DES.

RSA crypto-engines have been introduced, which regularly support 1024-bit key length and 2048-bit RSA on more sophisticated devices. More elaborate and advanced crypto-processors can be expected within the next five years as plain commodity products.

A higher-level support of software concepts is the provision of the so-called HW accelerators for selected operating systems or languages. The roadmaps of chip suppliers show HW accelerators for Java (HW byte code interpreter) to speed up Java code execution or to ease Java virtual machine integration by providing a large set of the basic virtual machine operations in hardware. Even operating systems like MULTOS could gain from the provision of individually provided HW accelerators. While Java language is a currently agreed demand on future smartcard systems it is not yet finally decided if Java is the open operating systems platform for long-term use on smartcards. At least it adapts well as an implementation, programming and interfacing language for scalable embedded devices for the next few years. In the long run even open source projects like LINUX could influence the smartcard industry. It depends on whether the development of the embedded LINUX version gets really pushed and stability and security issues are finally solved.

5.4.5   Software of security modules

The current state-of-the-art in software technology and operating systems is very similar to the UNIX consolidation 10 years ago, where some different implementations and many different branded products merged into a consolidated UNIX release, a newly rising OS windows NT and the upcoming LINUX hype later on.

While vendor-specific operating systems still dominate the smartcard market, the so-called open operating systems are now discussed for some years and products have been successfully launched recently. By open platform concepts the importance of standardised APIs is recognised as state-of-the-art and standardised APIs replace the individual vendor-specific APIs or clean up the existence of multiple API variants on smartcard platforms.

Today's commercial systems are based on symmetric key algorithms and there is a clear understanding of the security concept based on a mutually known secret called a secret key, which can be compromised independently by each of the both parties knowing the secret key by revealing the key willingly or by accident. Thus, with the symmetric key concept even the security model is symmetric and can be broken from sides knowing the secret.

More recent systems start to make use of asymmetric schemes using different keys for the different directions of the security relationship. This at least doubles the number of keys in use for centralised communications schemes, but exponentially increases the number of keys in use if peer-to-peer communication and additional security features like signing and authentication are also used separately. This leads to the so-called public key systems, storing all recipients' keys needed to send encrypted messages to them in a global public directory with various lookup mechanisms. The general assumption behind this is that using a public key does not allow for easy calculation on the private key needed for decryption.

PKI requires increased computational power of the SIM cards involved and has to store a larger amount of secret data on the card as well as to handle more dynamic cryptographic data interfacing with the crypto-unit. Therefore, PKI puts a load on storage, interface and computational power as well as on the clear and bullet-proof definition of a security model and an appropriate user interface.

Additionally, PKI allows for 'non-repudiation', which means that only a single person or entity can have performed a selected action (like a signature), since he is the only one knowing the key and nobody else holds a copy of it. This feature also has consequences for the smartcards production process to have the keys and PINs only stored on the card and nowhere else, while with symmetric approaches the keys have also to be stored outside the card, for example, to transfer the keys to the network operators. In contrast to the symmetric cryptography systems no second copy of a secret key exists in a PKI system, thus no one can misuse any copy of a secret key. But, on the other hand, PKI systems fully rely on the complexity of the underlying crypto-algorithms and result in larger key sizes to provide the same security level as symmetric crypto schemes.

5.4.6   *Tamper resistance of security modules*

One of the essential concepts and concerns in security device technology is tamper resistance. The general definition is that devices as well as their data and algorithmic behaviour cannot be manipulated, copied or revealed to or by any unauthorised person. Tamper attacks can be started from third parities as well as from inside the system.

The attack can be started at any one of the following access levels of a device:

- physical access to device with damaging analysis tools;

- physical access to device with non-damaging analysis tools;
- direct access to device during operation;
- passive access to device during operation;
- direct or passive access to external operations or protocols of a module.

Basic tamper attacks to security module devices by damaging analysis are, for example, done as physical attacks to the devices. This can happen in unwrapping any packaging or analysing the physical structure of the device by hardware manipulation, by tapping signal lines, by contacting areas on a chip, by analysing electromagnetic radiation or by willingly inducing effects or errors into the operating module. Results or any different behaviour after manipulation can give useful information for the analysis of the entire device. There exist examples of incidents where devices were analysed and manipulated by this approach. Most of these successful attacks are on offline systems like pay-TV descrambler or similar systems. Up-to-date technology of security modules provides means to successfully prevent such attacks.

Previously, attacks were based on timing analysis aimed at detecting timing differences for selected operations like checking wrong or correct PINs; these attacks were already knows from early mainframe architectures and have been successfully solved now. In recent years, a new class of attacks has shown up: Differential Power Analysis (DPA) and Simple Power Analysis (SPA). Both SPA and DPA try to analyse data and behaviour of algorithms by the detailed measurement of the power consumption of devices in use. They are based on the finding that different operations, different data sets and different algorithmic behaviours cause differences in the power consumption of a microprocessor executing any code. In measuring the detailed power consumption during operation one can easily detect loops and cycles as well as different kinds of memory accesses and differences if zero or non-zero values are written or read. A close differential analysis of the power consumption can even reveal secret keys of a device if no guards against such attacks are implemented in the hardware or software of the security module.

Today's technology of security modules targets to disallow the analysis of the modules to prevent the production of visual, functional and operational identical copy of the devices. While some of these features are reached by secured packaging including manipulation detectors, similar techniques are applied at the logical level in using secure operating systems on smartcards, or to design algorithms to make SPA/DPA impossible.

A secured operating system can, for example, repeatedly check for internal consistency and authenticate with and against external devices during any external communication, to detect potential manipulations, while carefully designed algorithms disallow SPA/DPA and hardware sensors prevent physical manipulations of the devices.

Besides the great efforts needed for producing and implementing tamper-resistant security devices any certification of a security device for tamper resistance is a time consuming and expensive process as well. Limiting the certifications process to a smaller device allows for an easier and less-complex verification process and for reuse of the security modules within multiple devices later on. Additionally, it also allows for easy adoption to local law requirements, which is essential for signature or encryption functionality in most countries.

Using a removable and exchangeable device allows to reuse certified technology within new or updated but not yet evaluated environments or devices. Additionally, removable tamper-resistant devices allow for secured transfer of secured information within possibly insecure environments by simply transferring the plug in module.

5.5 Technology roadmap

The technology roadmap can easily be calculated based on the currently used device and the known progress in high-end DRAM and CPU technology from the last few years. Any progress made in these technologies can be expected in the embedded mass market within the following year. Therefore, a confident roadmap of future technology data can be given today. Table 5.1 shows the current scenario in SIM card technology as found in shipping products today.

The prospective architecture of a security module is expected for two or five years. This is neither an official statement of any company involved nor a commitment on any development, but is based on the expectations from a technical point of view.

During the two-year time frame the improvements are expected to be similar to that shown in Table 5.2.

Besides the overall improvements in processor cycle time one can detect the progress of cryptographic coprocessors to the mass market as well as the slight increase in RAM memory and ROM storage available for security modules. For the five-year scenario the roadmap provides the performance characteristics as given in Table 5.3.

Table 5.1   Today's scenario

	Lowest cost	Mass market	Sophisticated
Technology	0.6 μm	0.35 μm	0.25 μm
MHz	1	1–5	10
RAM	1 kB	4 kB	6 kB
ROM	32 kB	64 kB	96 kB
EEPROM	16 kB	32 kB	96 kB
FERAM	None	None	None
Coprocessors	None	None	RSA, DES
External IF	Serial 9.6 kb	Serial 9.6 kb	Up to 115 kb
Storage cycle time	8 ms	4 ms	2 ms

Table 5.2   Two-year scenario

	Lowest cost	Mass market	Sophisticated
Technology	0.35 μm	0.25 μm	0.18 μm
MHz	1–5	12	66
RAM	2 kB	6 kB	12 kB
ROM/Flash EEPROM	32 kB	128 kB	256 kB
EEPROM	16 kB	64 kB	128 kB
FERAM	None	None	None
Coprocessors	DES, AES	DES, AES, RSA, ECC	DES, AES, RSA, ECC
External IF	Up to 115 kb	Up to 115 kb	Up to 115 kb/USB
Storage cycle time	3 ms	2 ms	2 ms

Table 5.3   Five-year scenario

	Lowest cost	Mass market	Sophisticated
Technology	0.25 μm	0.18 μm	0.12 μm
MHz	1–10	20–33	120
RAM	2 kB	8 kB	64 kB
ROM/Flash EEPROM	64 kB	256 kB	512 kB
EEPROM	32 kB	128 kB	256 kB
FERAM	None	None	1 MB
Coprocessors	AES, ECC	AES, RSA, ECC, Hash	AES, RSA, ECC, Hash
External IF	Up to 115 kb	Up to 115 kb/USB	Up to 115 kb/USB
Storage cycle time	2 ms	1 ms	1 ms to 100 ns

Current trends are to move towards MRAM and FERAM technology, to reduce the memory cycle time for long-term storage significantly and also to use it as a general RAM area, to expand the RAM storage available for the processor. While these technologies are promising, but not yet used in real products, the first flash memory based products are now ready to be used for smartcards. Compared to EEPROM cells less transistor cells are needed per bit, but the memory block size is increased resulting in new restrictions on the memory model.

5.6   Current usage of secure mobile tokens/prevalent tokens

Instead of a mobile phone being personalised to the subscriber, as in analogue systems, the handset is generic to all subscribers and is personalised by means of a smartcard

SIM. The SIM contains all the personalisation and encryption details relative to the specific end-user. A SIM contains a microprocessor capable of handling its own security and managing the flow of data within the chip and between the chip and the outside world. It requires an operating system to manage these functions, and this operating system must conform to specifications set by the international groups of ETSI, SCP and 3GPPT3.

In UMTS networks, subscribers are also personalised by smartcards like in GSM. This smartcard is called a USIM and is based on an UICC. The USIM contains new security algorithms and allows for mutual authentication of subscriber and serving network.

A WIM is used to provide security on the application layer in WAP. WIM implementation can be any tamper-resistant device, not necessarily a smartcard. Most likely WIM is integrated on a GSM SIM or on a USIM ICC.

5.6.1 GSM SIM

SIM plug-ins (ID-000) are derived from the well-know credit card-sized ID-1 format by a standardised punching around the embedded processor area, resulting in a 15 mm × 25 mm sized module. Figure 5.2 shows an ID-000 plug-in for a G&D UniverSIM card still contained in its ID-1 sized holder. The contact pads can be seen in Figure 5.2 and the integrated micro-controller is limited to a size of 25 mm². Typical SIM cards currently in use contain 2 kB RAM, 16 kB EEPROM and 32kB ROM. The characteristic features are shown in Table 5.4.

In Table 5.4, one can notice two bottlenecks: serial line speed 9.6 kb and EEPROM cycle time. Additionally, the $T = 0$ protocol limits the maximum data size to be transferred within a single command to 255 bytes.

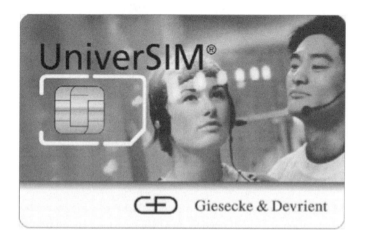

Figure 5.2   ID-000 plug-in on a UniverSIM card

Table 5.4   Characteristic features of SIM cards

	Lowest cost	Mass market	Sophisticated
Technology	0.6 μm	0.35 μm	0.25 μm
MHz	1	1–5	10
RAM	1 kB	4 kB	6 kB
ROM	32 kB	64 kB	96 kB
EEPROM	16 kB	32 kB	64 kB
FERAM	None	None	None
Coprocessors	None	None	RSA, DES
External IF	Serial 9.6 kb	Serial 9.6 kb	Up to 115 kb
Storage cycle time	8 ms	4 ms	2 ms

Table 5.5   Mandatory GSM files

File type	FID	Description
MF	3F00	Root directory
EF_{ICCID}	2FE2	Smart card ID
$DF_{Telecom}$	7F10	DF Telecom
EF_{ADN}	6F3A	Abreviated dial numbers
EF_{FDN}	6F3B	Fixed dial numbers
EF_{LND}	6F44	Last dialed numbers
EF_{SMSS}	6F43	Short message service status
EF_{SMSP}	6F42	Short message service parameters
EF_{SMS}	6F3C	Short messages
DF_{GSM}	7F20	DF GSM
EF_{LP}	6F05	Preferred language
EF_{KC}	6F20	Key Kc
EF_{SPN}	6F46	Service provider name
EF_{PUCT}	6F41	Price per unit
EF_{SST}	6F38	Sim Service Table
EF_{IMSI}	6F07	IMSI
EF_{LOCI}	6F7F	TMSI + location info
EF_{PHASE}	6FAE	GSM phase information

The SIM operating system has to conform to 3GPP specifications. It has to support the authentication algorithm needed for GSM networks and to implement a tree structured file system with two mandatory directories DF TELECOM and DF GSM. The basic files needed are shown in Table 5.5. GSM mobile phones make use of these files, which are usually read by the mobile during startup time. They are also used for storing long-term data like phone numbers and SMS data on the SIM card as well as to perform security functions like network authentication.

For network authentication, a symmetric key Ki is securely stored on the card and processed by the card's operating system, to deliver the response date to the mobile phone and to store the resulting key Kc (used for encrypting the voice channel). Therefore EF Ki are the most sensitive data on a SIM card and not directly accessible from the mobile via the file system and only used by the SIM card operating system itself. The same holds for the CHV or pin code date. They cannot be accessed from the mobile via the file system and are only used by the SIM card operating system, to compare with the PIN codes typed in by the user. Depending on the result access to the SIM card is granted or denied.

The basic GSM authentication procedure is based on a challenge protocol, where a random challenge is sent to the SIM card, some calculations done based on the symmetric key Ki and the results compared. If the results match on both sides access is granted to the network and for speech encryption the symmetric key Kc is used, which was derived on-the-fly during the authentication procedure.

Much more security and information data are managed and stored on a card than mentioned earlier, but the details are beyond the scope of this chapter. One can find most information in the ETSI and GSM specifications available from ETSI.

Figure 5.3 shows a screenshot of G&D's tool for SIM card production.

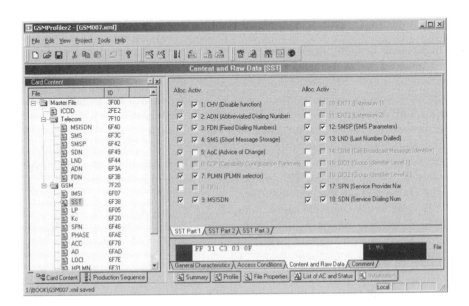

Figure 5.3   Snapshot of SIM production tool

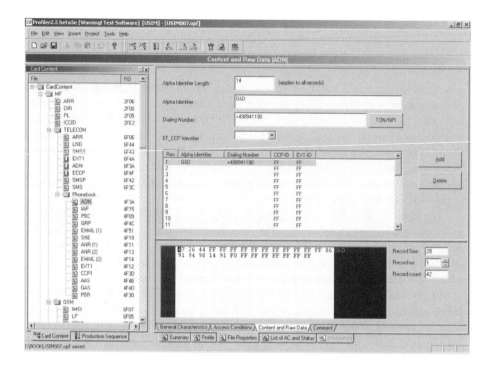

Figure 5.4   Snapshot of USIM production tool

5.6.2   UICC and USIM

The 3G authentication token is based on the UICC, which allows multiple applications (e.g. payment, signature, banking) to be hosted on a single ICC operating system. It can contain a GSM SIM application, a USIM environment and multiple proprietary applications or even customer-specific applications in case the UICC is based on Java technology. As a general restriction, 3G specifications requires not to have more than one network access application active on a single card at the same time. Generally, 3G specifications ensure backwards compatibility with 2G systems putting additional demands on UICC platforms in correctly interfacing both authentication environments if needed. The underlying UICC standards are resulting from the work of 3GPP T3 and ETSI Smart Card Platform (SCP) work.

Besides hosting multiple applications, the UICC provides a more sophisticated access control and file system based on an access control list, which allows individual access conditions to be defined for individual files and directories. An example of an USIM configuration is shown in Figure 5.4.

5.6.3   OMA WIM

OMA WIM (formally known as WAP WIM, defined by the WAP forum) defines a signature and key generation kernel based on RSA PKI in using a PKCS#15 file system. The PKCS#15 file system allows for implementation-independent usage of certificate and data storage on the WIM module. Up to now, WIM is the only standardised application for the SIM and USIM cards, which can be directly used by GSM mobile phones. WIM defines how WTLS and TLS session keys are derived and sent to the mobile for usage and defines how non-repudiation signatures work and are performed by the mobile in using the WIM functionality of SIM cards. The secret keys needed are securely stored on a tamper-proof SIM device and the interface to trigger an individual signature and how to handle the PIN codes are clearly defined. While PIN codes as well as private keys never leave the WIM card, the PKCS#15 file system allows access to public key, certificates and configuration information of the WIM and offers storage for new certificates.

Specifications such as PKCS#15 [5] allow for storage and access to data in a secure manner. The WIM is used in performing WTLS and application-level security functions. It stores and processes information needed for user identification and authentication. Sensitive information such as private keys are stored on the WIM and all operations where these keys are used or involved can take place on the WIM. For example, the WIM can be used to generate a WTLS session key where the WIM generates a random and takes the random generated from the server, applies a function and outputs a session key for encryption over a WTLS session (see Figure 5.5).

Other functions include SignText, which allows a user to sign any piece of text. A part of the text to be signed is passed to the WIM from the terminal, the WIM then uses the stored private key and applies the signing algorithm and signs the hash, and this is then passed back to the terminal; the private key never leaves the WIM.

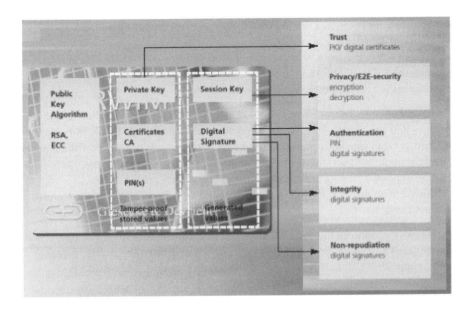

Figure 5.5 Structure of a WIM

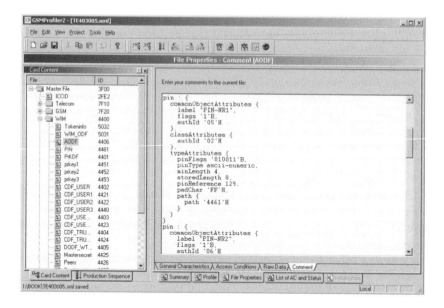

Figure 5.6   Example WIM configuration

The WIM follows the PKCS#15 file structure in order for it to carry out its requirements to store the following data (Figure 5.6):

- information on properties of the module: supported algorithms, etc.,
- key pairs for authentication, key establishment and digital signatures;
- own certificates for each key pair;
- trusted CA certificates;
- data related to WTLS sessions (including master secrets);
- information on the protection of data with PINs.

Cryptographic Token Information Format (PKCS#15) specifies how keys, certificates and application-specific data may be stored on an ISO/IEC 7816 compliant IC card or other media. It provides a hierarchical directory structure for the storage of information with certain directories having access control rights put upon them. PKCS#15 allows for a security module to present security credentials such as a digital certificate to an end application in a standardised way. PKCS #15 has the following distinct characteristics:

- allows multiple applications to reside on the card;
- supports storage of any type of objects (keys, certificates and data);
- support for multiple PINs whenever the security module supports it.

PKCS#15 defines a directory structure that allows for the storage of particular security information. It does this by defining directories such as PrKDF (private key directory file), PuKDF (public key directory file), etc. These directories then can be accessed in a standardised way by PKCS#15 compliant hardware. An important file in the structure is the Authentication Object Directory File (AODF), this is the directory where the authentication objects are kept (e.g. PINs). The WIM specification specifies at least one AODF must be present on a WIM. The first object in the AODF is considered as a general PIN (PIN-G), if not otherwise indicated, all relevant files (CDF, PrKDF) are protected with this PIN.

5.7   Personal security tokens – emerging use

5.7.1   EAP SIM/AKA

There is re-use of existing SIM and USIM technology for authentication and key-derivation on the emerging EAP authentication transport protocol. EAP is derived from early PPP protocols and available in latest consumer and server products (e.g. on CISCO and Microsoft). The basic idea behind EAPSIM/AKA is to transport GSM and 3G authentication data within EAP packets, thus integrating with existing AAA/RADIUS intra-structures.

5.7.2   ISIM

One natural way to expand functionality is to not only use the UICC for 3G network authentication, but also for other types of network authentication, user authentication or signature generation. Therefore, ISIM or USSD scenarios have already been proposed and other specifications are to be provided. It is also expected to provide basic identity services on UICC, to bootstrap one's identity and security environment from a security token, to get access to advanced services on any server on the net.

5.7.3   USB token

Security modules are not only available as ID-1 or ID-000 sized smartcard tokens, but can be used in other packaging as well. The most popular is the USB Token directly attached to the USB port of PCs or PDAs. The hot plug and play functionality allows for much more efficient usage compared to the serial or parallel interface technology available on standard PCs.

5.7.4   PSD

A personal security device is being used for bootstrapping one's identity. Based on successful authentication one can access local or remote security services or data. Finally, all authentication and identity management can be originated from such a device. This demands for more standardisation and integration work, but emerging standards like Microsoft Passport or Liberty Alliance already provide a infrastructure for federations, which could also be bootstrapped from personal security devices held by individual owners. The personal security device has to fulfil certain requirements to be accepted for federation concepts. Device issuer and manufacturer can provide the basic security assertions within their products (e.g. digitally sign the configuration) and personalise the PSD tokens with a unique identity and configuration for individual usage. The final device holder can add his personal data and register with existing infrastructures and federations. This is a future scenario, which needs more standardisation and implementation work during the next years, but again OMA WIM proposals exist to allow for on-board key generation and registration of generated key based on the device manufacturers' assertions or signatures. WIM key enrolment combines with basic operator authentication, thus turning the network operator into an identity provider position, providing identity and billing related services for telecommunications and other use, like Microsoft passport does for Microsoft-related activities at the moment.

5.7.5   Combined devices

MMC and FLASH disk devices can easily be turned into security devices, since these devices already contain embedded operating systems running on micro-controllers exactly the same as used in security module technology. Beside expanding the embedded operating system by security related functionality, the major task is to make the mobile devices accept the modules on the respective interfaces as well as to personalise and distribute the security modules. First flash disk devices already provide cryptographic services that are currently used to protect the stored data itself.

Compared to the regular SIM or USIM devices these devices are more expensive and not yet a commodity product, to be shipped with every mobile phone used in GSM. But combined with other user scenarios (like DRM, video, picture card) this functionality could make its way to products. These products still require to be imprinted and personalised, and certain assertions should be given on the quality and security of technology. Thus, the device manufacturers have to apply their knowledge in providing a tamper-proof and personalised device in the futures and to guarantee for security.

A future security token could be a minimal identity token to be used for bootstrapping the user's identity and security, with wired and wireless interfaces to make use of the basic functionality or integrate such a device into a new devices for emerging applications (like FLASH disks). Thus, a standardised security interface for re-use or co-use in other scenarios could be defined.

The requirements and characteristics of a future security module are:

- to be a personal device;
- to be based on a tamper-resistant device;
- to be a bootstrapping device for security and identity;
- to be transferable to other environments;
- to be exchangeable with other devices;
- to integrate into complex user scenarios;
- to authenticate with multiple networks and applications;
- to keeps private data secret;
- to encrypt and sign information and messages.

The technology for such products already exists, but specifications have to be developed and agreed on to integrate such devices into complex user scenarios and to make use of such technologies in standardised telecommunication and computational environments in the future. The essential role of a device manufacturer will be to guarantee for security and to personalise the individual devices before use.

Part II

Network security

The six chapters in Part II are concerned with provision of security for the mobile network itself.

In the brief history of mobile telecommunications, security has traditionally meant securing the radio path between the mobile phone and the local base station. The evolution of this security from the provisions in GSM to UMTS is covered in Chapter 6, *UMTS security*. This chapter presents a detailed description of the security facilities protecting the UMTS access network, within a context that explains the evolution from GSM. The chapter also describes the security provisions for UMTS internal network security, something not covered by previous mobile standards (notably GSM). This leads into a discussion of the use of IP security facilities for multimedia session control, exemplifying the growing convergence between mobile telecommunications and the Internet.

The issue of security of access to networks is continued in Chapters 7 and 8, which look at possible future developments in this area. Chapter 7 is concerned with a variety of architectural approaches to providing access security, whereas Chapter 8 focuses, in particular, on approaches based on public key cryptography. Both these chapters are based firmly on the results of the EU-funded project SHAMAN.

Work within SHAMAN has also supported the findings in Chapter 9, which considers security issues arising in the Personal Area Network. This is an area of rapidly growing importance as wireless enabled personal devices become increasingly common.

The notion of ad hoc networking, that is, networks with no centralised management function, is also rapidly gaining in importance. However, for these networks to function as they should, the security issues will need to be addressed, and Chapter 10 surveys recent work in this area.

Finally, Chapter 11 considers security issues for MobileIPv6. MobileIP permits mobile nodes to migrate from one access point to another, and clearly there are many major security issues with such mobility.

Chapter 6

UMTS security

Krister Boman, Günther Horn, Peter Howard and Valtteri Niemi

The Universal Mobile Telecommunications System (UMTS) is one of the new 'third-generation' mobile cellular communication systems. UMTS builds on the success of the 'second-generation' Global System for Mobile (GSM) system. One of the factors in the success of GSM has been its security features. New services introduced in UMTS require new security features to protect them. In addition, certain real and perceived shortcomings of GSM security need to be addressed in UMTS. This chapter surveys the major security features that are included in the first releases of the UMTS standards.

6.1 Introduction

The UMTS is one of the new 'third-generation' (3G) mobile cellular communication systems being developed within the framework defined by the ITU known as IMT-2000 [1]. UMTS builds on the capability of today's mobile technologies by providing increased capacity, data capability and a greater range of services using a new radio interface standard called UMTS Terrestrial Radio Access (UTRA) [2].

The basic radio, network and service parameters of the UMTS system were defined by the European Telecommunications Standards Institute (ETSI) in early 1998 [3]. ETSI developed the extremely successful second-generation GSM standard that is used by over 650 million customers world-wide and accounts for approximately 70 per cent of the wireless communications market [4]. An important characteristic of UMTS is that the new radio access network will be connected to an evolution of the GSM core network.

To help build on the global success of GSM, the UMTS standards work in ETSI was transferred in 1998 to a global partnership of regional standards bodies called 3GPP (3rd Generation Partnership Project) [5]. A separate partnership of standards bodies, known as 3GPP2 [6], are developing another 3G mobile cellular system based on a different 3G radio interface standard called CDMA2000 and a core network that is evolved from the North American ANSI-41 standard.

One of the aspects of GSM that has played a significant part in its global appeal is its set of security features. UMTS security builds on the success of GSM by providing new and enhanced security features. This chapter surveys the main security features that are included in the first releases of the UMTS standards.

6.2   Building on GSM security

Security has always been an issue for mobile phones. Many of the 'first-generation' analogue mobile phone systems were susceptible to abuse. For example, it was possible to eavesdrop the analogue radio path and thereby listen to other peoples' calls, or to reprogram the identities of mobile phones such that the cost of calls made using them appeared on another customer's bill. It was against this background of user traffic eavesdropping and cloning fraud that the 'second-generation' GSM system was designed.

GSM was designed to prevent cloning and to be no more vulnerable to eavesdropping than fixed phones. It addresses these goals by providing user-related security features for *authentication, confidentiality* and *anonymity* [7]. The authentication feature is intended to allow a GSM network operator to verify the identity of a user such that it is practically impossible for someone to make fraudulent calls by masquerading as a genuine user. Confidentiality protects the user's traffic, both voice and data, and sensitive signalling data, such as dialled telephone numbers, against eavesdropping on the radio path. The anonymity feature was designed to protect the user against someone who knows the user's international mobile subscriber identity (IMSI) from using this information to track the location of the user, or to identify calls made to or from the user by eavesdropping on the radio path.

The most novel feature of GSM security is the use of a smartcard, known as the subscriber identity module (SIM). The SIM contains all the identification and security-related data that the subscriber needs to make or receive a call. It is, in effect, a portable security module, personalised for the subscriber. The SIM can be used to access services in any network with which the subscriber's home network has a roaming agreement. During roaming, the subscriber's home network provides all the data needed by the serving network to operate the security features without revealing any of the sensitive security data stored in the subscriber's SIM.

The GSM security features have addressed to a very large extent the needs of operators and the aspirations of users. UMTS security builds on the success of GSM by retaining the security features that have proved to be needed and that are robust. As in GSM, a smartcard is used in UMTS to store all the identification and security-related data that the subscriber needs to make or receive a call. Although GSM security has been very successful, an objective of the UMTS security design was to improve on the security of second-generation systems like GSM by correcting real and perceived weaknesses. Some of the issues that have had an impact on the design of the UMTS access security architecture are listed below [8, 9].

- The currently used GSM cipher algorithms (used to provide confidentiality) are not published along with the bulk of the GSM standards. Instead, the GSM Association controls the distribution of the algorithm specifications. The decision not to make the algorithms available for peer review has received some criticism, with hindsight, from the academic world. However, it must be recognised that GSM security was designed at a time when the controls on the export and use of cryptography were much tighter. The regulatory situation was considerably relaxed in the late 1990s, which led 3GPP to adopt a more open approach to the design of the UMTS algorithms and to publish the algorithm specifications together with the rest of the UMTS standards.

- Unlike the cipher algorithm, the GSM and UMTS authentication algorithms do not need to be standardised and operators are free to design or select their own. In GSM, an example algorithm was not included in the standards. This resulted in some operators using an algorithm, known as COMP-128, that has been recognised to be vulnerable to cryptographic attack. After this attack was published on the Internet, the GSM Association made a replacement algorithm available. To help avoid inadequate algorithms being used in UMTS, an example algorithm called MILENAGE [10] has been included in the standards for use by operators who do not wish to design their own.

- The strength of the cipher algorithm depends, in part, on the length of the cipher key. In GSM, the cipher key is transported as a 64-bit structure. However, in practice the top 10 bits of the cipher key are set to zero to reduce the effective key length to 54 bits. This was due to the regulatory controls that were in force when GSM was designed. As these controls have been relaxed it is now possible for GSM to use full-length 64-bit keys. Creating a longer key than this for GSM is much more complex because it would require the ciphering algorithm to be replaced and the signalling protocols to be upgraded to support the longer key. UMTS required a new ciphering mechanism anyway so the opportunity was taken to increase the cipher key length to 128 bits, which should provide a good level of security for many years to come.

- GSM was not explicitly designed to protect against active attacks on the radio path, because they would require an attacker to masquerade as a GSM network (so-called 'false base station attacks'). These attacks were considered to be too expensive to mount compared to other attacks. However, as mobile cellular services become more widespread, the cost and availability of equipment that may be used to masquerade as a base station make such attacks more likely. Although GSM already provides some protection against certain types of false base station attack, a much more thorough threat analysis was performed during the UMTS design phase. This has led to the development of new security features, which are explicitly designed to counteract false base station attacks.

- For GSM circuit-switched services, user traffic and sensitive signalling information are protected on the GSM radio path between the mobile and the base station using a ciphering algorithm. While this protects communications on the most vulnerable radio path, an opportunity was taken in UMTS to extend ciphering further back into the network. This allows more links within the radio access network to be protected, including potentially vulnerable microwave links that may be used to connect base stations to the fixed part of the network.

6.3 UMTS access security

The access security features in UMTS are a superset of those provided in GSM. They are specified in TS 33.102 [11], which is included in the first major release of the 3GPP specifications, known as Release 99. The security features in UMTS, which are new compared to GSM, are introduced to correct the real and perceived weaknesses of GSM security explained in Section 6.2. The UMTS access security standards, in particular, the new authentication mechanism, are based on research work conducted by the EU-funded USECA project [12].

Entity authentication   UMTS provides mutual authentication between the UMTS subscriber, represented by a smartcard application known as the USIM (Universal Subscriber Identity Module), and the network in the following sense.

- *Subscriber authentication*: the serving network corroborates the identity of the subscriber.
- *Network authentication*: the subscriber corroborates that he/she is connected to a serving network that is authorised, by the subscriber's home network, to provide him/her with services; this includes the guarantee that this authorisation is recent.

It should be noted here that the concept of authentication, in general, has many subtle aspects [13, 14].

Signalling data integrity and origin authentication   The following security features are provided with respect to integrity of data on the network access link.

- *Integrity algorithm agreement*: the mobile station (MS) and the serving network (SN) can securely negotiate the integrity algorithm that they use.
- *Integrity key agreement*: the MS and the SN agree on an integrity key that they may use subsequently; this is realised as part of the protocol that also provides entity authentication.
- *Data integrity and origin authentication of signalling data*: the receiving entity (MS or SN) is able to verify that signalling data have not been modified in an unauthorised way since it was sent by the sending entity (SN or MS) and that the data origin of the signalling data received is indeed the one claimed. The use of the integrity feature for signalling data is mandatory.

This security feature has no equivalent in GSM. It provides protection against false base station attacks as the origin of signalling messages required to set up a communication with a mobile can now be authenticated by the mobile.

User traffic confidentiality   The following security features are provided with respect to confidentiality of data on the network access link.

- *Ciphering algorithm agreement*: the MS and the SN can securely negotiate the ciphering algorithm that they use.
- *Cipher key agreement*: the MS and the SN agree on a cipher key that they may use subsequently; this is realised as part of the protocol that also provides entity authentication.
- *Confidentiality of user and signalling data*: neither user data nor sensitive signalling data can be overheard on the radio access interface.

This security feature is the same as in GSM, but the entities between which protection is afforded are different. In UMTS, the protection extends to the Radio Network Controller (RNC) so that microwave links between the base stations and the RNC are also covered.

User identity confidentiality   The following security features related to user identity confidentiality are provided.

- *User identity confidentiality*: the permanent user identity (IMSI) of a user to whom services are delivered cannot be eavesdropped on the radio access link.
- *User location confidentiality*: the presence or the arrival of a user in a certain area cannot be determined by eavesdropping on the radio access link.
- *User untraceability*: an intruder cannot deduce whether different services are delivered to the same user by eavesdropping on the radio access link.

To achieve these objectives, the user is normally identified on the radio access link by a temporary identity by which he is known at the serving network. To avoid user traceability, which may lead to the compromise of user identity confidentiality, the user should not be identified for a long period by means of the same temporary identity. In addition, it is required that any signalling or user data that might reveal the user's identity is ciphered on the radio access link. These features are identical to those provided in GSM. They protect against passive attacks, but not against active attacks.

Mobile equipment identification   This feature is the same as in GSM. In certain cases, the serving network may request the mobile station to send its international mobile equipment identity (IMEI). Neither GSM nor UMTS provide a method for authenticating the mobile equipment identity. This is largely due to the complexity of designing and implementing a robust system. This means that any network features that are based on the IMEI, for example, the barring of stolen phones, relies on the terminal providing the genuine IMEI to the network. The standards, therefore, impose requirements on terminals to protect the integrity of the IMEI so that it cannot be tampered with or reprogrammed.

User-to-USIM authentication   This feature is the same as in GSM. It provides the property that access to the USIM is restricted until the USIM has authenticated the user. Thereby, it is ensured that access to the USIM can be restricted to an authorised user or to a number of authorised users. To accomplish this feature, user and USIM must share a secret (e.g. a personal identity number, or PIN) that is stored securely in the USIM. The user gets access to the USIM only if he/she proves knowledge of the secret.

USIM-terminal link   This feature ensures that access to a terminal or other user equipment can be restricted to an authorised USIM. To this end, the USIM and the terminal must share a secret that is stored securely in the USIM and the terminal. If a USIM fails to prove its knowledge of the secret, it will be denied access to the terminal. This feature is the same as for GSM. It is also known as 'SIM-lock'.

Secure messaging between the USIM and the network   The USIM Application Toolkit provides the capability for operators or third-party providers to create applications that are resident on the USIM (similar to SIM Application Toolkit in GSM). There exists a need to secure messages, which are transferred over the network to applications on the USIM, with the level of security chosen by the network operator or the application provider.

Visibility of security   Although, in general, the security features should be transparent to the user, for certain events and according to the user's concern, greater user visibility of the operation of security features should be provided. Therefore, in UMTS, a ciphering indicator is mandatory. It shows the user whether the confidentiality of user data is protected on the radio access link, in particular, when non-ciphered calls are set up. However, the ciphering indicator can be deactivated by the operator who issues the USIM by setting the appropriate bit in the USIM.

6.3.1   Mutual authentication and key agreement between user and network

The design of the authentication and key agreement (AKA) protocol for UMTS reflects the results of an analysis of the threats and risks in GSM. It was guided by the principle that the compatibility with GSM should be maximised and the migration from GSM to UMTS, and the handover between GSM and UMTS access networks, should be made as easy as possible. In particular, the changes to the GSM core network should be minimised.

The main changes with respect to the GSM authentication and key agreement protocol are as follows.

- The challenge is protected against replay by a sequence number and it is also 'signed' (integrity protected). This means that old authentication data intercepted by an attacker cannot be re-used.

- The AKA generates an integrity key in addition to a ciphering key. This integrity key is used to protect the integrity of the signalling data between the MS and the RNC.

In the following section, an overview of how the UMTS AKA protocol works is given.

6.3.1.1 Prerequisites

There are three parties communicating in the protocol: the authentication centre (AuC) in the home environment (HE) of the user, the Visitor Location Register (VLR) in the Serving Network (SN) and the user represented by his universal subscriber identity module (USIM). In the case of the packet-switched (PS) domain of UMTS, the role of the VLR is taken by the serving GPRS support node (SGSN).

The UMTS AKA protocol is of secret key type. One secret key, the authentication and key agreement key K, is shared by two parties, the AuC and the USIM.

The following trust relations are assumed: the user trusts the HE in all respects concerning this protocol. The HE trusts the SN to handle authentication information, sent by the HE to the SN, securely. The HE distributes authentication information only to SN entities it trusts to provide services to HE's users. The SN trusts the HE to send correct authentication information and pay for the services provided by the SN to the HE's users. The SN accepts authentication information only from entities it trusts.

It must be further assumed for the protocol to be secure that the intra-system interfaces linking the SN to the HE, and linking SNs, are adequately secure. This security is provided by network domain security, as described in Section 6.4.

6.3.1.2 Procedures

Authentication and key agreement (Figure 6.1) consists of two procedures. First, the HE distributes authentication information to the SN. Second, an authentication exchange is run between the user and the SN. The authentication information consists of the parameters necessary to carry out the authentication exchange and provide the agreed keys.

Figure 6.1 shows that, after receiving an authentication information request, the HE generates an ordered array of n authentication vectors. Each authentication vector consists of five components (and hence may be called a UMTS 'quintet' in analogy to GSM 'triplets'): a random number *RAND*, an expected response *XRES*, a cipher key *CK*, an integrity key *IK* and an authentication token *AUTN*. This array of n authentication vectors is then sent from the HE to the SN. It is good for n authentication exchanges between the SN and the USIM. In an authentication exchange the SN first selects the next (the ith) authentication vector from the array and sends the parameters *RAND*(i) and *AUTN*(i) to the user. The USIM checks whether *AUTN*(i) can be accepted and, if so, produces a response *RES*(i), which is sent back to the SN. *AUTN*(i) can only be accepted if the sequence number contained in this token is fresh. The USIM also computes *CK*(i) and *IK*(i). The SN compares the received *RES*(i) with *XRES*(i). If they match, the SN considers the authentication exchange to be successfully completed. The established keys *CK*(i) and *IK*(i) will then be

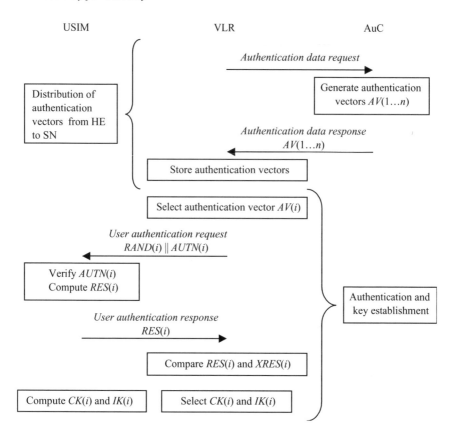

Figure 6.1   Overview of authentication and key agreement

transferred by the USIM to the mobile equipment and by the VLR (or SGSN) to the RNC; the keys are then used by the ciphering and integrity functions in the MS and in the RNC.

6.3.1.3   Authentication functions

No execution of cryptographic functions or storage of long-term secret keys is needed in the SN. This means that there is no need to standardise an authentication algorithm, and every operator is free to choose his own. However, for the reasons explained in Section 6.2, an example algorithm was included in the standards.

6.3.1.4   Compatibility with GSM security architecture

Especially in the initial years of UMTS, coverage will be provided only in isolated 'islands' so that handover between UMTS and GSM will be needed frequently. The UMTS AKA has been designed in such a way that roaming and handover between

GSM and UMTS works as smoothly as possible as far as security is concerned. This is facilitated by the similarity between UMTS quintets and GSM triplets. Conversion functions are specified to convert quintets into triplets and vice versa. (In the case of the conversion of a triplet into a quintet, of course, only GSM-grade security can be achieved.)

6.3.1.5   Compatibility with 3GPP2

As well as supporting roaming with second-generation GSM systems, it is also desirable for UMTS systems to support roaming to the 3G system being developed by 3GPP2. Rather than develop a different security architecture, 3GPP2 have adopted the 3GPP authentication and key agreement protocol as the basis for their security architecture to facilitate roaming between the two systems.

6.3.2   Integrity protection in the access network

6.3.2.1   Requirements for integrity protection

Integrity protection in UMTS prevents the insertion, modification, deletion and replay of signalling messages exchanged between the MS and the RNC. The reasons for supplementing the existing ciphering mechanism with a dedicated integrity mechanism in UMTS are summarised below.

- For various reasons UMTS networks must be able to instruct the MS to use an unciphered connection, that is, the use of ciphering cannot be made mandatory. Thus, an active man-in-the-middle attacker could potentially compromise user traffic confidentiality by masquerading as a network to establish an unciphered connection towards the user. Since integrity protection can be made mandatory, this attack can be prevented as the user can always verify the instruction from the network to establish an unciphered connection. In GSM, the instruction from the network to establish an unciphered connection is not integrity protected.

- The ability to integrity-protect ciphering algorithm negotiation messages provides protection against bidding-down attacks where an active attacker forces the use of an old ciphering algorithm that may, for instance, allow user traffic confidentiality to be compromised. This feature only becomes of interest when multiple algorithms are supported in the system, as is the case in GSM. In the first release of the 3GPP standards only one ciphering algorithm is available and all mobile stations must support this. However, it was considered desirable to design a future-proof system, which allowed new algorithms to be deployed in a way that protects against bidding down attacks.

- Although ciphering of signalling traffic provides some integrity protection and the ciphering of user traffic severely limits the usefulness of any successful compromise of signalling message integrity, the application of a dedicated integrity protection mechanism with its own integrity key increases the security margin of the system. This is seen as an important enhancement, which will ensure that 3G offers adequate protection against increasingly sophisticated active attackers.

- Although the application of user traffic ciphering is highly recommended not just for confidentiality but also for authentication and integrity purposes, there may be some exceptional cases where it is not applied. In these cases, integrity protection of signalling messages significantly increases the level of resistance against relatively unsophisticated attacks, which would have been effective had integrity protection not been provided.

With the above requirements in mind, the mechanisms that are provided for integrity protection by the 3GPP system will be reviewed.

6.3.2.2   Integrity protection of signalling traffic

Once an integrity key has been established as part of an authentication protocol run and once the available integrity protection algorithms in the MS are known, the network can start integrity protection. Integrity protection is applied in the mobile equipment (ME) at the user side and in the radio network controller (RNC) at the network side. A message authentication code function is applied to each individual signalling message at the radio resource control (RRC) layer in the UTRAN (UMTS Terrestrial Radio Access Network) protocol stack [15, 16].

After the RRC connection establishment and the execution of the security mode establishment procedure, most of the subsequent RRC signalling messages are integrity protected. This includes the RRC signalling messages themselves, plus the so-called RRC direct transfer messages, which contain protocol data units for higher layer dedicated signalling between the ME and the core network. Protection of direct transfer messages allows mobility management, call control and session management signalling to be protected.

Figure 6.2 illustrates the use of the integrity algorithm f9 to authenticate the data integrity of an RRC signalling message.

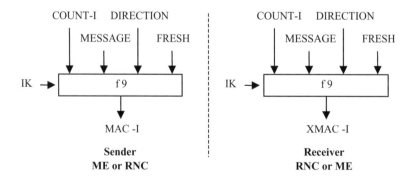

Figure 6.2   Integrity protection of signalling traffic

The input parameters to the algorithm are as follows.

- The integrity key IK, which is 128 bits long.
- An integrity sequence number (COUNT-I) and a random value generated by the radio network controller (FRESH). COUNT-I and FRESH are each 32 bits long. Together, they provide replay protection.
- A direction identifier (DIRECTION) to prevent the so-called reflection attacks.
- The RRC signalling message content (MESSAGE).

Based on these input parameters the sender computes the 32-bit message authentication code for data integrity (MAC-I) using the integrity algorithm f9. The MAC-I is then appended to the RRC message when sent over the radio access link. The receiver computes the expected MAC-I (XMAC-I) on the message received in the same way as the sender computed MAC-I on the message sent and verifies the data integrity of the message by comparing it to the received MAC-I.

The integrity feature also provides data origin authentication, so that the receiver of an integrity-protected signalling message can corroborate the identity of the sender. This allows an operator not to run the full authentication and key agreement protocol every time a connection is established.

6.3.2.3 Partial integrity protection of user traffic

UMTS also has a mechanism that prevents the insertion or deletion, but not the modification of user traffic. This feature is meant to prevent certain bandwidth hijacking attacks while avoiding the cost of full-blown integrity protection mechanisms for user data.

The procedure works by allowing the RNC to monitor the sequence numbers for integrity protection and ciphering associated with each radio bearer. The RNC may send an integrity-protected signalling message to the ME. The message contains the most significant parts of the counter values, which reflect the amount of data sent and received on each active radio bearer. On reception the ME checks that the counter values agree with the values maintained by the ME. If there is a difference then this is indicated in an appropriate response.

6.3.3 *Ciphering in the access network*

Ciphering, when applied, is performed in the RNC and the ME. The ciphering function is performed either in the radio link control (RLC) sub-layer (for non-transparent RLC mode) or in the medium access control (MAC) sub-layer (for transparent RLC mode) [16].

Figure 6.3 illustrates the use of the ciphering algorithm f8 to encrypt plaintext by applying a key stream using a bit per bit binary addition of the plaintext and the key stream. The plaintext may be recovered by generating the same key stream using the same input parameters and applying a bit per bit binary addition with the ciphertext.

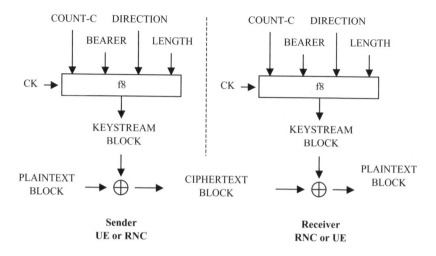

Figure 6.3   Ciphering of user and signalling traffic

The input parameters to the algorithm are:

- the cipher key CK, which is 128 bits long;
- the time dependent input COUNT-C of length 32 bits;
- the bearer identity BEARER;
- the direction of transmission DIRECTION; and
- the length of the required key stream LENGTH.

Based on these input parameters, the algorithm generates the output key stream block KEYSTREAM, which is used to encrypt the input plaintext block PLAINTEXT to produce the output ciphertext block CIPHERTEXT.

The input parameter LENGTH affects only the length of the KEYSTREAM BLOCK, not the actual bits in it.

6.3.4   Confidentiality and integrity algorithms

A common algorithm called KASUMI forms the basis for both the confidentiality algorithm f8 and the integrity algorithm f9, which are standardised for use in 3GPP systems. KASUMI is a block cipher [17] which is used in two different operating modes to construct f8 and f9 [18].

6.4   Network domain security

The term 'network domain security' in the 3GPP specifications covers security of the communication between network elements. In particular, the mobile station is not affected by network domain security. The two communicating network elements may both be in the same network administrated by a single mobile operator or they may belong to two different networks. The latter case, that is, internetwork communication, clearly requires standardised solutions, because otherwise each pair of two operators that are roaming partners would need to agree on a common solution. The intranetwork case also benefits from standardisation as many operators have network elements manufactured by several different vendors.

In the past, there have been no cryptographic security mechanisms available for internetwork communication. The security has been based on the fact that the so-called Signalling System Number 7 (SS7) network has been accessible only to a relatively small number of well-established institutions. It has been very difficult for an attacker to insert or manipulate SS7 messages. The situation is changing now for two reasons. First, the number of different operators and service providers that need to communicate with each other is increasing. Second, there is a trend to replace SS7 networks with Internet Protocol (IP) networks. The introduction of IP brings many benefits but it also means that a large number of hacking tools, some of which are available on the Internet, become applicable to the telecommunication networks. For instance, various denial of service attacks may be anticipated. For these reasons, the lack of cryptographic protection for internetwork signalling may increasingly become a security risk.

A major part of the 3GPP Release 99 specifications was devoted to the introduction of a completely new radio access technology while the core network part was an extension of the existing GSM specification set. This is the main reason why the protection mechanisms for core network signalling were not introduced in Release 99 but instead in later releases, starting from Release 4.

The mobile specific part of SS7 signalling is called the Mobile Application Part (MAP). In order to protect all communication in SS7 networks it is clearly not enough to protect only the MAP protocol. However, from the point of view of mobile communications, MAP is the essential part to be protected. For instance, the session keys for protecting the radio interface and authentication data are carried in MAP. On the other hand, the specification of a security protocol for SS7 would have been a big task unlikely to be completed in the required time frame. Mainly for these reasons, 3GPP has developed security mechanisms that are specific to MAP. The functional description of these mechanisms (stage 2) is given in the TS 33.200 [19] while the bit-level materialisation (stage 3) is described in the MAP specification itself TS 29.002 [20]. The whole feature is called MAPSEC and the first release in which it is included in 3GPP is Release 4. Note that the MAPSEC protocol protects MAP messages at the application layer. An overview of MAPSEC is given in Section 6.4.1.

Many different security mechanisms have been standardised by the Internet Engineering Task Force (IETF) for IP-based networks. Hence, there is no need to specify a completely new solution for 3GPP. On the other hand, it is still important to agree on how IETF protocols are used to protect IP-based communication in 3GPP networks. Specification TS 33.210 [21] is devoted to this task. It is included in the 3GPP Release 5 specification set. The main tool from the IETF used in 3GPP is the IPsec protocol suite [22].

Note that it is also specified in 3GPP how the MAP protocol can be run on top of IP. In this case, there are two alternative methods to protect MAP: either to use

MAPSEC or IPsec. The latter has the advantage that the protection covers also lower layer headers as it is done in the IP layer.

6.4.1   MAPSEC

The basic idea of MAPSEC can be described as follows. The plaintext MAP message is encrypted and the result is put into a 'container' in another MAP message. At the same time a cryptographic checksum, that is, a message authentication code covering the original message, is included in the new MAP message. To be able to use encryption and message authentication codes, keys are needed. MAPSEC has borrowed the notion of a security association (SA) from IPsec. The SA contains cryptographic keys but, in addition, it contains other relevant information such as key lifetimes and algorithm identifiers. Security associations of MAPSEC resemble IPsec SAs but the two are not identical.

3GPP Release 4 does not specify how SAs are to be exchanged between operators. In practice, this implies that the SAs are configured in the network elements manually. The plan at the time of writing is to later add automatic key management for MAPSEC (Figure 6.4) in the 3GPP specification set. The basic ingredient in MAPSEC automatic key management is a new element called a key administration centre (KAC). These KACs agree on SAs between themselves using the IETF Internet Key Exchange (IKE) protocol [23]. The KACs also distribute the SAs to the network elements. All elements in the same security domain, for example, elements in one operator's network, share the same SAs; they also share the policies about how to handle these SAs and incoming messages. The sharing of SAs is unavoidable as only networks, not individual network elements, can be addressed in MAP messages.

MAPSEC has three protection modes: no protection, integrity protection only and encryption with integrity protection. MAP messages in the last mode have the following structure: Security header || f6(Plaintext) || f7(Security Header || f6(Plaintext)), where f6 is the Advanced Encryption Standard (AES) algorithm in counter mode and f7 is AES in CBC-MAC mode. The security header contains information needed to

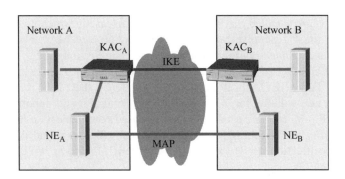

Figure 6.4   Automatic key management for MAPSEC

be able to process the message at the receiving end, such as the security parameters index, the sending network element identifier and the time variant parameter.

In MAPSEC only some of the MAP operations are protected, including the most critical operations, such as the authentication data transfer. This selective treatment of MAP operations is for performance reasons.

6.4.2 IPsec-based mechanisms

In the IPsec-based solution, all control plane IP communication towards external networks should go via a new element called a security gateway (SEG) (Figure 6.5). These gateways use the IKE protocol [23] to exchange IPsec SAs between themselves. An important conceptual distinction between a security gateway and a MAPSEC KAC is that the former uses the negotiated SAs while the latter distributes the negotiated SAs to other elements, which send and receive the actual MAPSEC messages. In 3GPP Release 5 the IKE is based on preshared secrets but in future releases support of a public key infrastructure (PKI) for key management is planned to be added.

A security gateway contains both an SA database and also a security policy database, which indicates how and when the SAs are used or have to be used. Naturally, the security gateway has to be physically secured. Typically, the security gateway could be combined with firewall functionality.

One obstacle in reaching full interoperability of IPsec is the great number of various options in the specifications [22]. In 3GPP, cutting down the number of options has solved this problem. The basic selections can be summarised as follows.

- Only Encapsulating Security Payload (ESP) [24] is used for protection of packets, while Authentication Header (AH) [25] is not used at all.
- ESP is always used in tunnel mode.
- 3DES is chosen as the mandatory encryption algorithm.
- HMAC with SHA-1 is the mandatory algorithm for integrity protection.
- IKE is used for key exchange in main mode phase 1 with preshared secrets.

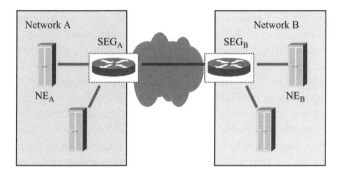

Figure 6.5   Network domain security for IP-based control messages

It should be noted, however, that operators might configure more options (e.g. transport mode) in their own networks. The specification TS 33.210 describes only the core part that guarantees interoperability between different security domains.

6.5   IP multimedia subsystem security

6.5.1   The use of SIP for multimedia session control in 3GPP

The IP multimedia subsystem (IMS) is a core network subsystem within UMTS. It is based on the use of the Session Initiation Protocol (SIP) [26] to initiate, terminate and modify multimedia sessions such as voice calls, video conferences, streaming and chat. SIP is specified by the Internet Engineering Task Force (IETF) [27]. IMS also uses the IETF Session Description Protocol (SDP) [28] to specify the session parameters and to negotiate the codecs to be used. SIP runs on top of different IP transport protocols such as the User Datagram Protocol (UDP) and the Transmission Control Protocol (TCP).

The IMS architecture is specified in TS 23.228 [29] and is built upon the UMTS packet domain. However, the architecture of the IMS is designed such that in future releases it may use mechanisms for IP connectivity other than those provided by the UMTS packet domain. This requirement is known as 'access network independence'. In order to meet this requirement the design relies to a large extent on IETF mechanisms. The 3GPP community has been very actively involved in work at the IETF to provide IETF standards that meet 3GPP requirements.

SIP itself is based on an IETF architecture that is very general and from which several trust models may be defined leading to hop-by-hop, end-to-middle and end-to-end security solutions. The IETF SIP working group has, therefore, defined several security mechanisms that can be applied to the different use cases of SIP. The mechanisms offer, for example, authentication, confidentiality and integrity of messages and replay protection. It should be stressed that SIP requests and responses cannot be fully encrypted or integrity protected on an end-to-end basis since parts of the messages by definition have to be available to proxies for routing purposes and for modification of messages.

At the lower transport and network layers, it is possible to use either Transport Layer Security (TLS) or IPsec to secure the whole SIP message. Although both UDP and TCP may be used in IMS, UDP is the default protocol and TCP is only used for large messages. As TLS can only be used on top of TCP, it has to be ruled out in the IMS. TLS can also be ruled out because it uses public key certificates.

At the application layer it is possible to use HTTP (HyperText Transfer Protocol) authentication and S/MIME (since SIP carries MIME bodies). S/MIME (Secure/Multipurpose Internet Mail Extensions) has the disadvantage that it is based on public key certificates, that is, it requires a public key infrastructure for the subscribers. It may also in certain cases generate very large messages (which has to be avoided over a wireless channel) and, therefore, in many instances it is recommended that TCP should be utilised. It should be mentioned that S/MIME cannot offer full protection of SIP since proxies need to have access to parts of the SIP messages. This also means that to some extent SIP proxies need to be trusted by the UAs.

A 3GPP IMS subscriber has one IP multimedia private identity (IMPI) and at least one IP multimedia public identity (IMPU). To participate in multimedia sessions, an IMS subscriber must register at least one IMPU with the IMS. The private identity is used only for authentication purposes.

There are four IMS entities relevant to the IMS security architecture.

UE: the user equipment (UE) contains the SIP user agent (UA) and the smartcard-based IMS subscriber identity module (ISIM), an application that contains the IMS security information. The ISIM can be a distinct application sharing no data and functions with the USIM, or it can share data and security functions with the USIM or it can be a reused USIM. There can only be one ISIM per IMPI.

P-CSCF: the proxy call session control function (P-CSCF) acts as an outbound SIP proxy. For the UA in the UE, it is the first contact point in the serving network. It forwards SIP requests towards the I-CSCF.

I-CSCF: the interrogating call session control function (I-CSCF) is the contact point in the home network and acts as an SIP proxy. It forwards SIP requests or responses towards an S-CSCF.

S-CSCF: the serving call session control function (S-CSCF) may behave as an SIP registrar, an SIP proxy server and an SIP UA. Before the UE can send an SIP INVITE message to set up a session it has first to register an IMPU at the S-CSCF. The registration of an IMPU is done by the UE by sending an SIP REGISTER message towards the home network.

Sessions are set up using INVITE messages. Figure 6.6 below describes a scenario where an INVITE is sent from one UE to another, both of which reside in a 3GPP

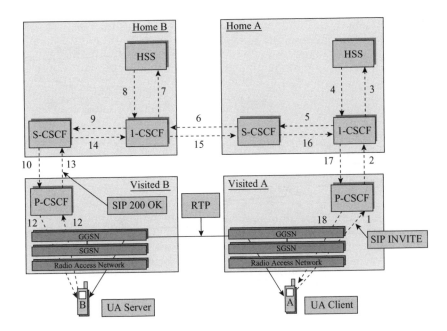

Figure 6.6   Session establishment in IMS

network. The INVITE from UE A in Home A first passes through a P-CSCF and then to an I-CSCF, which forwards the message to the home subscriber system (HSS), which looks up to which S-CSCF the user is registered. A similar process is done in Home B and the INVITE is finally terminated in UE B. The conversation may now start, for example, by using the IETF Realtime Transport Protocol (RTP) [30].

6.5.2 Security architecture for the IP multimedia subsystem

The 3GPP IMS security architecture is specified in TS 33.203 [31]. An IMS subscriber will have a private identity (IMPI), which is authenticated. All relevant subscriber data are stored in the HSS. During user registration, which will take place in the S-CSCF, the subscriber data are transferred from the HSS to the S-CSCF. Hence, upon request by a user the S-CSCF can match this request with the subscriber profile before access is granted, such that the home network can control access.

The registration process is shown in Figure 6.7. In SM1, the UE sends an unprotected REGISTER, which is forwarded towards the S-CSCF. The authentication and key agreement for IMS is based on the same mechanism as in UMTS, that is, the AKA protocol described in Section 6.3. The challenge is derived in the HSS when the UE REGISTERs for the first time. This challenge is forwarded by the S-CSCF towards the P-CSCF and the UE (SM4-SM6). The UE then checks that the challenge is authentic and sends the response back to the S-CSCF (SM7), who will authenticate the subscriber. This is somewhat different to the UMTS architecture where the

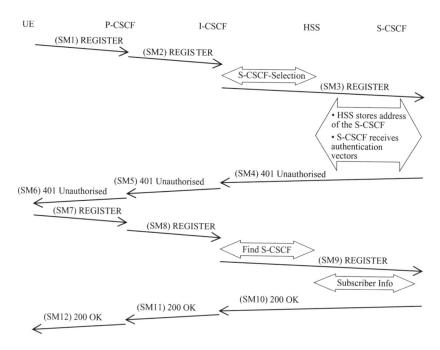

Figure 6.7 Registration process in IMS

authentication is delegated to the serving network. Hence, the trust in the P-CSCF is somewhat reduced. The S-CSCF in the home network can at any time require that the user is reauthenticated.

When the user has been successfully authenticated and a 200 OK has been received by the UE a security association (SA) is active between the UE and the P-CSCF for the protection of subsequent SIP messages between these two entities. The P-CSCF will obtain the integrity key for protecting SIP messages from the SM5 message sent by the S-CSCF towards the UE. Since the integrity key is passed from the S-CSCF to the P-CSCF this message has to be protected between the home network and the serving network. This is accomplished by using IPsec tunnels between the networks using the mechanisms specified in TS 33.210 [21] and described in Section 6.4.

Only integrity protection is used between the UE and the P-CSCF and this feature is important from a billing and charging perspective. No confidentiality protection is offered in IMS. Instead, an operator should use the confidentiality mechanism offered in UMTS at the link layer that is terminated in the RNC. For end-to-end security the IETF specifications SRTP [32] and MIKEY [33] can secure RTP and provide an appropriate key management mechanism.

The protocol, which is used for providing integrity protection of the SIP application control plane in the IMS, is IPsec ESP [24]. IPsec ESP requires security associations, which can be created using either IKE (Internet Key Exchange) [23] or 'manual keying'. An IKE implementation is quite complex and is, therefore, not feasible for mobile terminals. As a consequence, 'Manual Keying' is used in IMS. The term 'manual keying' is IPsec terminology and may be misleading here: in fact, key management is not manual, but automatically provided by the AKA protocol as described below. A local application programming interface (API, not specified in 3GPP) is used to transport the key and other information that is required by IPsec from the SIP layer to the IPsec layer.

6.5.3 Authentication using HTTP digest AKA

SIP is based on the HTTP framework and therefore SIP inherited the authentication framework of HTTP [34]. However, the IETF SIP working group has forbidden the use of HTTP Basic authentication because it is an insecure protocol due to the fact that the password is sent in clear. Hence, only HTTP Digest authentication is allowed within SIP. It should be noted that AKA is perfectly secure even though the parameter RES is sent in clear and hence the use of HTTP Basic together with 3GPP AKA would not cause any security weaknesses. However, 3GPP has aimed to be compliant with IETF specifications as far as possible and therefore HTTP Basic is not used.

The use of HTTP Digest together with 3GPP AKA [35] is described in the following. HTTP Digest is used for sending the authentication challenge from the S-CSCF towards the UE in a 401 Unauthorised response associated with the received REGISTER request. This message includes the WWW-Authenticate header, which transports, for example, the authentication challenge, the security mechanism identifier (which is called AKAv1-MD5) and the integrity key (IK) for the P-CSCF. Upon receiving the authentication challenge the UE first checks that the sequence number is in the correct range and that the message authentication code is correct. If these checks are successful the UE derives the authentication response RES based on the challenge and the long-term authentication key stored in the ISIM. The RES is treated as a password in the HTTP Digest framework and a response is calculated by using AKAv1-MD5 with the RES as the input. The S-CSCF has to make similar calculations based on the expected RES (XRES) before a check can be performed.

6.5.4   Integrity protection using IPsec ESP

IPsec Encapsulated Security Protocol (ESP) [24] is applied in IMS in transport mode between the UE and the P-CSCF. A pair of unidirectional IPsec security associations (SAs) between the UE and the P-CSCF is simultaneously established during authentication. The integrity key is the same for both security associations. In order to resist replay attacks the IPsec ESP anti-replay service is used. Furthermore, since the keys are the same in the two SAs, protection against reflection attacks is achieved by ensuring that the security parameters index (SPI) is different for each SA. It is specifically forbidden to use the NULL authentication algorithm since the SIP signalling originating from a user is the basis for charging and billing.

The authentication (i.e. integrity) algorithms that are used are either HMAC-MD5-96 [36] or HMAC-SHA-1-96 [37].

The SA is bound to the normal IPsec selectors, that is, source and destination IP addresses, and source and destination ports. The allowed transport protocols in IMS are UDP and TCP.

Not all messages can be integrity protected so the security architecture has to allow some unprotected messages to proceed (e.g. error messages). The P-CSCF shall accept unprotected messages on a port different from the port used for protected messages. A similar approach is applicable for the UE.

Since the keys generated during the HTTP Digest AKA procedures are used for integrity protecting the SIP messages with IPSec ESP, it means that the protection is tied with the authentication of the IMPI and hence the scheme is not vulnerable to protocol attacks like interleaving attacks.

The scheme for the agreement of Security Association (SA) makes it possible for a UE to receive INVITEs during a re-authentication and the set-up of a new Security Association. This is accomplished by allowing that INVITEs pass through the old SAs for a short period of time.

6.5.5   Security mode set-up and use of SIP security agreement

The mechanism for setting up SAs for IPsec ESP in the IMS is based on the use of three new SIP headers to negotiate different security mechanisms in a secure manner [38]. The mechanism defined for use within 3GPP is manually keyed IPsec without IKE, which is known as 'ipsec-man'. The negotiation procedure allows new mechanisms to be introduced in a way which is resistant to bidding down attacks. In SM1, in Figure 7, the UE sends the Integrity Algorithms it supports along with some other needed parameters such as the protected port. The P-CSCF stores the received parameters and based on the P-CSCF algorithm list ordered by priority the P-CSCF selects the algorithm with highest priority and which is supported by the UE. The P-CSCF then creates the SA, which is stored and activated. In SM6, the P-CSCF sends the algorithms list of the P-CSCF ordered in priority towards the UE along with other required parameters. Based on this list the UE selects the first algorithm, which is also shared by the UE and creates then the necessary SA. In SM7, the UE repeats the Algorithms List (and other parameters) received from the P-CSCF in SM6. The SA derived by the UE after receiving SM6 protects SM7. The P-CSCF shall upon receiving SM7 check that the included Algorithms List equals the list sent by the P-CSCF in SM6. This scheme will ensure that the UE and the P-CSCF can agree on an algorithm without making it possible for a man-in-the-middle to perform a bidding down attack.

6.6   Further developments in UMTS security

This chapter surveyed the major security features that are included in the first releases of the 3GPP standards.

Work on the next UMTS release has recently started. This will introduce new security features. Many of these features will be introduced to secure the new services that will be introduced, for example, presence services, push services and multicast/broadcast services.

Looking more into the future, mobile cellular systems will have to accommodate a variety of different radio access networks including short-range wireless technologies, connected to a common core network. On the user side the concept of a monolithic terminal, as we know it, is dissolving. Distributed terminal architectures are appearing whose components are interconnected by short-range radio links. These new developments represent a major challenge to the UMTS security architecture. A collaborative research project funded by the European Union and called SHAMAN (Security for Heterogeneous Access in Mobile Applications and Networks) [39] has recently tackled these issues.

A separate project has also identified research topics in the area of mobile communications as part of the European Union's 6th Framework Programme of sponsored collaborative research; the project is called PAMPAS (Pioneering Advanced Mobile Privacy and Security) [40].

References

1   http://www.itu-2000.org
2   RICHARDSON, K. W.: 'UMTS overview', *IEE Electronics & Communication Engineering Journal*, 2000, **12** (3), pp. 93–100
3   http://www.etsi.org
4   http://www.gsm.org
5   http://www.3gpp.org

6　http://www.3gpp2.org

7　GSM 03.20: 'Network related security features', http://www.3gpp.org

8　PIPER, F. and WALKER, M.: 'Cryptographic solutions for voice telephony and GSM', Proceedings of *COMPSEC'98* (Elsevier, Amsterdam, 1998)

9　WALKER, M. and WRIGHT, T.: 'Security aspects', in Hillebrand, F. (Ed.): 'GSM and UMTS: the creation of global mobile communication' (John Wiley & Sons, New York, 2002)

10　3GPP TS 35.205-35.208: 'Specification of the MILENAGE algorithm set; An example algorithm set for the 3GPP authentication and key generation functions f1, f1*, f2, f3, f4, f5 and f5*', http://www.3gpp.org

11　3GPP TS 33.102: 'Security architecture', http://www.3gpp.org

12　HORN, G. and HOWARD, P.: 'Review of third generation mobile system security architecture'. Proceedings of *ISSE'00*, Barcelona, 27–29 September 2000

13　GOLLMANN, D.: 'What do we mean by entity authentication', IEEE Symposium on *Security and privacy*, Oakland, CA, 6–8 May 1996

14　MENEZES, A., VAN OORSCHOT, P. and VANSTONE, S.: 'Handbook of applied cryptography' (CRC Press, Boca Raton, FL, 1997)

15　3GPP TS 25.301: 'Radio Interface Protocol Architecture' http://www.3gpp.org

16　3GPP TS 25.331: 'RRC Protocol Specification', http://www.3gpp.org

17　3GPP TS 35.202: 'Specification of confidentiality and integrity algorithms: KASUMI specifications', available under licence, http://www.3gpp.org

18　3GPP TS 35.201: 'Specification of confidentiality and integrity algorithms: f8 and f9 specifications', available under licence, http://www.3gpp.org

19　3GPP TS 33.200: 'Network domain security: MAP application layer security', http://www.3gpp.org

20　3GPP TS 29.002: 'Mobile Application Part (MAP) specification', http://www.3gpp.org

21　3GPP TS 33.210: 'Network domain security: IP network layer security', http://www.3gpp.org

22　IETF RFC 2401–2412 (1998): 'IPsec protocol suite', http://www.ietf.org

23　IETF RFC 2409 (1998): 'IKE: Internet Key Exchange', http://www.ietf.org

24　IETF RFC 2406 (1998): 'IP Encapsulating Security Payload (ESP)', http://www.ietf.org

25　IETF RFC 2402 (1998): 'IP Authentication Header', http://www.ietf.org

26　CAMARILLO, G.: 'SIP Demystified' (McGraw-Hill, London, 2002)

27　IETF RFC 3261 (2002): 'SIP: Session Initiation Protocol', http://www.ietf.org

28　IETF RFC 2327 (1998): 'SDP: Session description Protocol', http://www.ietf.org

29　3GPP TS 23.228: 'IP Multimedia Subsystem (IMS), Stage 2', http://www.3gpp.org

30　IETF RFC 1889 (1996): 'RTP: a transport protocol for real-time applications', http://www.ietf.org

31　3GPP TS 33.203: 'Access security for IP-based services', http://www.3gpp.org

32　IETF draft-ietf-avt-srtp-04 (2002): 'The Secure Real Time Transport Protocol', http://www.ietf.org

33 IETF draft-ietf-msec-mikey-01 (2002): 'MIKEY: Multimedia Internet KEYing', http://www.ietf.org

34 IETF RFC 2617 (1999): 'HTTP Authentication: Basic and Digest Access Authentication', http://www.ietf.org

35 IETF draft-ietf-sip-digest-aka-03 (2002): 'HTTP Digest Authentication Using AKA', http://www.ietf.org

36 IETF RFC 2403 (1998): 'The Use of HMAC-MD5-96 within ESP and AH', http://www.ietf.org

37 IETF RFC 2404 (1998): 'The Use of HMAC-SHA-1-96 within ESP and AH', http://www.ietf.org

38 IETF RFC 3329 (2003): 'Security mechanism agreement for SIP sessions', http://www.ietf.org

39 http://www.ist-shaman.org

40 http://www.pampas.eu.org

Chapter 7

Securing network access in future mobile systems[1]

Günther Horn, Valtteri Niemi, Kaisa Nyberg and Hannes Tschofenig

This chapter addresses security issues in future mobile communications systems, beyond the third generation currently being introduced in Europe under the name of Universal Mobile Telecommunications System (UMTS). A user in a future mobile communications system should be able to use services from anywhere in the system (global roaming), and when using these services, the particular access network technology should be transparent to the user. Finding a unique security solution that is largely independent of the potentially many different access technologies is a particular challenge that these future mobile systems pose. The security considerations presented in this article are based on the work of the EU-sponsored collaborative research project SHAMAN [1]. The results can be easily generalised to various types of future mobile systems. The concepts discussed in this chapter focus on the security features and mechanisms required to provide global IP connectivity and various forms of mobility to a globally roaming user in a future mobile system.

7.1 Introduction

We currently see the deployment of so-called third-generation (or '3G') mobile telecommunications systems. In Europe, the underlying technology is known as UMTS. But researchers and engineers are already working on the generation of mobile telecommunications systems 'beyond 3G'. While there are still many open questions concerning the specifics of mobile systems beyond 3G, a vision of such systems is taking shape, cf., for example, [2]: 'The concept of Ambient Intelligence provides a vision of the Information Society where the emphasis is on greater user-friendliness, more efficient services support, user-empowerment, and support for human interactions. People are surrounded by intelligent intuitive interfaces that are embedded in all kinds of objects and an environment that is capable of recognising and responding to the presence of different individuals in a seamless, unobtrusive and often invisible way.' Mobile systems beyond 3G will support us in every-day life in business, travel, entertainment, learning, health, commerce and other areas. A key requirement for such systems to work and be accepted by users is security. They must be secure against deliberate misuse, trustworthy and ensure the privacy of the parties involved. In this chapter, we present elements of a security architecture for mobile systems beyond 3G.

7.2   Outline of security architecture

Security is not meaningful as a stand-alone feature. While general security requirements from the stakeholders' (users', operators', regulators', etc.) points of view can be formulated without reference to a particular communications system, security features and mechanisms are only meaningful in the context of a system to which they are applied. This system need not be specified in detail, but its major technical characteristics need to be known.

The technology underpinning mobile systems beyond 3G is still mostly a subject of research, but certain key technological features are already emerging. There seems to be a common understanding that they will be characterised by the following features:

- the ubiquity of the IP-protocol, resulting in an all-IP-based core network and probably an extended use of IP up to the edge of the access network;
- the coexistence of heterogeneous radio access technologies including the GSM and UMTS radio access networks, Wireless local area networks, Bluetooth and broadband radio interfaces currently being specified;
- the transparency of the use of a particular access network technology for the user;
- the support for global roaming and a large variety of different forms of mobility.

There are various degrees of mobility that may be satisfied by different protocols and may also affect the security architecture. Mobile IP provides reachability under one global IP address. Nomadic mobility, which only requires that the user can connect to the network anywhere, does not require Mobile IP, but may still require a global authentication infrastructure. Another mobility requirement is that of seamless handover. For real-time services, seamless handover may imply that there is no time to restart a full authentication exchange with every movement and to establish a new security association with the new point of attachment. Therefore, there is a need for a transfer of the security context from the old to the new point of attachment. Another useful distinction is that between macro- and micromobility. Whereas macromobility enables global roaming through Mobile IP and/or a global authentication infrastructure, micromobility, in contrast, requires only mobility within a predefined local area. Micromobility may be used for optimal support of fast handover and minimisation of global signalling traffic.

Initiatives and research projects that have been working on mobile systems beyond 3G include the EU cluster beyond 3G [3], BRAIN [4] and MIND [5].

The basic components that need to be considered when designing a security architecture can be deduced from these activities. The resulting, very elementary reference architecture is depicted in Figure 7.1. It is briefly explained as follows: a mobile node MN connects to an access router AR (also called first hop router as it is the first router IP packets encounter on their way into the network). The MN may use different radio or fixed access networks to reach the AR, for example, the MN may use a Wireless LAN providing an access point AP or a UMTS radio access network including a radio network controller RNC. The key element of the security architecture is the well-known AAA (Authentication, Authorisation, Accounting) infrastructure, consisting of a local server AAAL in the visited network and a server AAAH in the home network. If Mobile IP is employed to support mobility then a Home Agent also will enter the picture.[2] If the decision on network access is based on the ability to pay online by, for example, an electronic purse or a credit card then a payment broker may be present in the architecture. Based on this very elementary reference architecture, it is already possible to start the work on the security mechanisms required for mobile systems beyond 3G.

There are many security issues relating to mobile systems beyond 3G. A distinction can be made between security for services (like web-browsing, e-commerce, etc.) and security for transport (e.g. IP connectivity, mobility management, Quality of Service, session control). The principle to separate transport from applications in

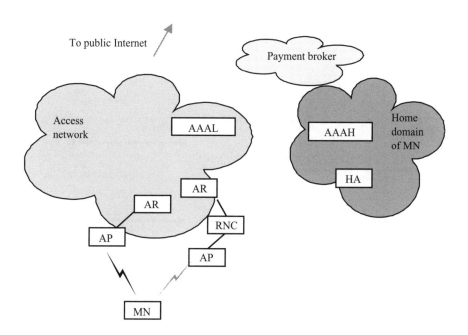

Figure 7.1 Security reference architecture

the design and implementation of the system seems to be generally accepted. The use of this principle will reduce the complexity of the overall system and allows for an independent evolution of transport networks and applications. This chapter focuses on security for the transport network; in particular, on the security features and mechanisms required to provide IP connectivity to a globally roaming user. A secondary focus is on security for Quality of Service procedures in a mobile system beyond 3G.

In order to cope with the uncertainties of the detailed architecture of future mobile systems and to further reduce the complexity of the work, a modular approach was chosen. Five basic functional building blocks have been identified that are likely to be required in any type of post-3G mobile system. The different building blocks were selected in such a way that a change in one building block would have a minimal affect on the other building blocks and that it should be possible to create the overall security architecture by combining these building blocks in a suitable way. The five main building blocks identified are:

• secure address configuration;
• authentication and security association establishment;
• IP layer security;
• link layer security;
• network domain security.

They correspond to the basic sequence of steps a mobile node performs when attaching to a (visited) access network. They are depicted in Figure 7.2. We will address them one by one in the remainder of this chapter.

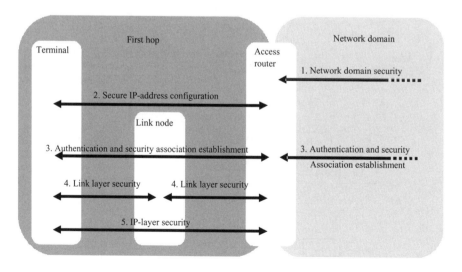

Figure 7.2   Basic building blocks

7.3 Secure address configuration

Since we consider the provisioning of IP connectivity to a roaming mobile node, the acquisition of an IP address is the first step a user has to perform. Currently, there exist two methods for an IPv6 node to automatically configure an IP address while attaching to a network: stateless and stateful address autoconfiguration. In the stateless address autoconfiguration, the IPv6 node forms an IP address by combining a prefix with topological significance and an interface identifier chosen by the node [6]. In the stateful address autoconfiguration, the IPv6 node obtains the IP address from a DHCP server along with other configuration information [7].

The mandatory duplicate address detection used with the stateless autoconfiguration may be exploited for denial-of-service attacks, and the common use of the link layer address to derive the interface identifier may betray the user's identity. The latter threat can be avoided when choosing the interface identifier at random. Furthermore, duplicate address detection may cause latency problems; in particular, during a handover because of the timeout values. But omitting duplicate address detection may be acceptable if the interface identifier is truly random, given its length of 64 bits, which makes collisions extremely unlikely.

A stateful autoconfiguration may also be susceptible to, among other things, denial-of-service attacks when an attacker tries to deplete the pool of IP addresses available at the DHCP server when no additional security measures are used. The security mechanism described in Reference 7 is not applicable in a mobility scenario as it assumes the pre-existence of a security association between the DHCP server and the mobile node. Stateful autoconfiguration seems to be generally less efficient as it adds another roundtrip between the mobile node and the DHCP server to the overall access procedure. Attempts to integrate the stateful address configuration with other parts of the access procedure (e.g. with the Mobile IP registration procedure) seem to go against the principle of modularity of the security architecture, and no such integrated solutions currently exist. The security problems associated with address configuration have been widely recognised and are currently being worked on in the SEND working group [8] of IETF (Internet Engineering Task Force).

7.4 Design alternatives for authentication and establishment of security association

7.4.1 Introduction

When a mobile node attaches to a visited network it has never been in contact with before, the network wants to be assured to get paid properly for the services granted to the mobile node and the mobile node wants to be assured that the access network does not tamper with any data the mobile node sends or receives via the access network. In accordance with an appropriate trust model, this requires some form of mutual authentication between the two, which can be carried out in a variety of ways.

The classical approach is the case where a subscription-like relation exists between the user of a mobile node and a home network, which also comprises the prior set-up of security information like keys, algorithms, etc. The cryptographic mechanisms used in the authentication procedure could be based on symmetric or asymmetric techniques. Whereas the former requires the involvement of the home network during the initial authentication process between the mobile node and the visited network, the latter allows for architectures that avoid an online involvement of the home network since the authentication may then be based on certificates. In this case, however, a public key infrastructure is required, which has to be involved online for the verification of the certificates.

Alternative approaches that do not rely on a subscription-like relation between the user of a mobile node and some home network may use credit cards, various forms of electronic money or other means of payment, leading to a quite different system and security architecture.

Besides the mere process of authentication, typically a session key and other parameters required for a security association such as ciphering algorithms are established between the mobile node and an entity in the access network during the initial access procedure. Such a security association could, for instance, be used to protect the first hop or create additional security associations between the mobile node and other entities in the access network (e.g. mobility agents, SIP servers, etc.). The establishment of the security association strongly depends on the specific architecture and could look quite different for public key and secret key based approaches. Note that there can also be several security associations for different purposes. For a detailed discussion of the relative merits of public key versus secret key based approaches see Reference 9.

Mutual authentication and key agreement can be established either in one step or in two steps. The two-step approach uses two separate authentication protocols, one for network authentication and the second for authentication of the MN. Typically, the network authentication protocol is executed first and it is used to create a protected tunnel through which the step-two authentication protocol is run. In particular, such a tunnel provides protection of user identity and other access negotiations at the initial access phase.

7.4.2   One-step approach

There is a large variety of protocols available to perform mutual authentication and key agreement in one step. In this section, we present an example protocol that has its origins in the UMTS authentication and key agreement protocol AKA [10], but is suitable for use in an Internet environment through the encapsulation of AKA in EAP, which has been specified in Reference 11. The Extensible Authentication Protocol (EAP) is an authentication protocol that supports multiple authentication mechanisms, the so-called EAP methods [12]. EAP typically runs directly over the link layer without requiring the IP. While EAP was originally developed for use with the Point-to-Point Protocol (PPP) [13] it is also in use with IEEE 802.1X [14]. EAP-AKA has recently become of interest in the

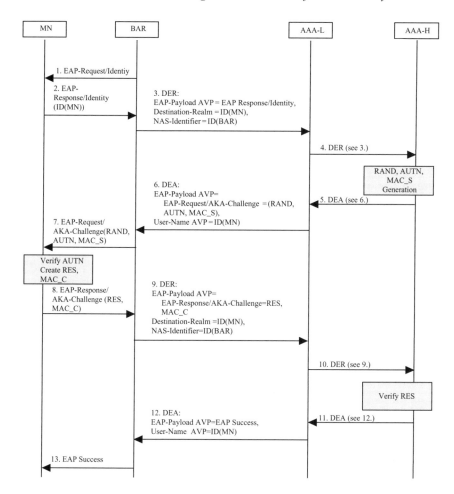

Figure 7.3   EAP-AKA authentication

context of interworking between 3G cellular and Wireless Access Networks (see Figure 7.3).

We assume that an AAAH server is able to obtain authentication information from a UMTS authentication centre, AuC, of the network to which the MN is subscribed. At the start of the protocol, the MN is prompted for its identity by means of the EAP-Request/Identity message (step 1). The identity is a Network Access Identifier (NAI) whereby the protocol permits the protection of the user identity against passive attacks through use of a pseudonym identifier.

The authentication of the MN to the home network is based on a challenge provided by the home network. Concerning the authentication of the home network to the MN, EAP-AKA does not use a challenge provided by the MN, but relies on sequence numbers (like the UMTS AKA protocol itself). Therefore, the home network starts the mutual authentication process by sending a challenge to the MN (RAND) as well as an authentication value (AUTN) calculated from a sequence number and the secret shared between MN and AuC (step 5). Both values, RAND and AUTN, which are generated by the AuC, are additionally protected by a message authentication code MAC_S, computed by the AAAH based on a key that is derived from the integrity and encryption keys (IK and CK). Both IK and CK are delivered to the AAAH in the authentication vector received from the AuC.

On receiving AUTN, RAND and MAC_S, the MN is able to authenticate the home network by verifying AUTN (based on the sequence number and secret shared with the AuC) and MAC_S (by deriving the required key from IK and CK. If the authentication succeeds the MN computes a response to the challenge RAND (RES) and a further message authentication code (MAC_C) and sends these values back to the AAAH. By verifying, in turn, the values RES and MAC_C the AAAH authenticates the MN and sends back an EAP success message.

7.4.3   Two-step approach

7.4.3.1   Initial authentication in two-step approach

The initial step in a two-step authentication protocol is to authenticate the network to the mobile client. Such protocols are typically implemented at the transport layer or network layer and will be discussed in this section. If the network authentication is implemented at the link layer only, the particular security requirements discussed in Sections 7.6 and 7.7.2.1 must be addressed.

7.4.3.1.1   Transport-layer protocols

The Protected EAP (PEAP) [15] claims to provide user anonymity and built-in support for key exchange. In Figure 7.4, the relationship between the EAP peer (client), NAS and backend authentication server are depicted. The EAP conversation 'passes through' the NAS on its way between the client and the backend authentication server. While the authentication conversation is between the EAP peer and backend authentication server, the NAS and backend authentication server need to have established trust for the conversation to proceed.

In PEAP, the conversation between the EAP peer and the backend server is encrypted and integrity protected within a TLS channel, and mutual authentication is required between the EAP peer and the backend server. This means that the NAS does not have knowledge of the TLS master secret derived between the EAP Peer and the backend authentication server, and cannot decrypt the PEAP conversation. In order to provide keying material for link layer cipher suites, however, the NAS does obtain the master session keys, which are derived from the TLS master secret via a key derivation function. The operation of PEAP consists of the following steps.

1. *Establish TLS connection.* The TLS record protocol provides a secure connection between the peer and the back-end authentication server.
2. *Authenticate TLS server.* The TLS handshake protocol is used for server authentication.

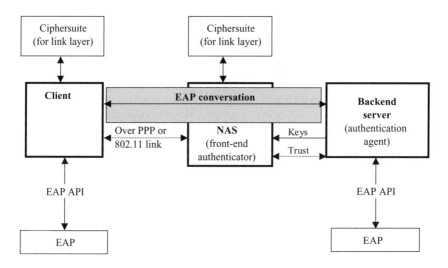

Figure 7.4   Relationship between EAP client, backend authentication server and NAS in PEAP

3. *Authenticate user.* The user of the peer authenticates by tunnelling another EAP mechanism inside the EAP-TLS connection. The back-end authentication server may have to contact another server to get the user authentication information validated.
4. *Generate session keys.* Using the TLS Pseudo-Random Function (PRF), the peer and the back-end server generate key materials for use between NAS and peer.
5. *Transport session keys.* The session key is transported from the server to the authenticator using, for example, Radius attributes and a secure connection.

In a subsequent (already expired) Internet draft, an application of PEAP to GSM authentication is proposed. In another proposal, EAP-TTLS [16], the architectural view is essentially the same as in PEAP. The purpose of EAP-TTLS is also to allow legacy password-based authentication protocols to be used against existing authentication databases, while protecting the security of these legacy protocols against eavesdropping, man-in-the-middle and other cryptographic attacks.

7.4.3.1.2   Network layer protocols

The PIC protocol [17] was a proposed method to bootstrap IPsec authentication via an Authentication Server (AS) and user authentication mechanisms (e.g. RADIUS). The client terminal communicates with the AS using a key exchange protocol where only the server is authenticated, and the derived keys are used to protect the user authentication. Once the user is authenticated, the client terminal obtains credentials from the AS that can be later used to authenticate the client in a standard IKE exchange with an IPsec-enabled security gateway. The PIC protocol proposed to secure this approach using simplified ISAKMP and IKE mechanisms. The protocol embeds EAP messages in ISAKMP payloads to support multiple forms of user authentication.

The work on PIC has now been stopped due to the development of the new version of Internet Key Exchange protocol, IKEv2 [18], which has inbuilt support for authentication using EAP methods based on RFC 2284 [12]. It is foreseen that the EAP methods to use with IKEv2 do not provide mutual authentication but are used for a client (initiator) authenticating to a server (responder). Such methods are also referred to as 'Legacy Authentication' mechanisms. The draft [18] mentions some of the most common EAP protocols expected to be used within IKEv2, but support for EAP methods is specified with the intent that new methods can be added in the future without updating the IKEv2 specification.

It is also important to note that the use of EAP is bound to the IKEv2 context in case the EAP method produces a shared key as a side-effect of authentication. More specifically, it is required that the shared key must be used by both parties to generate an AUTH payload (e.g. a MAC value) using the same procedure than what is specified for use with by IKEv2 in authentication based on shared secrets. Moreover, this shared key must not be used for any other purpose.

IKEv2 is a well-thought authentication protocol and offers many improvements over IKEv1. In the recent Internet draft [19], it is proposed to extend the applicability of IKEv2 beyond IP-based communication by defining an EAP type using IKEv2. If the IKEv2 uses EAP for client authentication the resulting protocol is a two-step authentication protocol that combines two EAP protocols in a very similar manner as PEAP does, but has some advantages. For example, EAP IKEv2 offers identity protection of the responder (authenticator) against passive attacks.

7.4.3.1.3   Summary

The purpose of PEAP, TTLS and PIC discussed above is to protect a possibly weak EAP protocol by encrypting the EAP protocol messages using a shared key between the network node and the user terminal (client). When establishing the tunnel, only the network is authenticated. This approach is depicted in Figure 7.5.

The secure legacy authentication in IKEv2 does not necessarily aim to protect a weak EAP method (although it can be used for this purpose also), but rather to combine two unilateral authentication methods that are most suitable to authenticate the parties when they are in an asymmetric position, and cannot be authenticated using a single method within one authentication domain.

In the scenario depicted in Figure 7.5, the client authentication protocol takes place between the client and the authentication server, to which the client is authenticated. The messages are transported within the protected tunnel between the client and the authentication agent. The secret key material needed for the protected tunnel is established as a result of the unilateral authentication protocol where the EAP authentication agent is authenticated to the client. In some cases, the authentication agent is the same network node as the authentication server. In some other cases, the authentication agent needs to communicate with a remote authentication server in the home network of the mobile client. It is assumed that this communication can be secured.

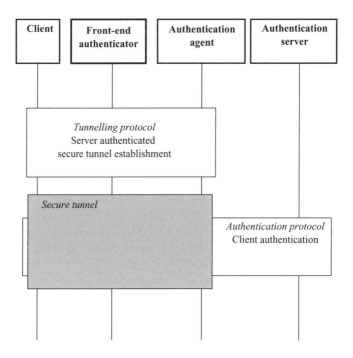

Figure 7.5 Initial network authentication and secure tunnel establishment

The result of the client authentication is communicated to the authentication agent. The authentication agent is also responsible for distributing session credentials and keys to the services and other network or link nodes. The network access point receives the session keys from the authentication agent through some fixed secured channel.

Such an approach also allows combining different authentication methods in situations like ad hoc or payment based network access. In such scenarios the network is typically authenticated using public key based methods, TLS or IKEv2. This requires the terminal to verify the authenticity of the public key of the network. Here different scenarios are possible. For example, the terminal can have a root key of the PKI system where the network belongs. In another, purely ad hoc scenario, the user obtains from some local service point, a secret password using which the terminal can verify the public key and the security context, which the network is using. For example, the MANA certificates described in this book ([20], Section 9.4.2.2) can be used for authentication of the network in ad hoc access. Server authentication is necessary in order to assure the user of the trustworthiness of the access network, considering that the user will send payment credentials to the access network later on in the access procedure. Moreover, a protected tunnel between the terminal and the network entity is set up, which is required in the payment-based access case in order to transmit confidential user and payment information.

7.4.3.2 Secondary authentication

The secondary protocol in a two-step authentication method is typically a legacy protocol, which is transported in EAP messages. It should be noted that a legacy protocol *per definitionem* originates from a different authentication environment, where it is used in a different manner. For example, plain SIM-based authentication is used for authentication in a GPRS network, while for WLAN access SIM-based authentication is transported in EAP messages.

The existing two-step (compound) authentication protocols have been specified to support any EAP type. A typical example would be the EAP/AKA protocol depicted in Figure 7.3 as the client authentication in the 802.1X framework. Then, the Home Subscriber Server plays the role of the authentication server in the UMTS system. Other typical, although completely different, examples are the simple password-based legacy EAP types: MD5-Challenge, One-Time Password (OTP) and Generic Token Card. The main difference is that EAP/AKA produces a strong shared secret between the parties, while such password-based client authentication methods do not provide a strong shared secret.

Also, several variants of client authentication exist in ad hoc access. As is explained in Reference 21, client access control is embedded in real time accounting and payment procedures. Hence, the client authentication may vary from a first check of payment credentials to virtually no client authentication at all. An approach using EAP offers sufficient flexibility to accommodate different needs for user authentication in ad hoc scenarios.

7.4.3.3 Man-in-the-middle attacks

The original drafts of PEAP [15] (up to version 0.5), TTLS [16], PIC [17], and the secure legacy authentication for IKEv2 [22] all shared the same security problem, which was discovered in Reference 23. While claiming support to all EAP types, it turned out that the security of some EAP methods was severely weakened if put in such a tunnelling framework. The problem is due to the fact that the security of the session credentials and keys depends solely on the same key material that is used to protect the tunnel between the authentication agent and the client. Hence, only the holder of the keys on the network side is authenticated, while the holder of the keys on the client side remains unauthenticated. Client authentication in the tunnel establishment is mentioned as an option in PEAP and EAP-TTLS, but is not mandated. PIC did not even offer this option.

The flaw opens up a backdoor, which allows a man-in-the-middle (MitM) unrightful access to resources that are granted to a legitimate and successfully authenticated client. The attack can be initiated if it is possible to run the legacy EAP protocol without the specified protected tunnel. A concrete example of the MitM attack is given in Figure 7.6. The protocol is a two-step protocol PEAP with EAP/AKA as the client authentication protocol.

The attack proceeds as follows.

1. MitM waits for a legitimate device to enter an unprotected client authentication protocol and captures the initial message sent by the legitimate client.

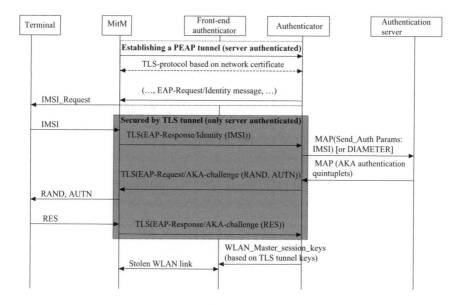

Figure 7.6   Man-in-the-middle in PEAP

2. MitM initiates a protected authentication protocol with the authentication agent.
3. After the protected tunnel is set up between the MitM and the authentication agent, the MitM starts forwarding the legitimate client's EAP messages through the tunnel.
4. MitM unwraps (removes the tunnel protection from) the EAP messages received through the tunnel and forwards them to the legitimate client.
5. After the client authentication ended successfully, the MitM derives the session keys from the same keys it is using for the protected tunnel.

A closer analysis reveals the following reasons for the failure. First, the client authentication protocol is a legacy protocol (EAP/SIM, EAP/AKA) that is typically used also without TLS tunnelling, and also independently without any tunnelling. The MitM can set up a legacy authentication setting, for example, a false cellular base station in GSM, to request a victim's identity (e.g. IMSI) and, subsequently, the required authentication data (e.g. RES). The second reason for the failure is that the session keys are derived from the Master Key generated using the tunnel protocol (the same key as used to create the tunnel). Keys derived in the EAP protocol (e.g. UMTS AKA Master Keys) are not used.

It is concluded in Reference 23 that the tunnelled authentication protocols as originally proposed are secure only if the client authenticates the network properly, and if one of the following conditions is satisfied.

1. Both ends of the tunnel are authenticated, that is, also the client is authenticated to the authentication agent.

2. The client authentication key is specified to be used only when the client
    authentication method is run with the specified two-step authentication protocol.

These conditions are in many practical cases either unrealistic or pose an undesirable
hurdle. The first condition is not true in the typical situation. The second condition does not hold for the common application areas such as integrated terminals
(e.g. GPRS/WLAN), and situations where the same general-purpose authentication
token is used with separate single-purpose terminals (e.g. WLAN, GPRS). Binding
the preliminary and the secondary authentications together in a secure manner can
solve the problem. Some examples of protocol binding methods are described in
Section 7.4.3.4.

Such a binding is not possible to achieve if the secondary authentication protocol
does not provide a strong shared secret, in which case the second condition remains
valid.

7.4.3.4 Binding methods

In this section, two basic approaches are presented for corroboration of the authenticity
of the entity that holds the client's copy of the session keys and credentials. Such
methods were first proposed in Reference 23. Analysis of the MitM problem and
possible solutions are also discussed in Reference 24. Note that a binding method is
been specified in the IKEv2 [18] for authentication of the IKE security association in
case an EAP method is used that produces a strong shared secret.

Recall that the MitM attack becomes possible if it is possible to run the legacy
client protocol in a different context outside the specific tunnel. To prevent this attack
it is sufficient to bind the shared secret produced by the EAP method to the tunnel
context of the preliminary authentication method. There the following approaches
can be taken [23].

1. A shared secret session key between the client and a network entity is derived
    that depends on both first- and second-step authentication protocols in a secure
    manner.
2. An additional step is added to the protocol to verify that the same client and the
    same network entity has taken part in both authentication protocols. This can be
    implemented using a MAC, which depends on both authentication protocols.

For example, the binding method specified for IKEv2 using EAP follows essentially
the second approach. The tunnel information consists of the shared Diffie–Hellman
secret g^{ir} and the context information, which is the ASCII coded string of the text 'Key
Pad for IKEv2'. The MAC value is computed with the context information as input
using the shared secret derived from the EAP method as the key for the MAC. Then,
the shared session keys (various SK values) are derived from the Diffie–Hellman
value g^{ir}.

7.4.4 Secure initial access and access control

This section discusses the security aspects of the initial communication over the first
hop between the terminal and the first packet router AR. Within the generic network

access model depicted in Figure 7.1, a wireless link connects the mobile terminal to an access point AP, which is a link node connected to the AR via a fixed net link.

The security requirements for the initial access are twofold and partially conflicting each other. First, there is the basic requirement from the network side of giving only minimum amount of resources to an unauthenticated, potentially unauthorised mobile node. The objective of access authorisation is in correct accounting but also in prevention of denial of service attacks.

The second set of requirements comes from the side of the mobile node. Protection of anonymity and location privacy has become a standard requirement, for example, in the case where the purpose is to provide access over the Internet to a corporate network. Also, the threat of false access networks, or false base stations, as it became known in the context of GSM networks, requires preventive countermeasures.

7.4.4.1 Phases of initial access

The initial access of the terminal to the network service has the following phases, where security procedures can be implemented.

1. Link layer access to the underlying wireless service (e.g. GSM, UTRAN, HIPERLAN/2, IEEE 802.11) and establishment of a link layer connection to the AP.
2. Establishing a network layer connection to the AR and access to AAA infrastructure to carry out AAA procedures [25, Section 3].
3. Set-up of network layer security association between the terminal and AR (or ANP) that ensures integrity and confidentiality of user data on the first hop. The security context for this connection is derived from authentication of the terminal or of the user. The authentication of the terminal or user may involve communication with the home authentication server or a payment broker (ad hoc access).

Security measures may not be required in phases 1 or 2 of the initial access, depending on the security requirements and the deployment scenario. It they are required they may be implemented in phase 1 or 2, or both. The security context for protecting link or network layer communication during phase 1 or 2 of the initial access must be available without online communication between the terminal and the home authentication server.

In two-step authentication (Section 7.4.3), phases 1 and 2 are separated from phase 3. Some one-step authentication methods may provide security functionality also at phase 1 or 2, for example, to protect anonymity of the mobile terminal.

7.4.4.2 Protecting access privacy

Often it is necessary to protect the identity of the MN in initial access. In addition to the identity also some other sensitive information may be necessary to send at an early stage in the connection set-up, and should be protected. The protection may be implemented on the link or network layer.

To implement anonymity protection the mobile node must share a secret key between a link or network node. Often distinction is made between passive and active attacks against access privacy. The reason for this is that passive wiretapping attacks can be prevented without any previous security association using a secret key that is derived from a key exchange protocol based on asymmetric (public key) techniques. A typical example is an anonymous Diffie–Hellman key exchange protocol that is included as an option in TLS. If the identity of MN is required to be protected against active attacks, such as MitM, then it is essential that the MN authenticates the link or network node and the secret key is derived from this authentication.

At this phase (phase 1 or 2) of the initial access it is not possible to have online communication with any back-end authentication server. Hence, the authentication of the network or link node, and hence protection against active attacks must be based on some previously established authentication information. For example, if the MN has a valid TLS certificate of the network, or if it is able to validate it, then anonymity protection can be performed using TLS security mechanisms and a TLS authenticated secret key. Also locally distributed authentication information such as the MANA certificates [20, Section 9.4.1.2] can be used.

If access authentication is based solely on a shared secret key between the MN and the authentication server, then the anonymity of the MN cannot be protected at the very first time the MN connects to the network. Later, however, protection against passive attacks is possible in various forms of pseudonyms or similar methods. An active attack by simulating the initial access where the MN must send its true identity still remains possible. An example of such an attack is the 'IMSI catcher' in a GSM network.

Anonymity protection of the MN can be extended all the way to the home network. More typically, anonymity protection terminates at some network or link node, where true identities are needed. In any case, protection of user identity may open the network to a denial of service attack performed by sending garbage data in place of encrypted identities.

7.4.4.3   Link layer access control

Unauthorised access to IEEE 802 LAN infrastructure is controlled by IEEE 802.1X port-based access control [14]. IEEE 802.1X distinguishes between controlled and uncontrolled ports. It hereby provides a means of authenticating and authorising devices attached to an IEEE 802 LAN port that has point-to-point connection characteristics, and of preventing access to that port in cases where the authentication and authorisation process fails. It has become a standard authentication framework for wireless LANs and is being adopted by other wireless link layer technologies.

Access to the controlled port is controlled by an authenticator, which is the network entity that wishes to enforce authentication in the link layer. Within the generic access model of Figure 7.1, this functionality is placed in the first link node, AP. The port that wishes to access the services offered by the authenticator's system adopts the role of the supplicant. The MN takes this role.

The authenticator and the authentication server can be co-located, thus eliminating the need for communication with an external server. Similarly, a port can adopt the role of the supplicant in some authentication exchanges, and the role of the authenticator in others.

However, when aiming at a security solution for which the particular access technology is transparent then the use of WLAN-specific security measures such as IEEE 802.1X may not be the right choice unless it is adopted by other link layer access technologies.

7.4.4.4   Network layer access control

Similar control mechanisms as specified by the 802.1X framework can also be implemented on the network layer. In Reference 25 this idea is realised by assigning two functions to an access router: first, a packet filter that looks at each packet and lets pass only packets of senders that were granted the necessary authorisations before and, second, an attendant that relays packets requesting authentication to a AAA server in the local access network. Only after a successful authentication and authorisation the packet filter will update the filter rules to allow the relaying of packets of the newly authorised sender, see Figure 7.7.

The policies are configured in a user profile database. Depending on the policies, also higher layer control functions may be needed to restrict access to specific services or applications.

Source Spoofing (e.g. spoofing the IP address of the MN) can be a threat to access security, even after authentication. An attacker might hijack the IP address of an authenticated client. Enforcing an IPsec security association between MN and

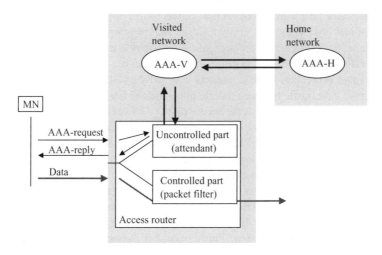

Figure 7.7   Implementing secure initial access according to Reference 25

AR can prevent this threat. Other threats applicable in this context are described Reference 26.

7.4.5　*Session key derivation and key transport*

Protocols for authentication and key agreement/key transport allow a master key (MSK) to be established as part of the protocol execution. For integrity and/or confidentiality protection of data traffic or signalling messages, a security association is required which includes cipher suites, integrity algorithms and various parameters in addition to a number of session keys (integrity, encryption session keys, IVs, etc.).

Session keys are established as part of the initial authentication and key agreement procedure. In case of subscription-based environments, the security association between the MN and the AAAH is typically based on symmetric cryptography. For alternative means of access, asymmetric cryptography is used between the mobile node and the local AAA server (AAAL). In case of network layer protection an IPsec security association has to be established with the first-hop router (access router). Issues related to the establishment of an IPsec SA are elaborated in Section 7.4.6. If a link layer security association has to be established then the necessary parameters have to be delivered to the link layer device (access point) closest to the mobile node. An architectural overview is given in this section.

As a result of the authentication and key exchange process a master session key is established between two EAP end points, whereby the two end points typically are either the MN and the AAAH (in case of traditional subscription-based environments) or the MN and the AAAL (in case of alternative means of access scenarios). However, the goal of the network access authentication procedure is to allow access to network resources based on the successful authentication and authorisation. To prevent network access by unauthorised nodes some security mechanisms have to be enabled. Packet filters are based on the source address of the mobile node to filter packets of unauthorised nodes. However, a non-cryptographic protection cannot prevent certain attacks such as address spoofing. To provide ironclad security protection against unauthorised access data origin authentication has to be provided. To enable cryptographic protection of data packets a security association is required for the protection of data traffic between mobile node and the first device in the visited network. At the mobile node and at the access point (AP) some functionality is required to derive link layer keys from a master session key and an API is necessary for the communication between the different protocol stacks and software components. The term master session key is taken from Reference 27. Since the master session key is not available at access point, some key transport protocol has to be used.

The master session key can be securely and reliably delivered to the AP using the AAA infrastructure. Protection of the master session key is provided by CMS as defined in Reference 28 or by IPsec/TLS in a hop-by-hop fashion. CMS protection is clearly the preferred approach. Furthermore, it is necessary to ensure that the distributed master key is fresh and unique as demanded in Reference 29. Figure 7.8 shows the protocol interaction.

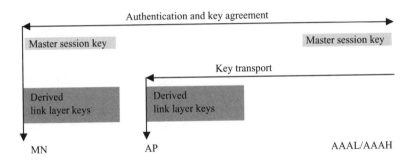

Figure 7.8   Derivation of link layer keys from the master key

The master session key is, according to Figure 7.8, established between MN and the AAAL/AAAH. Then the master key is securely transported to the AP. In the AP, the master session key is taken to the corresponding software module, where the appropriate link layer keys are derived. The same key derivation functions are executed at the MN.

Which link layer security associations have to be established for a particular link layer technology is described in Section 7.6.

7.4.6   IPsec security association negotiation protocol

7.4.6.1   Introduction

In most mobile environments, devices must use an authentication and key agreement protocol (e.g. AAA) to authenticate to the network to transmit and receive data and to subsequently access services within the network. EAP (together with an AAA protocol) is a protocol that enables the mobile node and a node in the network (e.g. the first-hop router) to establish a master session key. A derived session key might be used to secure data traffic over the wireless link using IPsec. The AAA protocol distributes a unique and fresh session key but unfortunately not enough information to establish an IPsec security association (SA) immediately after the initial network access authentication protocol execution. Seamless mobility with its performance constraints requires efficient means to establish an IPsec SA.

Currently, IKE with pre-shared secrets authentication is a standard IETF protocol, which is widely available and, therefore, the most obvious choice for establishing an IPsec SA after a finished network access authentication protocol run. Once IKEv2 is finished it can also be used to establish IPsec SAs since a pre-shared secret key authentication mode is also supported. A session key derived from the master session key (or even subsequently derived keys) can thereby serve as the pre-shared secret for IKE. This protocol interaction is, however, inefficient with regard to the number of roundtrips, bandwidth and computational resources. IKE implementations are large in size and tend to be complex. Low-performance devices might, in some cases, not even be possible to support a full-size implementation of IKE. IKEv2, which is currently under development at the IETF, is less complex than IKE, but its implementation may still pose a heavy burden on some small-scale mobile devices. Hence there seems to be a need for an efficient IPsec Security Association Negotiation protocol that takes a previous authentication and key establishment protocol (e.g. PANA [30]) into consideration.

Section 4.6.2 briefly describes existing IPsec SA establishment protocols. It will be explained why they are not perfectly suitable for network access authentication environments. Subsequently, requirements for an efficient IPsec SA negotiation protocol are listed. A new IPsec SA Negotiation protocol cannot be described in this chapter. However, based on the description, on the listed requirements and on the scenarios the reader will be able to see the basic idea. For a full specification, the reader is referred to Reference 1.

7.4.6.2   Existing IPsec SA establishment protocols

7.4.6.2.1   *The Internet key exchange protocol*

IKE provides authentication and key agreement during IKE phase I whereas IPsec SAs are negotiated in IKE Phase II (Quick Mode). IKE in main or aggressive mode requires a large number of messages and compulsorily executes Diffie–Hellman to provide perfect forward secrecy. IKE phase II requires two messages and offers an optional Diffie–Hellman exchange. In addition, IKE is known to be difficult to implement in terms of size and complexity. Furthermore, it has been argued that the negotiation of algorithms (SA payloads) suffers from an exponential explosion [31].

A number of authentication modes are offered in IKE Phase I: authentication with pre-shared secrets, authentication with public key encryption, a revised mode of public key encryption, a signature-based authentication mode and some other authentication modes that have been submitted as individual IETF draft contributions. Each of them has different security properties and different computational requirements. Based on a fresh and unique session key of the initial authentication exchange the pre-shared secret key authentication mechanism seems to be best suited to establish an IPsec SA.

7.4.6.2.2   *Kerberised Internet negotiation of keys (KINK)*

KINK [32] allows establishing an IPsec SA whereby authentication and key distribution is provided by Kerberos. From an abstract point of view KINK changes IKE Phase I with the simpler and more efficient Kerberos authentication and key distribution mechanism. The payloads of the Quick Mode exchange are used to negotiate the parameters and to establish an IPsec security association. KINK thereby reduces the number of message exchanges to two (in the optimum case) or three. Additionally, new features like dead-peer detection were added to the protocol. Similar to IKE, KINK is a standalone protocol. It heavily relies on Kerberos, which restricts its usage to certain environments. Finally, KINK heavily relies on the IKE payloads and, therefore, inherits some of the limitations of IKE (e.g. the inflexibility of the ID payload to provide more fine-gained traffic selectors).

KINK, however, provides some advantages: It is efficient in terms of exchanged protocol messages and the computational requirements. KINK can be run without executing asymmetric crypto-operations. It should be noted that the message flows required to request the ticket granting ticket and the service ticket adds some additional message roundtrips and their exact number heavily depends on the network architecture.

7.4.6.2.3   IKEv2

With IKEv2 the IETF IPsec Working Group is creating an IKE successor. To provide more information on the design rationale, Radia Perlman published a separate design document [33]. Secure legacy authentication (SLA) is supported by IKEv2. Based on SLA it is possible to execute EAP exchanges secured with an IKE-SA based on a unilateral authentication (responder to imitator). The EAP exchange will then provide the client authentication.

IKEv2 contains a number of attractive features, such as:

- optional DoS protection,
- reduced number of roundtrips (by piggy-backing IPsec SA negotiation on top of an IKE exchange),
- more powerful Traffic Selector negotiation,
- NAT handling,
- payloads to configure a client in a remote access scenario,
- reduced number of authentication mechanisms (pre-shared secret and public key based authentication only).

IKEv2 provides a major improvement to IKE with regard to complexity and functionality. However, for the special case of network access authentication still a number of improvements are possible (e.g. to completely avoid public key computations).

To summarise, the following properties of the above-described protocols can be identified.

- The listed protocols do not allow a separation between authentication and key exchange and IPsec SA Negotiation into separate protocols. In order to establish an IPsec SA the entire protocol has to be executed.
- None of the protocols provides third-party IPsec SA Negotiation. Both IKE and IKEv2 protocols are two-party protocols where only these two parties interact directly.
- None of the above protocols provides in-band IPsec SA negotiation. The term in-band refers to an IPsec SA negotiation protocol that is embedded into another protocol. IKE, IKEv2 and KINK are standalone protocols.
- Restrictions for Traffic Selector negotiation exist. Furthermore, it is not possible to update or delete security association and security policy entries in all protocols in a sufficiently reliable and flexible manner.

7.4.6.3   Requirements

The main requirements for an IPsec SA negotiation protocol as used in the area of network access authentication are: session key freshness and uniqueness and mutual authentication. Once such an initial authentication and key exchange procedure is finished and these requirements are fulfilled it is possible to start with an IPsec SA negotiation.

The requirements can be described in more detail for a particular application. The following description elaborates the AAA case in more detail, which is strongly influenced by fact that the AAA exchange already provided authentication and fresh keys to the involved parties. Furthermore, there is a need for efficiency in terms of bandwidth and performance consumption, features that are also required by, for example, IKEv2 but to a lower degree because of the different assumptions.

The discussed protocol assumes that a previous authentication and key exchange protocol run is completed and that the following properties are satisfied.

- A protocol interaction with mutual authentication was finished. Mutual authentication, therefore, refers mainly to the EAP end points. Additionally, authentication takes place between the entities participating in the AAA exchange.
- The distributed session key is fresh and unique.

Additional requirements have to be fulfilled by the usage of CMS encrypted key transport. These requirements are also important but of minor interest for our investigation.

For network access in a mobility environment the following properties are of interest.

- The mobile user is authenticated to his home network.
- The home network is authenticated to the user.
- Based on the trust relationship and on the protocol exchange between the visited and the home network the visited network is guaranteed that the established financial settlement allows proper charging of the resources consumed by the mobile user.

Note that the properties described above address a traditional subscriber-based authentication procedure between the mobile node and his home network. Other non-traditional authentication procedures may also be used but they are not directly addressed in this section. Requirements for such a new IPsec SA Negotiation protocol would, therefore, be the following.

- The protocol MUST be efficient.
- The protocol MUST be extensible.
- The protocol SHOULD allow both parties to initiate the protocol run.
- The protocol MUST be independent of the underlying transport protocol.
- The IPsec SA negotiation protocol MUST be integrity protected.
- The IPsec SA negotiation protocol MUST be replay protected.
- The IPsec SA negotiation protocol SHOULD provide proper error handling.
- The IPsec SA negotiation protocol SHOULD NOT introduce new denial of service attacks.
- The protocol SHOULD NOT transmit user identities in the clear.
- The protocol MAY use a Diffie–Hellman exchange.
- The protocol MUST specify how to derive the keys required for the IPsec SAs.
- The protocol SHOULD exchange information about traffic selectors.

- The protocol MAY support the establishment of multiple IPsec SAs within a single message exchange.
- The protocol SHOULD provide lifetime parameters for an SA.
- The protocol SHOULD provide parameters to differentiate IPsec AH/ESP SAs.
- The protocol MAY provide negotiation of IPComp SAs.
- The protocol SHOULD provide a functionality to delete an IPsec SA.
- The protocol SHOULD support parameters to differentiate Tunnel and Transport Mode.

Based on these requirements it is possible to envision such an IPsec SA Negotiation protocol, which is similar to IKE Phase II (Quick Mode) from the payload structure with a simple messaging structure (e.g. create, delete, update). To provide flexibility for the security policy database either the Traffic Selector structure of IKEv2 or the structure proposed in Reference 34 should be used. To illustrate its usefulness we will describe two message flows where such a protocol can be utilised. Other scenarios are third-party IPsec SA establishment (e.g. in the area of Mobile IP to establish an IPsec SA between the MN and the HA, etc.).

7.4.6.3.1 First-hop IPsec SA establishment

The first scenario combines AAA for IPv6 [25] ideas with an IPsec SA Negotiation protocol to establish an IPsec SA between the MN and the AR in a single inter-domain roundtrip. This approach represents an optimal solution for mobility in terms of bandwidth consumption, latency and number of roundtrips. The main idea of the described procedure is to use the AAA infrastructure and their trust relationship to establish IPsec SA in-band between two nodes. IKE does not allow other parties to negotiate security associations on behalf of someone.

In Figure 7.9, the mobile node receives a Router Advertisement message (possibly with a flag indicating the support for the described fast IPsec SA set-up), which triggers the depicted message flow. Using message (2) the mobile node adds an IPsec SA Negotiation payload. The IPsec SA Negotiation payload consists of a header payload structure (HDR), a security association proposal of the mobile node (SAii) and a Traffic Selector (TS) payload if required. By omitting the TS payload an IPsec SA is established with an IPsec Security Policy Database (SPD) entry, which protects the entire data traffic inside a single security association. For QoS data traffic it is sometimes desired to have several different IPsec SA: one for each QoS class. In Figure 7.9 protocol payloads of the IPsec SA Negotiation protocol are, for editorial reasons, formatted in bold to highlight the important parts.

To prevent downgrading attacks the included payloads and attributes are integrity protected with the long-term security association between the mobile node and its AAAH server. This mechanism reuses the security properties of the AAA for IPv6 security scheme. For replay protection the AAA for IPv6 specific challenge/response scheme is used in this case. The mobile node includes a local challenge and a home challenge value. The home challenge value is a random number received with the previous AAA exchange. The local challenge provides a freshness guarantee for the mobile node.

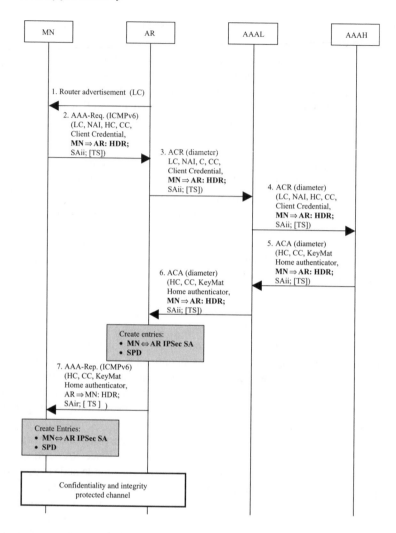

Figure 7.9   IPv6 network access using AAA

The inclusion of the IPsec SA parameters and their protection is no deviation from the ideas of IPv6 Network Access using AAA protocols. The only modification to the existing mechanism is the inclusion of the IPsec SA payload and to include these payloads as input to the keyed message digest computation. The BAR is not able to cryptographically verify message (2) according to the AAA exchange procedure. Hence, the AR is also unable to immediately verify the received IPsec SA Negotiation payloads, since they are protected with the security association shared with the AAAH server. They have to be forwarded to the AAAL server in message (3) and to the AAAH server (4). In the above scenario, we assume that the AR later selects the required IPsec SA parameters itself. It would also be possible to allow the local AAA server to select the parameters for the AR and to allow the home AAA to assist in the IPsec SA selection process.

Furthermore, we assume that the AAA protocol distributes keys. These keys are then used for IPsec SA establishment. When the AR receives message (6) it is able to verify the received parameters and to determine which IPsec SA parameters and Traffic Selector attributes to select. If the BAR was able to select an IPsec SA then the AR is able to create the corresponding security association database (SADB) and the security policy database (SPD) entries. Message (7) must not be protected by the established IPsec SA since the mobile node has not installed its SAs and has not created SPD entries before receiving message (7). If no Traffic Selector is specified by the mobile node then protection for the entire data traffic between the mobile node and the BAR is assumed (usually traffic which is not addressed to the AR itself in transport mode). In some cases, a more fine grain specification of the Traffic Selector might be desired, which is then indicated in the TS attribute.

Note that AAA-Req./AAA-Rep. messages as used in messages (2) and (7) must not experience IPsec protection even after the establishment of the IPsec security association since they represent the control channel for the IPsec SA set-up. In the above cases (AAA for IPv6), ICMP is used for the communication between the mobile node and the AR. To use the ICMP protocol for message transport might have been chosen because of the small message size. Other transport protocols might, however, be more applicable. The fact that the IPsec SA negotiation message payload is small and the short distance between these two nodes does not seem to cause any particular problems. For roaming within the administrative domain the established IPsec SA is transferred from one AR to another using an IPsec Context Transfer protocol.

To provide a transport mechanism for information exchanged between the mobile node and the AAA attendant the PANA working group was founded. The message flow in Figure 7.9 uses ICMP for this purpose. A more abstract message flow is shown in Figure 7.10. It uses an EAP for the purpose of network access authentication. Subsequently, IPsec SA Negotiation is used to establish an IPsec SA between the MN and the AR. This protocol exchange makes a clearer separation between the authentication and key establishment procedure and the IPsec SA Negotiation protocol. As a difference to Figure 7.9 and other application scenarios of the IPsec SA Negotiation protocol a secure channel is required between the MN and the AR to protect the negotiation after the EAP authentication is finished. This secure channel would, for example, use an HMAC-SHA1 to secure the negotiation with the EAP provided session key. This session key is obtained as an output of the session key derivation procedure.

The message flow in Figure 7.10 has the following properties.

- It aims to keep different protocol steps separate.
- It allows a modular approach by optionally decoupling the IPsec SA negotiation procedure.
- It does not state how the communication between the MN and the BAR is accomplished. Any transport protocol (such as TCP, SCTP, etc.) might be used. Even

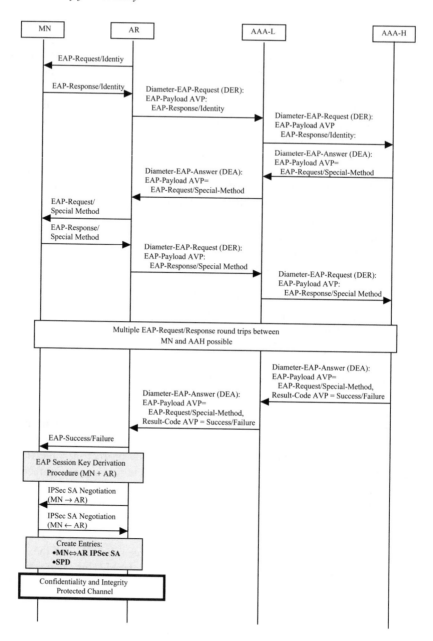

Figure 7.10   Network access using EAP to establish an IPsec SA

EAP over IEEE 802.1x is a possibility. Deciding about a transport mechanism is a religious issue where no best answer exists.

- User identity protection is provided if a tunnelled TLS variant is used which is executed within EAP itself. Other approaches for user identity confidentiality can also be provided (i.e. pseudonyms, temporal identities, public key encrypted identities, etc.)
- The number of roundtrips is slightly larger than the message flow in Figure 7.9.
- Security for the IPsec SA negotiation has to provide independently. The fresh and unique EAP session key is used as input.

To prevent replay protection during the IPsec SA Negotiation a binding to the previous EAP exchange should be used. The fact that the session key is fresh and unique with each EAP exchange provides a guarantee that the IPsec SA Negotiation immediately following the exchange is also fresh. Thus, replay protection could be based on a cryptographically generated random number, which uses the session key and some parameters (such as the identities of both end points as input) to generate a start value.

7.5 IP layer security

7.5.1 Introduction

In this section, we assume that network layer protection is provided with IPsec, since it is the 'natural' candidate to protect IP traffic. IPsec is a well-known and widely used protocol, which supports data origin authentication, connectionless integrity, (windowed) replay protection, optionally confidentiality protection (with ESP [35]) and limited traffic flow confidentiality. In this case, an IPsec security association has to be set up between the mobile node and an appropriate entity in the access network in the course of the initial access procedure. IPsec offers two modes: transport and tunnel mode. For our description the tunnel mode is of interest since IP traffic which experiences protection does not terminate at the access router.

If network layer protection has to be enabled between the mobile node and the access router then there is strong relationship with the authentication and key exchange building block described in Section 7.4. Methods to establish an IPsec security association based on a pre-shared secret were described in Section 7.4.6.

There are various ways by which mobility can be handled and the specific design has a strong influence on the security issues involved. If we assume that the protection of the first hop is implemented using IPsec and that in the course of a handover the mobile node changes its point of attachment to the access network, the IPsec security association has to be newly established between the mobile node and the new point of attachment. This could be done by either running a new negotiation or by transferring the IPsec security context from the previous point of attachment to the new one. Since the former is assumed to be too time consuming, we focus on the latter approach, which is discussed in the next section.

7.5.2   IPsec context transfer

Members of the Seamoby working group in the IETF work on IPsec context transfer. Note that in this section we assume that the IPsec tunnel is terminated at the AR although other configurations are possible and described in the Seamoby working group. Furthermore, it has to be mentioned that we focus on the security aspects of the Context Transfer protocol and not on a generic handover procedure whereby Quality of Service, Header Compression state and other information is transferred.

The following assumptions were specified that allow such an IPsec context transfer to be accomplished.

1. The IPsec Context Transfer takes place between ARs of the same administrative domain. Hence, there is a strong trust assumption between these nodes.
2. To secure the transfer of sensitive information (e.g. keys and SA parameters) between the two ARs a security association is required to provide authentication and authorisation, integrity, confidentiality and replay protection.
3. In order to initiate the handover procedure and to trigger the IPsec Context Transfer it is required to share a security association between the mobile node and the previous AR. We assume that network access authentication produced such a security association compatible with the Context Transfer protocol.

Mobility is a broad topic and, therefore, it makes sense to talk about two different categories.

1. The mobile node roams and the entity with which the mobile node shares its security association changes. This scenario is described above with the mobile node roaming from one AR to another. The mobile node can modify its own IPsec security association whereas a 'signalling protocol' is required to trigger the transfer (and the adaptation) from one node to another. The Seamoby working group provides the signalling protocol for the Context Transfer case. Their protocol provides more functionality than a simple Context Transfer since micro-mobility issues and forwarding of in-flight packets should also be addressed. Their work is, however, not finished yet.
2. The mobile node roams but the entity with which the mobile node shares its security association does not change. This is a non-typical context transfer scenario since there is only a need to 'patch' the existing security association at both the mobile node and at the other end. Patching the security association refers to the fact that the modified IP address of the mobile node affects the security policy database (SPD) with their Traffic Selectors of the both entities. Without updating the Traffic Selector entries packets forwarded to the mobile node (or sent by the mobile node) are not recognised as data traffic that requires appropriate protection. Typical cases where such situations arise are VPN scenarios where the mobile node shares a security association with its home domain, that is, with the security gateway of the home domain and scenarios with IPsec SAs that are used to protect the traffic between the mobile node and some network servers

within the visited network. It is worth noting that using a home address instead of the care-of address may change the need for such a processing but this is not always an option.

In the Seamoby working group, technical issues with the transfer of an IPsec SA have been discussed which relate to the types of attributes of IPsec SA. Static and dynamic attributes have to be transferred between the ARs. Static attributes are fixed at the time of IPsec SA establishment whereas dynamic attributes need to be modified during the SA lifetime. The following list of attributes gives information that attributes to transfer between two nodes.

Static attributes
- SPI value,
- IPsec Mode (ESP/AH),
- Authentication and Encryption Algorithm,
- Key Length,
- Key Rounds,
- Keys,
- Replay Window Size,
- Path MTU,
- SA Lifetime Type and SA Duration (seconds/kilobytes),
- Traffic Selector Fields specifying which traffic must be secured.

Note that IKE does not use the term Traffic Selector; instead, the Traffic Selector information is encoded in the Identity Payloads.

Furthermore, compression state information needs to be stored along with the security relevant attributes if IPComp SAs are used like

- Compression Dictionary Size,
- Compression Private Algorithm,
- IP Compression Algorithms.

Dynamic attributes
- Accumulated kilobytes relevant for the SA Lifetime,
- Highest Sequence Number stored,
- Flags which packets within the windows have already been received.

We assume that the established IPsec SA is set up in tunnel mode since the protected data traffic should not terminate at the AR (which is true for nearly all messages). In Reference 36, the authors additionally allow more than a single security association to be transferred between the previous BAR and the new BAR. This is required whenever protection is provided at a more fine-grained granularity other than in a 'protect everything between the MN and the AR' policy style of policy.

The context transfer can, however, lead to problems. These exceptions are listed below.

SPI collision

The SPI is chosen locally and could, therefore, conflict with an already used SPI value at the AR when transferred. If the SPI value of the transmitted IPsec SA already exists at the new AR then various alternative decisions are possible ranging from rejecting the transfer to re-negotiation of a new SPI value.

SA conflict

In Section 6.4 of Reference 36, an additional error case is mentioned that could occur during a context transfer. If the new AR does not support the algorithms of the transferred SA then a so-called SA Conflict occurs. Current proposals cover only the standard case where both ARs use the same set of algorithms and no such SA conflict can occur. To cover the generic case (e.g. in an environment where devices of different vendors and with different capabilities are used) a protocol is required to handle issues regarding the negotiation of an IPsec SA between the new AR and the mobile node by including the old AR in the communication.

Note that none of the existing proposals transfer other SAs like RSVP SAs or ISAKMP SAs as part of the context transfer procedure. Hence, without an ISAKMP SA no subsequent re-keying is able to take place without re-executing the IPsec SA generation mechanism. Similarly, without transferring non-IPsec SAs (e.g. SAs required for quality of service signalling protocols like RSVP or future NSIS protocols) performance and latency problems are caused and re-creating them may be the consequence. The fact that SAs (except for IPsec SAs) are not separated from the protocol-specific data causes practical problems when transferring SA information since no generic API is able to dump the content of this database. Hence, it would be useful to store SAs of other protocols within a unified key storage (possibly within the kernel-space) and to provide access by privileged application based on a standardised API like PF_KEY [37].

Additionally, it would be useful to delete old SAs at the previous AR after a successful context transfer, otherwise they remain active for their entire lifetime and consume the memory of the corresponding AR.

7.6   Link layer security

For the security mechanisms to protect confidentiality and integrity of the first hop in our security architecture, see Figure 7.1, basically three different approaches exist.

1. The terminating points of the security mechanisms are identical to the endpoints of the first hop, that is, MN and AR.
2. The terminating points of the security mechanisms are identical to the endpoints of the wireless link, that is, MN and AP.

3. The network termination point of the security mechanisms is located between the AP and the AR in a radio network controller (RNC), where the protection of control data and user data may be terminated in different physical entities.

In the first approach, the security mechanisms are located in Layer 3. In the second and third approaches, the native wireless link layer security mechanisms may be used. It should be noted, however, that the trust model may imply that layer 3 security is involved in the latter two approaches as well.

In this section, we first briefly review the security mechanisms offered by existing link layer technologies. Then, we continue by studying the most important issues with link layer security in our security architecture. Finally, we list problems faced in case link layer security is totally disabled.

7.6.1 Existing link layer technologies

In GSM, some signalling information elements are considered sensitive and may be protected (which in GSM means encryption). To ensure identity confidentiality, the Temporary Mobile Subscriber Identity must be transferred in a protected mode at allocation time and at other times when the signalling procedures permit it. For both signalling and user traffic, the layer 1 data flow is ciphered by a bit per bit or stream cipher, that is, the encrypted data flow on the radio path is obtained by the bit per bit binary addition of the plaintext data flow and a ciphering bit stream, generated by algorithm A5. For more information, see Reference 38.

In GPRS, encryption is performed at the LLC level [38] between the MN and the SGSN. The plaintext at the LLC level encryption is the data consisting of the payload of the LLC frame (i.e. the information field) and the cyclic redundancy check. This means that the header part of the LLC frame is not ciphered.

In UMTS [10, 39], two security mechanisms for the radio link are provided. An integrity protection mechanism is provided at the RRC level to protect the integrity of the signalling. The confidentiality protection mechanism is performed at a lower level, either at RLC or MAC layer. The end points in both protection mechanisms are the ME and the RNC.

The Bluetooth wireless link layer communication system [40] provides an authentication mechanism for peer device authentication. At the authentication mechanism, a key for an encryption mechanism is derived. The Bluetooth encryption mechanism is applied to the payload of Baseband packets. Signalling and user information is transmitted in the packet payload, which can be encrypted, while the access code and the packet header are never encrypted. Authentication and encryption are based on a bond between two Bluetooth devices, which is created by establishing a common link key between the devices. The Bluetooth specifications provide a method, called as pairing procedure, through which the two devices can establish a common link key from a passkey entered by the users to the devices. It is also possible to import a common link key to the devices from a higher layer.

IEEE 802.11 [41] provides link level authentication between stations. The purpose of IEEE 802.11 authentication is to bring the wireless link up to the assumed physical security of a wired link. For the same purpose, IEEE 802.11 provides the ability to encrypt the contents of messages. This functionality is provided by the privacy service: an optional privacy algorithm, *Wired Equivalent Privacy* (WEP), is specified.

It should be noted that after the publication of some of these security mechanisms they were demonstrated to have fatal vulnerabilities. In particular, the WEP authentication protocol is not secure. Later also the WEP algorithm itself was shown to have weaknesses in its key schedule. Furthermore, integrity was (claimed to be) protected only by means of a CRC checksum, which can be manipulated. The IEEE 802 group has withdrawn the WEP algorithm and initiated work for a replacement. For an overview of the recent development in 802.11 security, see Reference 20.

The main purpose of the security in Hiperlan/2 is to provide the possibility to secure the air interface [42]. The system provides authentication and encryption mechanisms. In Hiperlan/2, authentication has two options. It is based either on a pre-shared secret or public keys of the terminals. In the case of public key authentication, both parties may have to trust a third party in order to verify the other party.

In Hiperlan/2, encryption is activated before authentication. The shared secret encryption key is established in the MN and the AP using anonymous Diffie–Hellman protocol. The encryption is used to protect the identity information of the MN when it is sent to the AP at the initial access phase. Clearly, this protection is only against passive attacks. It does not offer any protection at all against active attacks launched by false malicious access points or malicious terminals. Only after successful authentication, in which the public Diffie–Hellman tokens are authenticated, is secure encryption possible. Note that message authentication is not provided at all, which leaves sensitive data open for manipulation.

7.6.2   Link layer SA negotiation

A set of basic parameters and keys (which together form an SA) are needed to secure the link layer communication. Negotiation of these SAs, that is, the security set-up, is not necessarily executed at the link layer. For instance, Hiperlan/2 uses a layer 2 mechanism while Bluetooth can utilise higher layer mechanisms to bootstrap the link layer security. Scenarios where the SA negotiation is done completely at a higher layer will require that either the SA negotiation protocol is aware of the underlying link layer it negotiates for, or that there exist a common set of security parameters that can be negotiated in order to be able to create the full SA. In the latter case, this would require that all the link layer technologies are fully aligned with a common set of parameters that can be negotiated by the higher layer irrespective of the actual link layer mechanism used.

7.6.3   Link layer context transfer

When a mobile node moves between different APs or RNCs, while using link layer security to secure the path, a link layer SA has to be established between the mobile node and the new point of attachment. Instead of running the full link layer SA negotiation protocol again, we could transfer the security context from the old point of attachment to the new. Security context transfer for the link layer exists in some systems. Also, some systems have only specified the parameters that need to be transferred but have left out the transfer itself from the standard (e.g. Hiperlan/2).

In our model, three different types of security context transfer can be identified.

- Security context transfer between nodes of the same type belonging to the same AR.
- Security context transfer between nodes (possibly also of different types) within the same AN (e.g. between Hiperlan/2 and WLAN 802.11).
- Security context transfer with nodes outside the AN.

In the first case, the context transfer could be handled between the APs or RNCs directly. However, this might require that quite a few SAs are set up in between them. Another option is to do the context transfer directly between the nodes, but always through the AR. Now the link between the AP and the AR must be protected if keying material for the link layer security is sent down.

In some situations it might be practical to be able to have context transfer between different types of nodes. However, big differences in security functionality make it hard to define a mapping from one system to another (one of the most extreme cases would be between GSM and Bluetooth). By letting the AR take on the responsibility of handling the context transfer for the link layer, one can achieve a common mechanism that is independent of whether the MN moves between APs/RNCs belonging to the same AR or belonging to different ARs. It will then be possible to rely on the network domain security (described later) for the communications that is done between different ARs within the AN.

Context transfer between different ANs would require higher layer signalling and possibility for the context transfer node in one AN to communicate (directly or indirectly) with the context transfer node in the other AN. Such possibility may lead to new threats: for example, tampering a node from which the security context transfer is initiated would also make the receiving AN vulnerable for attacks. Therefore, a re-authentication would, in general, be necessary when executing a handover to a different network (especially when it is done to a network with a different operator).

7.6.4 Problems with disabled link layer security

It is clear that signalling messages related to the control and management of the wireless communication link itself cannot be protected at a higher layer; potential vulnerabilities must be protected using security mechanisms at the link layer. Therefore, when studying the importance of protection at the link layer, the vulnerabilities and threats for link layer signalling messages are essential. Anonymity protection is not considered in this context, as it creates a problem of its own. If anonymity is to be achieved, the protection should start before the first authentication over the network is performed. Also, user identity consists of several pieces of information, such as MAC addresses, IP headers, channel access codes, etc., which are distributed between different communication layers.

An unprotected wireless link can be tampered in different ways causing undesirable damage to the victim mobile devices and victim networks. At least the following threats are opened by disabling security mechanisms for the signalling and control messages on the wireless link layer:

- false network access points;
- mobile station gaining unjustified high QoS level;
- tampering of signalling data causing channel deterioration;
- manipulation of link layer mobility functions;
- tampering of power control functions;
- link hijacking by a mobile station;
- link hijacking by a false network (e.g. garbage advertising and spamming).

It should be mentioned, however, that some of these threats (e.g. link hijacking) could be mitigated by network layer security mechanisms so that the gain of an attacker is very limited.

It should be noted that some link layer messages cannot be protected. This affects all information that needs to be unrestrictedly available:

- broadcast messages, for example, network information;
- paging messages directed to a larger set of mobiles;
- initial communication between MN and AP/RNC before protection is negotiated.

7.7   First hop security options

In this section, we investigate various approaches for first hop security. Particular attention is paid to the role of link and network layer security protection mechanisms on the wireless connection between the mobile station and the access point. For this purpose, it is useful to make a distinction between the following types of data transmitted over the wireless link:

- user data;
- network layer signalling messages, for example, IP headers and ICMP-messages, that are not protected by network layer security mechanisms;
- link layer signalling messages that can be protected at the link layer;
- link layer signalling messages that cannot be protected (e.g. broadcast messages).

In the following we describe different types of security threats for the first hop among MN, AP/RNC and AR. We assume that the link between the MN and the AP is wireless and there is a wired connection between AP/RNC and AR.

7.7.1   Threats

On the physical layer, an attacker may hijack the connection (channel) between the MN and the AP or take over the wire between the AP and the AR. Furthermore, the connection can be disturbed by radio noise or by manipulation of the wire, which may result in degradation of service quality or complete service interruption. Since there may be a large number of distributed APs, a physical attack or a complete takeover cannot be ruled out.

There are a variety of threats that would be produced by disabling link layer protection. These were already discussed in Section 7.6. A number of threats can arise from manipulations and attacks at the network layer. Since the IP protocol provides a cornerstone for our architecture, network layer vulnerabilities must, hence, be considered particularly dangerous. These threats include at least the following:

- eavesdropping of user data;
- manipulation of user data;
- denial of service attacks against the network router (AR) and other network components (e.g. AAA server);
- attacks against any IP based service of the AR;
- denial of service attacks against the mobile station (MN);
- attacks against any IP based service on the MN;
- unauthenticated users get access to the network;
- manipulation of quality of service attributes;
- spoofing or manipulations of the sender's and/or receiver's IP address;
- eavesdropping of the mobile's IP address.

7.7.2 Security options

We described earlier that the different layers have their specific security threats. In this section, we discuss what kind of *combination* of security mechanisms meets the requirements. A basic observation is that user data can be protected both at the network layer and at the link layer. Since user data protection would represent a minimal requirement, three different options (focusing on the wireless link) remain to be studied.

7.7.2.1 Link layer protection only

The first option uses only link layer protection between the MN and the AR. This is the only possibility if the mobile node is not capable of performing security mechanisms at the network layer. In this case all transmitted data will be decrypted and/or integrity checked at the AP/RNC, depending on the functionality of the link layer security mechanism. If the access point is physically separated from the access router, then appropriate mechanisms need to be performed when transferring data over a fixed net link between these two nodes. The protection can be at the network layer or at the link layer. In this scenario, the links between the MN and the AP/RNC and between the AP/RNC and the AR are protected separately and the hop(s) in the middle (AP/RNC) need to be trusted (see Figure 7.11).

This option may offer some advantages if the mobile node has limited resources and if the access point and the access router are physically united.

In this option, the complete IP packet is protected as part of link layer frame encryption and integrity protection. But since the complete IP packet is treated as payload, the AP/RNC is, in general, not able to analyse the IP header. Since key agreement and authentication traffic needs to reach the AR, the AP or RNC would not be able to perform access control unless port-based access control is implemented.

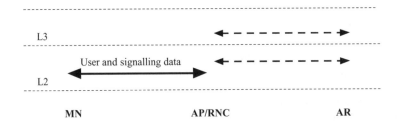

Figure 7.11   Protection only at link layer

Assuming that there is no additional network layer protection, the threats listed in Section 7.7.1 persist.

It should be noted that a mobile terminal that provides a variety of link layer technologies (e.g. GERAN, UTRAN, Bluetooth and IEEE 802.11) has to implement all the different protection mechanisms and encryption algorithms. Hence, mobile terminals for heterogeneous access have to provide a variety of different protection mechanisms and encryption/integrity algorithms.

Peer entity authentication at the link layer, as implemented, for example, for Bluetooth, where it also includes mechanisms for the derivation of the session keys, may be a very useful mechanism. This is the case particularly in situations, where the session keys, for example, for link layer message authentication, need to be refreshed during the session. Such a situation may occur in access point roaming when changing to a new access point. On the other hand, it may be argued that performing authentication at the middle of a session may just cause unnecessary delay in session re-establishment, and should be avoided.

7.7.2.2   Network layer protection only

Next, we describe our second option in which user data and part of the network and higher layer signalling is protected using IPsec encapsulated security payload (ESP) over the first hop. The rest of the network layer signalling data and the link layer signalling data remain unprotected (see Figure 7.12). The link is vulnerable to all attacks identified in Section 7.7.2.1 (mainly for denial or degradation of service).

Only the threats of hijacking can to some extent be prevented using the network layer protection. While the wireless link can be hijacked by a malicious device or by a malicious network, the attempts to use the hijacked link for transferring user data will be detected at the network layer. IPsec protects the user data against manipulation and eavesdropping, ensures data integrity, prevents unauthorised access and protects IP-based services on both the MN and the AR.

Apart from the above-mentioned link layer threats, the following network layer threats persist.

- Denial-of-service attacks against IPsec by forcing the AR and/or the MN into resource-consuming cryptographic computations with malicious packets (note, however, that this threat is fairly general, i.e. it applies to use of IPsec in any environment where no additional security is provided).

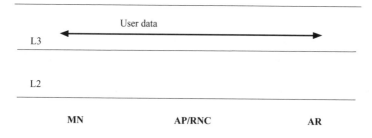

Figure 7.12   Protection only at network layer

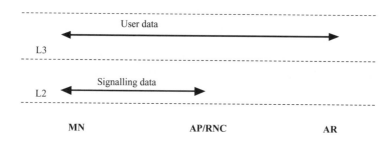

Figure 7.13   Protection at network and link layers

- Manipulation of the outer header's IP addresses (note here that there is not necessarily a big damage that can be caused by this).
- Eavesdropping of the mobile's IP address (only outer address).

The authentication, key agreement and IPsec security association negotiation protocols need to be carefully designed to provide the required level of security. In particular, protection against active attacks and user identity protection is a challenging issue.

7.7.2.3   Both link and network layer protection

Now we move to the third option that can be used for the first hop protection. It was shown above that the 'network layer only' option does not protect tampering of the link layer signalling messages, which may cause serious problems at the link layer, depending on the deployment scenario. To prevent such problems occuring it seems advisable to protect at least the authenticity and integrity of the link layer signalling messages at the link layer for certain scenarios.

Furthermore, some remaining threats from Section 7.7.2.2 (mainly IP-based DoS attacks) can be prevented by combined link and network layer security. With link layer protection enforced, an attacker is no longer able to insert any IP packets that would have to be processed by IPsec (at least not after authentication and key agreement) (see Figure 7.13).

The properties of the current wireless link technologies vary significantly with respect to providing authenticity and integrity of the signalling data. In UTRAN, it is made possible to treat RRC signalling in a different way than user data. UTRAN provides specific procedures for RRC signalling, which forms the essential part of the link layer signalling. This is not possible in Bluetooth, where all different communication channels are mapped on the packet payload, and encrypted. This means that in this option, user data is subject to double encryption, which may significantly increase the processing overhead required by security mechanisms, if IPsec encryption is used. Basically, there are the following sub-options.

- Encryption of all data on the link and the network layer (implies double encryption of the data).
- Selective encryption and integrity protection of signalling data on the link layer (may not be supported by all radio technologies), encryption of user data on the network layer.
- Encryption of all data on the link layer and only authentication and integrity protection on the network layer (with IPsec AH or ESP Null Encryption). This would, for example, be sufficient for quality of service (QoS) protection. It would require additional protection for the wired links.

Which of these sub-options should be used could also depend on the link layer technology. If it already offers strong encryption (e.g. UTRAN) the third sub-option may be the best. If, on the other hand, the encryption is weak and selective encryption is available, then the second option may be the preferred one. If the link layer technology does not support selective encryption, only provides weak encryption or encryption does not consume considerable resources, then the first sub-option seems preferable. This also holds if the link layer is a common standard or mandatory for some other reason.

7.7.2.4   Summary

We have now presented three different strategies for providing security for the first hop. The optimal performance of the wireless communication cannot be guaranteed without protecting the basic signalling messages at the wireless link. On the other hand, authentication, integrity protection, QoS and access control issues are most efficiently dealt with at the network layer. This shows that a combination of link and network layer security mechanisms seems to satisfy best all of the relevant security requirements.

But this choice may not be optimal with regard to processing and protocol overhead. It remains to be seen whether double encryption (at the link and network layers) is feasible with limited devices. Multiple encryption may also interfere with compression schemes. On the other hand, the processing power of mobile devices is continuously increasing and implementations of security algorithms are improved so that processing power limitations may not be a problem in the future. Higher layers may add additional encryption mechanisms anyhow and from an architectural point of view, a clear separation of the individual layers seems preferable.

7.8 Network domain security

Network domain security is about securing the communication between network internal nodes, that is, the communication beyond the first hop. It is just as important as access security. Fortunately, the problem seems easier to solve.

It is assumed that the network domain of a mobile system beyond 3G is entirely IP based. The basic approach proposed here is to secure all traffic in the network domain (as far as protection measures are required) at the network layer by using IPsec. This has the big advantage to make the security independent of any particular protocols at the transport layer or above. Therefore, the approach used in 3GPP for securing signalling traffic in the UMTS core network seems a promising candidate for network domain security also in mobile systems beyond 3G. For a description of the 3GPP approach see Reference 39.

Some cautionary remarks regarding the application of 3GPP concepts to network domain security in mobile systems beyond 3G seem in order. In particular, the trust model for mobile systems beyond 3G may turn out to be fundamentally different from that in 3G systems. This may affect the question whether hop-by-hop protection is still adequate in all cases. Furthermore, protocols used for certificate management may change. But, in general, a PKI-based approach to security association management for network domain security in post-3G systems seems the way to go.

7.9 Security for QoS signalling

7.9.1 Introduction

Signalling QoS information in an end-to-end fashion allows end hosts to specify a certain behaviour from the network, which differs from the standard 'best-effort' service offered today. By introducing this service differentiation a number of challenges have to be dealt with. It is fairly obvious that better service for some users means that there must be a 'penalty' for not reserving resources such as additional costs. Without restriction any user would just use better QoS, which leads to no improvement at all. With mobility and telecommunication networks today authorisation can (or should) be seen in an abstract form, as 'Is one of the signalling participants able to pay for the reservation?'

In this section, we mainly focus on the Resource Reservation Protocol (RSVP) [44], which is currently the only standardised QoS signalling protocol in the IETF. It is worth noting that the IETF NSIS working group analyses signalling protocols and tries to create a signalling protocol for a broader usage as generic signalling; a signalling protocol that signals information other than only QoS such as firewall and NAT information, active networking code and other applications that need the path coupled signalling functionality.

RSVP is currently only deployed in closed environments such as enterprise networks or in intradomain environments. In such an environment authorisation often means role-based access control based on group membership or special rights to use a service. Users are typically not charged directly for their generated QoS traffic

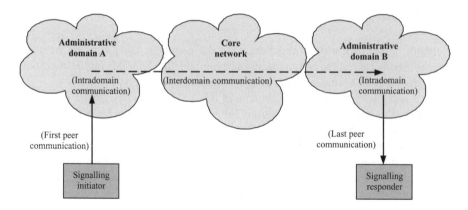

Figure 7.14   Networking parts in an end-to-end signalling environment

or for QoS reservations. If the signalling messages (and thereby the QoS reservation) travel beyond the administrative domain, then the enterprise network is charged and not the individual end-user directly. This is based on a peering relationship and business contracts between the neighbouring networks. An analysis of RSVP and its security properties can be found in Reference 44. Issues about authorisation of QoS reservation requests are discussed in Reference 45.

Figure 7.14 shows the different network parts involved in an end-to-end signalling message environment. A characterising feature of path-coupled end-to-end QoS signalling is that state information is stored at intermediate nodes along the path between the sender and the receiver. In the case of RSVP this state is the soft-state and times out if refresh messages are not provided. This state information is both for the QoS reservation itself and also signalling state, which ensures that messages can be delivered back and forth along the path. Note that RSVP depends on routing but does not modify routing.

Although Figure 7.14 lists a number of different networking parts, which have different characteristics and deserve their own analysis we will focus on the edge communication only. First and last peer communication is interesting from a security point of view due to the large number of scenarios, environments and architectures where such a QoS signalling protocol is used. As such it is difficult to agree on one particular security mechanism (e.g. Kerberos). Additionally, it is the end host that triggers the communication and is also finally charged for the consumed resources.

RSVP signalling thereby typically starts with a PATH message transmitted from the initiator to the responder. The PATH message is addressed to the responder and has a router alert option attached to it. Every RSVP aware router along the path intercepts the signalling message. After processing it and performing the desired actions the message is forwarded to the responder. The responder returns an RESV signalling message in a hop-by-hop fashion by using state installed with the PATH message to route the signalling message back to the initiator using the same path (and the same

RSVP nodes). The PATH message itself mainly serves as a 'path discovery' and to pin the route for the RESV message. The RESV message finally installs the state.

7.9.2 Threats

Knowing the communication patterns for a QoS signalling protocol it is reasonable to address threats. A more detailed discussion of security threats in NSIS can be found in Reference 46.

- A number of threats are related to authentication and the establishment of a security association for securing signalling messages. Mitm attacks caused by missing or unilateral authentication are realistic threats.
- Furthermore, it is possible to attack the signalling message exchange itself by replay, signalling message injection, by eavesdropping signalling messages, modifying content of messages, performing denial of service attacks and other more sophisticated attacks (e.g. fraud).

A problem of RSVP is that signalling message delivery is combined with discovery. In a generic environment it is, therefore, not possible to secure the messages in a proper way. However, in a typical environment the edge routers are the first QoS aware nodes and as such a mobile node already knows the first RSVP aware node. This is, among other things, an architectural assumption, which might not hold in all environments. Furthermore, we will later make a decision for signalling message protection, which is again an assumption that cannot be made in general. For mobile environments, RSVP does not provide the functionality to learn the necessary information required to start RSVP in an off-the-shelf manner (e.g. negotiation of the security mechanisms which has to be used).

7.9.3 Requirements

To address these threats we list a few security requirements and we will not separate them into requirements for a signalling application (such as QoS signalling) and requirements for signalling message transport as done in the NSIS working group:

- mechanisms must exist which enable strong authentication;
- state installation (e.g. for a QoS resource reservation) must be authorized;
- signalling messages must be integrity and replay protected;
- as a minimal requirement peer-to-peer security must be provided;
- denial of service attacks must be prevented to some possible extent;
- user identity confidentiality and topology hiding should be provided;
- confidentiality of signalling messages should be provided.

Other requirements analysed within the NSIS working group can be found in Reference 47. A description of the security requirements can also be found in the same document.

Note that mobility specific requirements and threats are not addressed in this section.

7.9.4   Framework

In order to protect protocols it is common practice to split the protocol execution into two phases. The first phase provides authentication and key exchange whereas the second phase ensures that the established security association is used to protect the protocol itself. RSVP provides a mechanism to protect the signalling messages with the help of the RSVP Integrity object as defined in Reference 48. Furthermore, user authentication was introduced in Reference 49, based on Kerberos and on public key based authentication. We will not make use of the latter one since network access authentication was already assumed to be present before a QoS signalling protocol is executed. RSVP does not provide key management for the RSVP Integrity object, which allows us two possible solutions.

1. We establish the necessary security parameters for RSVP based on the previously executed network access authentication procedure. To obtain this solution we need to derive session keys from the master key provided with the network access authentication procedure. Both the mobile node and the access router need to derive the keys in the same way and make it available to the RSVP daemon. This is an architectural decision and needs to be specified accordingly. Furthermore, it is necessary to specify how replay protection in RSVP is handled (i.e. the starting sequence number) and which algorithms are used for computing the keyed message digest routine. A similar approach is used in the PacketCable environment.

2. As a second alternative we could use the fact that all messages experience IPsec protection to the access router, which is also assumed to be the first RSVP aware router in the access network. Every incoming packet at the access router is only processed or forwarded if cryptographic verification with IPsec was successful. This ensures the access router that intercepted RSVP signalling messages are also cryptographically bound to a previous authentication and key exchange protocol. Furthermore, the IP address used by the mobile node cannot be modified and can, therefore, serve as an identity that can be mapped to a user identity. This process has to be done anyway to associate consumed resources to a particular entity. Policy based admission control procedures still have to be executed to ensure that the requested resources are available and that the requesting entity is allowed to make a reservation.

By using either one of these two approaches a number of security requirements can be fulfilled. The separation of the protocols allows requirements specific to the authentication and key exchange protocol be provided by a number of network access authentication protocols discussed in this chapter. Requirements for strong authentication and key exchange, denial of service protection and user identity confidentiality are requirements that are typically addressed to these protocols. For the signalling message protection itself IPsec might additionally provide confidentiality protection (by using IPsec ESP SAs) whereas RSVP does not provide confidentiality protection. Providing fresh and unique session keys with a sufficient length make attacks against the signalling protocol practically impossible.

7.9.5   Conclusion

We can conclude that the design possibility for an architecture or framework it is possible to specify an interworking between different protocols to combine individual building blocks in a way that makes protocol operation much simpler. In order to protect QoS signalling the split between the authentication and key exchange protocol and the actual signalling message protection allows to introduce flexibility and to tackle the huge number of scenarios in an efficient way. Together with the capability of Context Transfer even a more advanced QoS management is possible. The ability to make such architectural assumption, which are justified by the way how QoS works, speaks against defining custom-security protocols (such as those introduced in Reference 49) although they have been used in Windows 2000 (see Reference 50, where Kerberos and RSVP interworking is described).

7.10   Conclusions and future work

This chapter discussed functional building blocks for security architecture of future mobile systems expected to follow the third-generation UMTS systems. To support heterogeneous access the security architecture uses a modular approach. Five basic functional building blocks were identified which are likely to be required in any type of future mobile system:

- secure address configuration,
- authentication and session key establishment,
- IP layer security,
- link layer security, and
- network domain security.

Within these building blocks the primary focus was on the security features and mechanisms required to provide global IP connectivity and various forms of mobility to a globally roaming user. A secondary focus was put on security for Quality of Service procedures in such a system.

For address configuration, both stateless and stateful address autoconfigurations were considered. Both methods may be subject to attacks if used unprotected. However, for both methods security features and extensions exist which offer a sufficient security level. The special demands of access privacy protection were investigated for both stateful and stateless cases.

In Section 7.4, authentication and key agreement protocols for heterogeneous access in future mobility architectures were discussed. The EAP offers a standard framework for carrying the authentication exchanges. The authentication and key agreement protocols may be combined in a sequence of protocols, *an authentication sequence*, where each sub-protocol of the sequence involves different entities at different communication layers. The special cases of one-step and two-step approaches were discussed in more detail. The two-step approach to authentication and key agreement can provide a possible solution for identity protection against active attacks and for the support of legacy authentication protocols. However, to become functional, it requires an efficient key management infrastructure to provide the roaming users with the credentials of the local network. Also the security of the two-step protocols was analysed. It is shown that if the authentication of the mobile terminal uses the same long-term authentication key in different environments then a two-step protocol is exposed to an active MitM attack to impersonate the MN. One solution would be to implement a cryptographic binding between the first-step and the second-step protocols. It would also allow a more flexible usage for existing strong EAP methods.

In order to establish IPsec SA a secure channel and an IPsec SA negotiation protocol is required. Section 7.4.6 gave a detailed problem analysis, highlighting existing protocols that provide such a functionality and finally some requirements for a future protocol. When a mobile node roams into a network (interdomain mobility) an authentication and key establishment process is triggered. As part of this procedure the mobile node is authenticated and authorized to grant access. Additionally, an IPsec security association is established. Mobility within the same administrative domain (intradomain mobility) is significantly improved if the already established SA is moved from the old access router to the new one and to adopt some attributes. Section 7.5.2 described the procedure of IPsec context transfer and listed both static and dynamic IPsec fields that have to be securely transmitted between the routers. Furthermore, the impact of QoS protected data traffic was investigated. Whenever IPsec in tunnel mode is applied the size of every IP packet is increased due to encapsulation and double encryption leads to performance degradation. Double encryption can be avoided by a tricky selection of Traffic Selectors in combination with IPsec AH and ESP.

The lower the protection that takes place in the protocol stack the more communication can be protected. Specific link layer signalling can obviously not be protected at higher layers. Current radio link layer security systems include mechanisms for peer entity authentication and session key derivation during the set-up procedures at the network layer. Mechanisms for confidentiality protection of the traffic have been specified for all link layer systems. However, message authentication (integrity protection) is only provided in UMTS, with a clear purpose of providing integrity protection to the radio control signalling messages only. What is the optimal selection for the cryptographic mechanisms to be used to protect the link layer remains for further study.

If the keys for these security mechanisms are derived at the network level authentication procedure when the network access connection is set up, then the protection of the link level can be provided only after the initial authentication. Set-up procedures that would allow starting link layer protection earlier remain to be studied. Further, a general framework for link layer SA negotiation might be desired as well as a standard way of doing context transfer.

Three different strategies for providing security for the first hop were studied. It is most apparent that optimal performance of the wireless communication cannot be guaranteed without protecting the basic signalling messages at the wireless link. On the other hand, authentication, integrity protection, QoS and access control issues are most efficiently dealt with at the network layer. This shows that a combination of link and network layer security mechanisms seems to satisfy best the relevant security requirements. However, this optimal solution cannot always be achieved and must be traded off with processing and protocol overheads. It may not even be required, depending on the deployment scenario and business case. Therefore, approaches relying either on network layer security or on link layer security were also investigated.

Considering network domain security it was found that the solutions that are currently being developed for IP-based core networks in current mobile systems are also expected to be suitable for future systems.

Finally, it was concluded that without security a specific QoS level (except best effort) cannot be guaranteed. It is not only important to protect QoS signalling, but also the QoS reserved flows so that these cannot be, for example, hijacked or in some other way misused. The main requirement that needs to be fulfilled to create a good level of protection is that integrity protection on both the QoS signalling and the QoS reserved flows must be applied.

Notes

1 The work described in this chapter has been supported by the European Commission through the IST Programme under Contract IST-2000–25350. The information in this document is provided as is, and no guarantee or warranty is given or implied that the information is fit for any particular purpose. The user thereof uses the information at his/her sole risk and liability.
2 We assume Mobile IPv6, so no Foreign Agent shows in our pictures.

References

1 IST-2000-25350 SHAMAN, Security for Heterogeneous Access in Mobile Applications and Networks: 'Deliverable D13 Annex 1', http://www.ist-shaman.org
2 ISTAG Scenarios for Ambient Intelligence in 2010: European Commission Community Research, February 2001, www.cordis.lu/ist/istag.htm
3 A vision on systems beyond 3G, European Commission Community Research, February 2002, ftp://ftp.cordis.lu/pub/ist/docs/ka4/mb_sb3g-vision2001.zip
4 IST-1999-10050 Broadband Radio Access for IP based Networks (BRAIN) Deliverable D2.2: 'BRAIN architecture specifications and models, BRAIN functionality and protocol specification', March 2001, http://www.ist-brain.org/
5 IST-2000-28584: 'Mobile IP based Network Developments (MIND)', http://www.ist-mind.org/
6 THOMSON, S. and NARTEN, T.: 'IPv6 Stateless Address Autoconfiguration', RFC 2462, December 1998
7 DROMS, R. (Ed.): 'Dynamic Host Configuration Protocol for IPv6 (DHCPv6)', IETF draft, draft-ietf-dhc-dhcpv6-28.txt (work in progress), November 2002
8 Securing Neighbor Discovery (SEND), IETF Working Group: http://www.ietf.org/html.charters/send-charter.html

9 SCHWIDERSKI-GROSCHE, S. and KNOSPE, H.: 'Public key based network access', Chapter 8, this book

10 3GPP TS 33.102 V4.3.0 (2001–12) Technical Specification: '3rd Generation Partnership Project; Technical Specification Group Services and System Aspects; 3G Security; Security Architecture (Release 4)'; http://www.3gpp.org

11 ARKKO, J. and HAVERINEN, H.: 'EAP AKA Authentication', Internet draft (work in progress), draft-arkko-pppext-eap-aka-09.txt, February 2003

12 BLUNK, L. and VOLLBRECHT, J.: 'PPP Extensible Authentication Protocol (EAP)', Internet RFC 2284

13 SIMPSON, W.: 'The Point-to-Point Protocol (PPP)', STD51, RFC 1661, July 1994

14 IEEE Standard 802.1X-2001: 'Standards for Local and Metropolitan Area Networks: Port Based Access Control', June 2001

15 ANDERSSON, H., JOSEFSSON, S., ZORN, G., SIMON, D. and PALEKAR, A.: 'Protected EAP Protocol (PEAP)', Internet draft (work in progress), draft-josefsson-pppext-eap-tls-eap-06.txt, March 2003

16 FUNK, P. and BLAKE-WILSON, S.: 'EAP Tunneled TLS Authentication Protocol (EAP-TTLS)', Internet draft (work in progress), draft-ietf-pppext-eap-ttls-02.txt, November 2002

17 SHEFFER, Y., KRAWCZYK, H. and ABOBA, B.: 'PIC: A Pre-IKE Credential Provisioning Protocol', IETF draft (work in progress), draft-ietf-ipsra-pic-06.txt, October 2002

18 KAUFMAN, C. (Ed.): 'Internet Key Exchange (IKEv2) Protocol', IETF ipsec working group draft (work in progress), draft-ietf-ipsec-ikev2–07.txt, April 2003

19 TSCHOFENIG, H. and KROESELBERG, D.: 'EAP IKEv2 Method (EAP-IKEv2)', Internet draft, Internet Engineering Task Force, April 2003

20 GEHRMANN, C. and NYBERG, K.: 'Security in personal area networks', Chapter 9, this book

21 KNOSPE, H. and SCHWIDERSKI-GROSCHE, S.: 'Secure m-commerce', Chapter 14, this book

22 HARKINS, D., PIPER, D. and HOFFMAN, P.: 'Secure Legacy Authentication (SLA) for IKEv2', IETF personal draft, draft-hoffman-sla-00.txt, December 2002

23 ASOKAN, N., NIEMI, V. and NYBERG, K.: 'Man-in-the-Middle in Tunnelled Authentication Protocols', http://eprint.iacr.org/2002/163/, October 2002. A revised version of the paper was presented at the International Workshop on Security Protocols, 2–4 April 2003, Cambridge, England, to appear in the Workshop Proceedings

24 PUTHENKULAM, J., LORTZ, V., PALEKAR, A., SIMON, D. and ABOBA, B.: 'The compound authentication binding problem', IETF personal draft (work in progress), draft-puthenkulam-eap-binding-02.txt, March 2003

25 FLYKT, P., PERKINS, C. and EKLUND, T.: 'AAA for IPv6 network access', Internet draft (work in progress), Internet Engineering Task Force, May 2003

26 PARTHASARATHY, M.: 'PANA threat analysis and security requirements', Internet draft (work in progress), Internet Engineering Task Force, April 2003

27 ABOBA, B. and SIMON, D.: 'EAP keying framework', Internet draft (work in progress), draft-aboba-pppext-key-problem-06.txt, March 2003

28 CALHOUN, P., FARRELL, S. and BULLEY, W.: 'Diameter CMS security application', Internet draft (work in progress), Internet Engineering Task Force, March 2002

29 WALKER, J., HOUSLEY, R. and CAM-WINGET, N.:'AAA key distribution', Internet draft (work in progress), Internet Engineering Task Force, April 2002

30 FORSBERG, D., OHBA, Y., PATIL, B., TSCHOFENIG, H. and YEGIN., A.: 'Protocol for Carrying Authentication for Network Access (PANA)', Internet draft (work in progress), Internet Engineering Task Force, April 2003

31 HARKINS, D., KAUFMAN, C. and PERLMAN, R.: 'Overview of IKEv2', *in* 'Proceedings of the Fifty-second Internet Engineering Task Force', Salt Lake City, Utah, USA, 9–14 December 2001, available at <http://www.ietf.org/proceedings/01dec/slides/ipsec-10.pdf>

32 THOMAS, M. and VILHUBER, J.: 'Kerberized Internet Negotiation of Keys (KINK)', Internet draft, Internet Engineering Task Force, January 2003

33 PERLMAN, R.: 'Understanding IKEv2: tutorial, and rationale for decisions', Internet draft (work in progress), Internet Engineering Task Force, February 2003

34 SRISURESH, P. and VILHUBER, J.: 'IKE extensions to support dynamic policies', Internet draft (expired), draft-srisuresh-ike-policy-extensions-00.txt, January 2001

35 KENT, S. and ATKINSON, R.: 'IP Encapsulating Security Payload (ESP)', RFC 2406, November 1998

36 HAMER, L.-N., HAZY, P., GOPAL, R., KRISHNAMURTHI, G. and SENGODAN, S.: 'Issues in IPSec Context Transfer', IETF draft (expired), draft-gopal-semoby-ipsecctxt-issues-01.txt, February 2002

37 McDONALD, D., METZ, C. and PHAN, B.: 'PF_KEY key management API, version 2', RFC 2367, Internet Engineering Task Force, July 1998

38 3GPP TS 03.20 V8.1.0 (2000-10) Technical Specification: '3rd Generation Partnership Project; Digital cellular telecommunications system (Phase 2+)', Security related network functions (Release 1999)

39 BOMAN, K., HORN, G., HOWARD, P. and NIEMI, V.: 'UMTS security', Chapter 6, this book

40 Bluetooth SIG: 'Specification of the Bluetooth System', Volume 1, Part B – Baseband Specification, v 1.1, 2001

41 IEEE Std. 802.11, Part 11: 'Wireless LAN Medium Access Control (MAC) and Physical Layer (PHY) Specifications', 1999 edition

42 ETSI TR 101 683 v1.1.1: 'Access Networks (BRAN); HIPERLAN Type 2', Broadband Radio System Overview

43 BRADEN, R., ZHANG, L., BERSON, S., HERZOG, S. and JAMIN, S.: 'Resource ReSerVation protocol (RSVP)–version 1 functional specification', RFC 2205, Internet Engineering Task Force, September 1997

44 TSCHOFENIG, H.: 'RSVP security properties', Internet draft (work in progress), Internet Engineering Task Force, March 2003

45 TSCHOFENIG, H., BUECHLI, M., VAN DEN BOSCH., S. and SCHULZRINNE., H.: 'NSIS authentication, authorization and accounting issues', <draft-tschofenig-nsis-aaa-issues-01.txt> (work in progress), March 2003

46 TSCHOFENIG, H. and KROESELBERG, D.: 'Security threats for NSIS', Internet draft (work in progress), Internet Engineering Task Force, March 2003

47 BRUNNER, M.: 'Requirements for QoS signaling protocols', Internet draft (work in progress), Internet Engineering Task Force, March 2003

48 BAKER, F., LINDELL, B. and TALWAR, M.: 'RSVP cryptographic authentication', RC 2747, Internet Engineering Task Force, January 2000

49 YADAV, S., YAVATKAR, R., PABBATI, R., FORD, P., MOORE, T., HERZOG, S. and HESS, R.: 'Identity representation for RSVP', RFC 3182, Internet Engineering Task Force, October 2001

50 Whitepaper: 'Windows 2000 RSVP Kerberos User Authentication Interoperability', January 2000, available at: <http://www.microsoft.com/windows2000/techinfo/howitworks/communications/trafficmgmt/rsvp.asp>, July 2001

51 BUCKLEY, A., SATARASINGHE, P., ALPEROVICH, V., PUTHENKULAM, J., WALKER, J. and LORTZ, V.: 'EAP SIM GMM Authentication', Internet draft (work in progress), draft-buckley-pppext-eap-sim-gmm-00.txt, August 2002

52 DIERKS, T. and ALLEN, C.: 'The TLS Protocol Version 1.0', RFC 2246, January 1999

53 MISHRA, A. and ARBAUGH, W.: 'An initial security analsyis of the IEEE 802.1X Standard', UMIACS-TR-20020-10, University of Maryland, February 2002

Chapter 8

Public key based network access

Scarlet Schwiderski-Grosche and Heiko Knospe

The security of access procedures to mobile networks is very critical, because wireless communication can be easily compromised. So far, network access procedures to wireless networks have primarily been based on secret key techniques. In this chapter, we want to survey current secret key approaches, motivate public key approaches and present major public key protocols for network access.

8.1 Introduction

Access to wired networks raises less security concerns than access to wireless networks. The reason is that wired networks offer some inherent level of security. There is a physical link between users and networks that can only be compromised by gaining physical access to network equipment. In wireless networks, however, the security requirements are much higher because communication over the wireless air interface can be easily intercepted. Strong encryption to prevent eavesdropping and tampering, and mutual authentication of user and network to ensure that sensitive information is transmitted only to the legitimate network are, therefore, necessary [1].

Network access security constitutes a group of security features providing mobile users with secure access to a serving network. The wireless part of the connection, the so-called radio access link, requires particular protection. For example, in Universal Mobile Telecommunications System (UMTS) the network access security comprises the following features.

- Mutual authentication of user and serving network.
- Confidentiality – user traffic and signalling data needs to be encrypted over the radio link.

- Anonymity – confidentiality of the mobile user's identity over the radio link. In UMTS, this is achieved with temporary identities, called Temporary Mobile Subscriber Identities (TMSIs).
- Signalling data integrity and origin authentication – the security of the signalling data is mandatory in UMTS and protects against man-in-the-middle attacks.

Today, access procedures to mobile networks are based on secret key techniques. If a mobile user seeks access to his/her home network, both communication partners share a common secret through which the mobile user and the home network mutually authenticate and establish confidential communications. If a mobile user seeks access to a foreign access network, access is granted by exploiting the trust relationship between the mobile user and the home network, and a roaming agreement between the foreign access network and the home network. Hence, in the roaming scenario, a triangular relationship between mobile user, home network and access network is utilised.

In the design of 3G (third-generation) UMTS networks, public key techniques for network access were considered, but were abandoned later on for reasons of complexity [2]. This was mainly due to the fact that some of the key advantages offered by public key techniques were not seen as sufficient to justify their deployment. After all, changing from symmetric to asymmetric techniques in the transition from 2G to 3G would have required new and more powerful smartcards in the user equipment and would have ruled out any option for 3G mechanisms to be backward compatible [3]. However, given the fact that post-3G systems may have different architectural and business assumptions, public key techniques seem to be worth a new consideration. The security architecture of current and potential future mobile systems has been studied in the IST SHAMAN project [4].

If network access procedures are based on secret key techniques, there are imminent problems with user identity confidentiality and non-repudiation. In Global Systems for Mobile (GSM) communications, user identity confidentiality is achieved by means of temporary identities. However, this approach is open to active attacks. Moreover, secret key techniques cannot offer solutions for non-repudiation.

The major advantages of a public key based approach to network access are as follows.

- Online connection to the mobile user's home network is not required for authentication purposes.
- User identity confidentiality can be achieved.
- Non-repudiation can be accomplished, which means that neither the mobile user nor the access network are able to repudiate a connection after its termination.
- A public key based system may be reusable for other purposes, such as electronic payment or digital signatures.

In References 5 and 6, we present an approach to 'alternative access' where the mobile user does not have a home network or the home network is not accessible. Instead, access to foreign access networks is granted by means of ad hoc electronic payment. In this case, public key based techniques to network access are the only feasible option, since there is no home network with which the mobile user shares a secret key and through which billing of services is made.

In this chapter, we want to give an overview of techniques for public key based network access. In Section 8.2, we start by surveying current secret key based approaches. Section 8.3 gives a background on public key solutions. Different public key protocols for network access are introduced in Section 8.4.

8.2 Existing approaches to network access

8.2.1 History

A few decades ago, all cryptographic algorithms and security protocols were based on secret keys that the communication partners needed to share. It was even assumed that the exchange of secret keys over a secure channel was a prerequisite for establishing a protected communication over an unsecured channel. Symmetric-key encryption is characterised by a close relationship (or even identity) between encryption and decryption keys and algorithms [7]. Similarly, a security protocol (e.g. for authentication, key agreement) is called symmetric key protocol if it is based on a secret key that the adversaries share.

It was only in the 1970s when Diffie and Hellmann invented[1] a method for secret key exchange over unsecured channels [8] and Rivest, Shamir and Adleman found a method (RSA) to encrypt messages with a public key [9] that public key cryptography emerged.

8.2.2 Secret key protocols in mobile telecommunication

Before the introduction of GSM, telecommunication was mostly unprotected. Neither fixed line systems nor analog wireless systems offered standardised encryption or authentication mechanisms. Today, security protocols are an essential element in modern telecommunication systems. The growing demand for secure communication is also caused by the popularity of mobile services: security is regarded as particularly important when users communicate over a wireless air interface.

In wireline networks, physical connections between users and network equipment offer an inherent level of security. However, in wireless networks, communication can be easily intercepted in transit using wireless network analysis software. Hence, in order to avoid eavesdropping and man-in-the-middle attacks, strong security mechanisms are required.

Modern mobile systems offer authentication mechanisms and protection against unauthorised access, eavesdropping and tampering. The security mechanisms usually cover the air interface but end in the operator's network, which is often assumed to be protected. The mobile user may either trust the transit networks or negotiate a separate security protocol (on the network, transport or application layer) with the target system or application.

Despite the availability of public key mechanisms, secret key protocols still play a substantial role in the protection of mobile communication. Current wireless technologies offer physical or link layer security mechanisms, as listed in Table 8.1.

Table 8.1 Security protocols for mobile technologies

Mobile technology	Authentication and key agreement protocols	Integrity protection and encryption protocols
GSM (Global System for Mobile Communications)	A3/A8 (operator specifiable)	A5/1, 2, 3
GPRS (General Packet Radio Service)	A3/A8 (operator specifiable)	GEA/1,...,7
UMTS (Universal Mobile Telecommunications Systems)	f1, f1*, f2, f3, f4, f5, f5*	KASUMI f8, f9
DECT (Digital Enhanced Cordless Telecommunications)	DSAA	DSC
TETRA (Terrestrial Trunked Radio)	TAA1	TEA1,...,4
IEEE 802.11 (Wireless Local Area Networks)	WEP	CRC, WEP
Bluetooth	E1, E2, E3	E0

In fact, all these protocols are secret key protocols. In the following, we discuss some characteristic properties of current security protocols.

For public mobile telecommunication networks using GSM, General Packet Radio Service (GPRS) or UMTS technology, the secret key Ki (resp. K) is stored in the operator's authentication centre (AuC) and on a smartcard (the SIM (Subscriber Identity Module) and USIM (UMTS Subscriber Identity Module), respectively), which the user inserts into his/her mobile phone [10]. This method requires a previously established relationship (subscription or roaming agreement) between the mobile operator and the user. Authentication is based on a challenge–response protocol using the secret key Ki. Both the network and the mobile user derive secret session key(s) from Ki and user data are then encrypted with these keys using symmetric protocols (A5 resp. f8). The key Ki itself is never sent over the network.

The use of secret keys requires additional communication between the home and the visiting network in the roaming scenario. Since the visiting network does not possess the user's secret master key, it needs to contact the user's home domain for authentication data and keying material. A trust relationship between the user, respectively, home domain and the visiting network is necessary in this scenario. The access procedure is depicted in Figure 8.1.

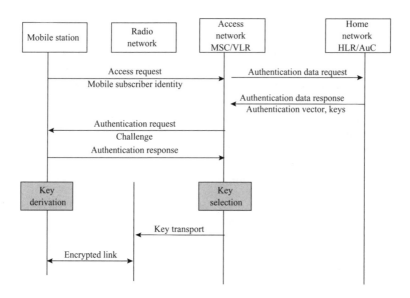

Figure 8.1 Main steps in authentication and key agreement for GSM and UMTS

Other local wireless technologies (e.g. IEEE 802.11 Wireless LAN, Bluetooth) require symmetric configuration of a secret key in both the user device and the access point (or the peer device). The Wireless LAN security protocol Wired Equivalent Privacy (WEP) employs the stream cipher RC4 for authentication and encryption. Since WEP has serious weaknesses [11], the IEEE will replace it by the new security standard IEEE 802.11i. Authentication and key agreement will then probably be based on IEEE 802.1X [12] and on Extensible Authentication Protocol (EAP) [13], which can employ public key methods (e.g. Transport Layer Security (TLS) [14]; see Section 8.4.2.1). The forthcoming 802.11i standard will also define new encryption and integrity protection protocols.

Bluetooth was mainly designed for spontaneous communication between peers and within personal area networks. Security mechanisms are optional but if link level security is enforced then the Bluetooth peers employ a symmetric link key that they establish in a pairing procedure. This procedure (which may be regarded as vulnerable) is based on a shared passkey (PIN), which is either typed in or set to a factory default. The subsequent authentication (with a challenge–response protocol) and the generation of a secret encryption key use the link key.

Higher layer security protocols (network, transport and application layers) are often realised as hybrid protocols where public key techniques are used to authenticate and to establish a symmetric session key, which is then used for the encryption of bulk data.

An example is the network layer encryption protocol Encapsulating Security Payload (ESP) [15] within the IPsec suite. ESP uses a secret key for encryption (e.g. with the 3DES cipher), which can be negotiated by a preceding public key protocol, for example, with Internet Key Exchange (IKE) [16], see also Section 8.4.1 for the IKE-replacement protocol. Similarly, the popular transport layer security protocol SSL/TLS [14] employs a symmetric key for data encryption, but public key methods are used for authentication (often only unilateral server authentication) and for encryption of the symmetric key. Also, the public key crypto application Pretty Good Privacy (PGP) [17] encrypts bulk data with a symmetric cipher (IDEA). Here, public key mechanisms are used to protect the symmetric key (using RSA or Diffie–Hellmann algorithms) and to carry out digital signatures (with RSA or the Digital Signature Standard, DSS).

8.2.3   Advantages and disadvantages of secret key protocols

Symmetric key cryptography is well established, thoroughly investigated and available in hard- and software. The main advantages are [7]:

- symmetric key ciphers can be designed for high throughput;
- symmetric keys are relatively short compared to public/private key pairs;
- symmetric ciphers can be composed to produce stronger ciphers with longer keys;
- symmetric ciphers are easier and cheaper to implement on smartcards than public key ciphers (bit operations versus multiplications).

But there are some well known disadvantages (cf. Reference 7 for some points and the discussion in Section 8.3.3):

- the secret keys need to be agreed (either over a secure channel or with a separate protocol);
- the keys must remain secret at both ends;
- in a large network, a considerable number of keys ($\frac{1}{2}n(n-1)$ in a fully meshed network with n peers, or n keys in a centralised network with n clients) have to be managed;
- a frequent change of secret keys would be recommended but may turn out to be organisationally difficult;
- non-repudiation of user, respectively, network data or actions cannot be achieved;
- it is more difficult to keep the user identity secret (GSM and UMTS use temporary identities here).

Many current cryptographic systems and protocols exploit the strength of both secret and public key algorithms (hybrid cryptographic systems). Symmetric key mechanisms can be efficiently employed for encryption and integrity protection of bulk data but there are some limitations when they are used for entity authentication and key agreement.

8.3   Public key based network access

8.3.1   Public key infrastructure

In order to support a public key based approach to network access, a public key infrastructure (PKI) needs to be established that takes the particularities of the mobile environment (such as limited processing power, memory and display size of a mobile device and limited bandwidth on the air interface) into account. The following components are needed to form a public key infrastructure:

- (public key) users, in this case mobile users, access and home networks that need to communicate securely;
- certification authorities (CAs) that issue certificates;
- certificate servers that store valid certificates and certificate revocation lists.

An in-depth description on PKI in a mobile environment is given in Reference 18 (Chapter 2).

8.3.2 Motivation

So far, public key based techniques have only played a subsidiary role in network access. However, in future mobile telecommunication systems, this may change. The following paragraphs give reasons why public key based techniques for network access could be a means of choice in future mobile systems.

The number of access networks may grow. If future telecommunication systems support roaming to networks that provide hotspot coverage, then the total number of operators and, hence, roaming agreements between operators may grow significantly. If the number of operators reaches a critical level then bilateral agreements will be inefficient, especially for small operators. In GSM, small operators often use roaming brokers such that they do not have to establish bilateral roaming agreements with all other operators. Instead, they establish a single roaming agreement with the broker, which, in turn, has roaming agreements with lots of other operators. The roaming broker acts as a signalling gateway during roaming for the exchange of user profiles, authentication data, real-time charging information for pre-pay subscribers, etc. It also acts as a clearinghouse for billing information for contract (i.e. post-pay) subscribers.

Stronger requirements for user privacy. Requirements for user privacy seem to be getting stronger. Public key authentication offers more possibilities for providing strong user identity confidentiality. The (permanent or temporary) user's identity does not need to be transmitted in plaintext over the air interface. The protocol starts anonymously and the identity can be sent encrypted after network authentication using a public key. In GSM and UMTS, 'International Mobile Subscriber Identity (IMSI) catching' is possible and the network can always require the mobile node to send the user's identity in plaintext. Because of the need to provide for recovery in the event of loss of synchrony for temporary user identities between a mobile and the network, the level of user privacy that can be supported using only secret key cryptography is inevitably imperfect.

Public key techniques could be re-used for other purposes. Public key mechanisms for network access could be re-used for application security, electronic payment and digital signatures. In the past, it was claimed that the use of public key cryptography for network access could stimulate the growth of a mobile PKI, which could then be used as an enabler for m-commerce. However, perhaps things will work in the other direction. As PKI systems for m-commerce are deployed, it might become more attractive to re-use that infrastructure for basic network access authentication.

No online connection to the home network for the purpose of authentication is required.　In GSM, the triplets used for authentication are generated within the home network (i.e. HLR/AuC) and must be retrieved from there. Public key based techniques would allow the access network to authenticate the user without involvement of the home network.

The user is unable to deny that he or she actually used the network.　This is known as non-repudiation, and may save network operators from losing revenue when a user claims that calls were never made. Non-repudiation also ensures that the network operator cannot deny receiving particular information from the user.

8.3.3　Characteristics of symmetric versus asymmetric solutions for network access

Choosing between symmetric and asymmetric techniques for network access is not an obvious task. There are advantages and disadvantages in both solutions. The following sections discuss some of the characteristics that may play a role in the decision process.

8.3.3.1　Computational burden

Public key algorithms are typically more computationally demanding than secret key algorithms given the current stage of development of cryptography; indeed, there is no reason to believe that this balance will change in the foreseeable future. In the past, this was usually given as one of the main reasons for not considering public key solutions for network access security where the latency and cost of authentication and key agreement mechanisms must be kept small. However, the development of less computationally demanding algorithms (e.g. those based on elliptic curves) and, in particular, the availability of more powerful low-cost processing technology (particularly at the user end) has meant that public key based solutions are much more feasible for future systems.

Possible asymmetric load distributions can help to mitigate the computational load problems arising from the use of public key cryptography. Protocols can be designed, for instance, where the public key cryptographic operations at the user end are much less demanding than the corresponding operations required at the network end.

Furthermore, when hybrid cryptosystems (cf. Section 8.2.2) are used, the computational burden for the public key part is rather small and the ciphering can be efficiently realised even on limited devices or smartcards.

Typically, public key based schemes require less online interaction between the access network and the home network than secret key based schemes. This may increase the level of interactivity required at the authentication point in the access network, and may require more complex processing and storage capabilities. This will be significant since the network access security functionality in the access network will typically be distributed among nodes at some point near the edge of the network.

8.3.3.2 Cryptographic parameter size

Key sizes for public key algorithms (e.g. 1024 bits for RSA) are generally much larger than those required for secret key algorithms (e.g. 128 bits for AES). Only elliptic curve algorithms allow smaller key sizes (e.g. a minimum of 160 bits for the Elliptic Curve Digital Signature Algorithm, ECDSA). Furthermore, when cryptographic mechanisms are required to provide data origin and integrity protection, digital signatures created using public key techniques are larger than corresponding secret key based message authentication codes (MACs). Both of these characteristics have a particular bearing on network access security because of the typically limited storage at the user end and the nature of the transmission channels over the air interface.

8.3.3.3 Algorithm availability

It has always been claimed that there are a limited number of generally accepted public key algorithms, and devising secure new schemes appears to be difficult. On the other hand, the methods needed to produce new symmetric algorithms are relatively well understood. The choice in asymmetric algorithms may also be limited due to the fact that a greater proportion of public key algorithms are patented.

However, this issue is probably not of any great significance today. In terms of standardised techniques, the choices for symmetric and asymmetric techniques appear roughly equally wide.

8.3.3.4 Frequency of communication between access and home network

While in secret key techniques the home network must be involved during the initial authentication procedure, the use of public key techniques can limit the need for home network involvement. Such interaction may increase the latency and cost of individual authentication processes, particularly when the access network and home network are very remote. Indeed, even if the user is roaming on an access network, which is closely associated with the home network, the need for online communication with the home network may still be a significant cost.

The advantage of less interaction with the home network might also be particularly attractive in future systems where there is a trend towards smaller access networks, which also has the effect of increasing the mobility of users between different access network authentication points.

Furthermore, the online interaction requires high availability of the home network's authentication infrastructure and short response times (e.g. within fraction of a second) since the user's access is usually blocked before completion of set-up procedures.

A disadvantage of having less interaction with the home network is that it is generally more difficult for the home network to authorise individual service requests without complex revocation mechanisms. However, this may also be considered to be a problem in secret key based schemes where re-authentication is often possible without explicit reference to the home network.

If public key cryptography is used then there may no longer be a need for real-time communications between access network and home network for authentication purposes. However, the advantage of not needing online interaction with the home network for authentication may be partly counteracted by the fact that access networks may need to interact with the home network anyway to support mobility management and the delivery of real-time charging and user-status information. Although, the fact remains that these interactions can occur in parallel with MN authentication and will thus not necessarily increase access latency.

It should also be noted that the access network may need to contact entities in a PKI (i.e. a Certificate Server) for verifying user certificates. Depending on the configuration, this can be time consuming and might eat up some or all of the time savings achieved by not contacting the home network during the initial authentication.

8.3.3.5   User identity confidentiality

With secret key based schemes it is not always possible for the user to be able to send his identity or certificate to the access network in encrypted form. Temporary identities are, therefore, typically used to provide identity and location confidentiality. These mechanisms often require a facility to allow the network to request the user identity in plaintext so that it can recover from temporary identity mismatch. In general, it seems difficult to protect against active attacks on the user identity where the attacker spoofs this message in order to obtain the user identity in the clear (even the 3GPP mechanism is not completely secure against such attacks). The asymmetric nature of public key techniques, however, allows us to overcome these problems by providing a mechanism whereby the user's identity can always be securely sent to the network by encrypting it using an authentic copy of the access network's public encryption key.

8.3.3.6   Non-repudiation of network access

Non-repudiation of network access means that the user is unable to deny that he or she used the network. The use of public key techniques for network access security can make it much easier to achieve non-repudiation of network access through the use of digital signature systems (in combination with appropriate time-stamping). Non-repudiation of network access is important if the trust relationships are such that the access network needs to be able to present strong evidence to prove a particular user's chargeable actions to the home network, for example, in order to resolve accounting disputes, or if the home network needs to be able to prove a user's chargeable actions to an arbitrator in order to resolve a billing dispute. This may be the case if larger numbers of smaller access networks come into existence.

8.3.3.7   Re-use of security infrastructure for end-to-end security

An advantage of using public key techniques to provide network access security is that the supporting infrastructure may then be available to provide end-to-end security and other value-added security services where public key mechanisms are essential. Examples include electronic payment, where non-repudiation is important, or some end-to-end security scenarios where key management in a network where there are many-to-many relationships is much easier to support if public key techniques are used.

8.3.3.8 Migration

Current solutions to network access are based on symmetric techniques (cf. Section 8.2.2). Promoting asymmetric techniques means that migration from a current symmetric to a new asymmetric-based system requires a redesign of the security architecture; the secret key infrastructure needs to be complemented by a PKI; a different set of protocols needs to be supported at the MN, the access network and the home network; public key certificates have to be issued at a trusted source; certificate revocation has to be implemented. In order to smooth the transition from the subscriber's point of view, the operators could, for example, issue SIM cards with additional public key functionality. The subscriber would use traditional symmetric key protocols for access to GSM or UMTS networks and public key mechanisms for other access networks (e.g. Wireless LAN) or to achieve end-to-end security. So far, operators and manufacturers have avoided this challenging task. However, this development may be unavoidable in future mobile telecommunications systems.

8.4 Protocols

In Sections 8.4.1 and 8.4.2, we present two important public key based protocols for network access: IKEv2 and EAP-TLS.

8.4.1 Son-of-IKE

IPsec represents a series of standards that have been developed by the IP Security (IPsec) Working Group of the IETF [19]. Several public key based key set-up and key agreement protocols have been proposed, and one, Internet Key Exchange (IKE) is the current standard [16]. IKE has a number of deficiencies, the three most important being that the number of rounds is high, that it is vulnerable to denial-of-service attacks and that the specification is complex. Currently, an IKE-replacement protocol, the son-of-IKE, is being developed by the IETF. In this section, we give an overview of the son-of-IKE, called IKEv2 [20]. IKE performs mutual authentication and establishes an IKE security association (SA) that can be used to efficiently establish SAs for IPsec's ESP (Encapsulated Security Payload, cf. Reference 15) and/or AH (Authentication Header).

IKEv2 uses Diffie–Hellman key agreement to provide unauthenticated key agreement [7]; two parties that have never met in advance establish a shared secret by exchanging messages over an open channel. Basic Diffie–Hellman works as follows: A and B own private keys a and b. A and B negotiate a Diffie–Hellmann group, that is, public parameters g and p. The corresponding public keys are g^a and g^b. The shared secret is calculated as $(g^b)^a = (g^a)^b = g^{ab} \bmod p$. In the protocol shown in Table 8.2, the nonces N_A and N_B are used to ensure key freshness.

Table 8.2   IKEv2 phase 1 between endpoints A and B

	A		B
(1) IKE_SA_init_request	$\dots SA_{A1}, g^a, N_A$	$\rightarrow$	
(2) IKE_SA_init_response		$\leftarrow$	$\dots [CERTREQ.] SA_{B1}, g^b, N_B$
(3) IKE_auth_request	$\dots SK\{ID_A,$	$\rightarrow$	
	$[CERT,],$		
	$[CERTREQ,],$		
	$[ID_B,] AUTH \dots\}$		
(4) IKE_auth_response		$\leftarrow$	$\dots SK\{ID_B, [CERT,] AUTH\dots\}$

In the following, we explain the initial exchange of IKEv2, called 'phase 1' exchange, consisting of four messages (two request/response pairs). Subsequent IKEv2 exchanges are called 'phase 2' exchanges and are not described further in this document. 'Phase 2' exchanges may be used to establish additional CHILD-SAs between the same authenticated pair of endpoints and to perform housekeeping functions [20]. Please note that for simplicity reasons, we have left out all those details of the protocol that are not essential in understanding its basic cryptographic features.

Messages (1) and (2) in IKEv2 exchange Diffie–Hellman exponentials (g^a and g^b, respectively) and random numbers, the nonces N_A and N_B. Nonces achieve freshness in the key derivation procedure. Moreover, the SA payloads contain a choice of cryptographic algorithms that A and B support. Neither message (1) nor message (2) are encrypted or authenticated. That means, that at this stage, A and B do not have any proof of the real identity of their peers.

At this point in the negotiation (which means, after message (2)), each party can generate SKEYSEED from which all keys for encryption and authentication are derived. SKEYSEED is computed as follows:

$$SKEYSEED = prf(N_A|N_B, g^{ab})$$

where prf is a pseudo-random function, N_A and N_B are the nonces exchanged during the IKE_SA_init exchange and g^{ab} is the Diffie–Hellman shared secret established during the exchange. Each party, A or B, computes a corresponding set of keys for encryption and authentication that are denoted SK_{Ae} and SK_{Be}, and SK_{Aa} and SK_{Ba}, respectively. Please note that SK_d is used for deriving new keys for CHILD-SAs and is not relevant in the initial exchange:

$$\{SK_d, SK_{Aa}, SK_{Ba}, SK_{Ae}, SK_{Be}\}$$

$$= prf + \{SKEYSEED, g^{ab}|N_A|N_B|CKY_A|CKY_B\}$$

where prf+ describes the function that outputs a pseudo-random stream (by using prf iteratively) and CKY_A and CKY_B are A's and B's cookies.[2] SK_d, SK_{Aa}, SK_{Ba}, SK_{Ae}, and SK_{Be} are taken in order from the generated bits of prf+. The notation $SK\{\dots\}$

in messages (3) and (4) indicates that the corresponding payloads are encrypted and integrity protected using that direction's keys, SK_e and SK_a, respectively.

In message (3), A asserts his/her identity with ID_A and the AUTH payload authenticates ID_A and integrity protects the contents of the first two messages. Moreover, A may send his/her certificate in the CERT payload and a list of his/her trusted anchors for verifying B's certificate in certificate request CERTREQ payload(s). In ID_B, A can specify which of B's identities she/he wants to talk to, because B may use multiple identities at the same IP address.

B's response in message (4) has a similar structure. B asserts his/her identity with ID_B and the AUTH payload authenticates ID_B and integrity protects the contents of the first two messages. Again a certificate or certificate chain CERT may be included to provide evidence that the key used to compute the digital signature belongs to the corresponding identity.

Mutual authentication of A and B can be based on either of the following.

- Digital signatures (using public key cryptography)
  - $SIG_A(\ldots SA_{A1}, g^a, N_A, N_B)$
  - $SIG_B(\ldots [CERTREQ,]SA_{B1}, g^b, N_B, N_A)$
- MACs (using a pre-shared secret key)
  - prf (shared secret | 'key pad for IKEv2', <message bytes>) where 'key pad for IKEv2' is added if the shared secret is derived from a password, in order to make sure that this exchange will not compromise the use of this password in other protocols.

Please note that all payloads of messages (1) and (2) are included under the signature or MAC in order to guarantee the integrity of the information sent in the first two messages.

8.4.2 Public key based EAP methods

EAP is an authentication protocol that supports multiple authentication mechanisms, the so-called EAP methods [13]. EAP can run directly over the link layer without requiring IP. While EAP was originally developed for use with the Point-to-Point Protocol (PPP) [21], it is also in use with IEEE 802.1X [12,13] and can be regarded as a general framework for various authentication and key agreement protocols.

Although EAP supports a bunch of authentication methods, only few are commonly used today and among them, three are public key based, namely the following.

- EAP-Transport Layer Security (EAP-TLS) [22] – for public key certificate based mutual authentication.
- Tunnelled Transport Layer Security (TTLS) [23] – for public key certificate based server authentication; the client uses another authentication method.
- Protected EAP (PEAP) [24] – for public key certificate based server authentication; an EAP exchange is run over the TLS tunnel.

In EAP-TLS, digital certificates are used to authenticate clients and servers. Although server authentication is well established in current TLS implementations, client authentication is rare and requires the roll-out of a complete PKI on the client side, providing certificate generation, distribution and revocation. This represents an enormous overhead to organisations wanting to employ EAP-TLS. TTLS and PEAP were developed to ease the PKI problems of EAP-TLS. Here, unilateral server authentication is combined with well-established, older methods of client authentication that do not impose the management overhead of running a complete PKI. Another advantage of TTLS and PEAP is that, as opposed to TLS, the EAP exchange and, in particular, the user identity is protected by a TLS tunnel. In TLS, the user identity is sent in the clear, before certificates are exchanged. Recently, a flaw in the two-step protocols TTLS and PEAP has been discovered [25], which allows a man-in-the-middle to access resources that are granted to a legitimate and successfully authenticated client. The three protocols are introduced in the subsequent sections.

According to Reference 13, the following entities take part in an EAP exchange.

- *Authenticator.*   The end of the link requiring authentication.
- *Peer.*   The other end of the link that is being authenticated by the authenticator.
- *Authentication server* (also called EAP server).   An entity that provides an authentication service to an authenticator. This service verifies the credentials provided by the peer.

The authentication server acts as a 'back-end' server that implements the corresponding authentication mechanisms, while the authenticator passes through the authentication exchange and performs access control (i.e. blocks all other traffic for unauthenticated peers).

8.4.2.1   EAP-TLS

EAP-TLS is a public key based EAP method that has been standardised by the IETF [22]. It is based on the SSL/TLS protocol [14] that is commonly employed for securing Internet traffic on the transport layer.

TLS authentication within EAP works as follows: each message in a TLS session establishment dialogue between the peer and the authentication server is packed into an EAP-TLS packet. When the TLS authentication dialogue succeeds, the authenticator is informed and access to the network is granted.

EAP-TLS sets up an encrypted channel between the peer and the authentication server. However, the peer wants to establish secure communications with the authenticator, not with the authentication server. Hence, the keying material created during the EAP-TLS exchange is transferred securely from the authentication server to the authenticator.

In common Internet applications, SSL/TLS is employed with unilateral server-side authentication only. This means the client does not send a public key certificate to the server for authentication (for the mere reason that client certificates are usually not installed in the end-user's web browsers). However, mutual authentication is desired in EAP-TLS in order to avoid man-in-the-middle attacks. The fact that peers have to be equipped with public key based certificates to exploit the full capabilities of EAP-TLS represents a complication in the employment of the protocol. Many organisations are not ready for that (see discussion on migration in Section 8.3.3.8).

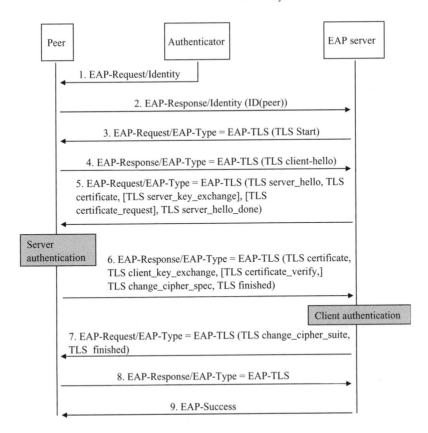

Figure 8.2 EAP-TLS exchange

As a result of the EAP-TLS conversation, the EAP end-points mutually authenticate, negotiate a cipher suite and derive a session key.

The EAP-TLS conversation typically begins with the authenticator and the peer negotiating EAP. Then, the authenticator sends an *EAP-Request/Identity* packet to the peer (step 1 in Figure 8.2), and the peer responds with an *EAP-Response/Identity* packet to the authenticator, containing the peer's user ID (step 2). From this point forward, while nominally the EAP conversation occurs between the authenticator and the peer, the authenticator may act as a pass-through device, with the EAP packets received from the peer being encapsulated for transmission to an authen-tication server (called here the EAP server). Having received the peer's identity, the EAP server responds with an EAP *TLS Start* packet (step 3). The peer answers with a *client_hello* handshake message (step 4), followed by the EAP server's TLS *server_hello* handshake message (step 5). At this point, the peer authenticates the EAP server (server authentication). The next message (step 6) contains, among other things, a *client_key_exchange* message, which completes the exchange of a shared master secret between the peer and the EAP server. If the EAP server sent a *certificate_request* message in the preceding EAP-Request packet, then the peer must send, in addition, *certificate* and *certificate_verify* handshake messages. The former contains a certificate for the peer's signature public key, while the latter contains the peer's signed authentication response to the EAP server. After receiving this packet, the EAP server verifies the peer's certificate and digital signature, if requested (client authentication). If the peer authenticates successfully, the EAP server sends a response containing a *finished_handshake* message (step 7). If the EAP server is correctly authenticated, the peer must send an *EAP-Response* packet of *EAP-Type=EAP-TLS*, and no data (step 8). The EAP server must then respond with an *EAP-Success* message (step 9).

Figure 8.2 illustrates the protocol steps in the case where the EAP-TLS mutual authentication is successful. For further examples of protocol flows (i.e. where fragmentation is required, where client and server authentication respectively fail, where a previously established session is being resumed, etc.) see Reference 22.

8.4.2.2   EAP-TTLS

Tunnelled TLS (TTLS) is a two-step protocol that extends EAP-TLS [23]. EAP-TTLS is suitable for those users who want the security of TLS, but have a legacy authentication mechanism on the client side. Hence, it combines unilateral server authentication (the server is authenticated by the peer) using EAP-TLS with a legacy authentication mechanism for client authentication (the peer is authenticated by the server). In EAP-TTLS, a secure connection established by the TLS handshake, the so-called TLS tunnel, is used to run another authentication protocol for authenticating the user. This other authentication protocol could be based on clear-text passwords, challenge–response passwords or token-based authentication. The second step is based on the exchange of 'attribute-value-pairs' (AVPs). Thus, EAP-TTLS allows legacy-based authentication protocols to be used against existing authentication databases, while protecting the security of the legacy protocol against eavesdropping, man-in-the-middle and other cryptographic attacks [23].

One advantage of EAP-TTLS in comparison to EAP-TLS is that the user identity is only sent after the TLS tunnel is established. Hence, the user identity is protected during transmission.

According to Reference 23, the EAP-TTLS negotiation comprises two phases:

* phase 1 – TLS handshake phase;
* phase 2 – TLS tunnel phase.

In phase 1, the TTLS server is authenticated to the client and, optionally, the client is authenticated to the TTLS server (see Figure 8.3). The TTLS server is an AAA server which implements EAP-TTLS. At the end of phase 1, a secure tunnel exists between the client and the TTLS server ('secure password authentication tunnel' in Figure 8.3). It should be noted that the client name should not be sent in the clear until the secure tunnel is established between the client and the TTLS server. This way, the user identity is protected.

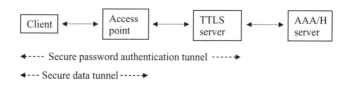

Figure 8.3   EAP-TTLS network architecture [23]

In phase 2, information can be tunnelled between the client and the TTLS server. This information might include not only user authentication, but also key distribution, communication of accounting information and a number of other functions. The user authentication may itself be EAP or a legacy authentication protocol. The user authentication information is forwarded to the AAA/H server, which is an AAA server in the user's home domain and responsible for authentication and authorisation for the specific user. If the user authentication between client and AAA/H server is successful, the TTLS server issues an '*EAP-Success*' message and distributes shared keying information to the client and access point 'to permit encryption and validation of the wireless data connection subsequent to authentication, to secure it against eavesdroppers, and to prevent channel hijacking' [23] ('secure data tunnel' in Figure 8.3). It should be noted that user authentication is not necessary in phase 2, if the user has already been authenticated via the mutual authentication option of TLS.

8.4.2.3   PEAP

PEAP [24] is another two-step protocol that establishes a server authenticated TLS tunnel in the first step and then uses this tunnel to protect the client authentication in the second step. Similar to EAP-TTLS, PEAP protects the EAP negotiation that addresses a number of deficiencies of plain EAP. The main difference to EAP-TTLS is that the second step is not based on attribute-value-pairs but on a separate (and hence protected) EAP exchange.

8.5   Conclusions

In this chapter, we have discussed current approaches to network access that are mainly secret key based, and have given a motivation for public key based and hybrid approaches in future network access, especially in access to mobile networks. We have presented two important public key based protocols that provide mutual authentication of client and server and session key establishment, the son-of-IKE protocol IKEv2 and EAP-TLS. Moreover, we have sketched two extensions of EAP-TLS, namely TTLS and PEAP. Whereas EAP-TLS requires the rollout of a full-fledged PKI, TTLS and PEAP combine the security of unilateral server authentication in basic TLS with the convenience of employing a legacy authentication protocol for the client, for example, based on passwords or secret key approaches.

Acknowledgements

The authors would like to thank Hannes Tschofenig for his valuable comments on an earlier version of this chapter.

Notes

1 Ellis, Cocks and Williamson who worked at the British GCHQ organisation probably found public key mechanisms some years before but the results were not
published at that time.
2 In IVEv2, cookies are used to defeat Denial-of-Service (DoS) attacks. Cookies
are included in the IKEv2 message header. In this paper, we will not explain the
concept of cookies further.

References

1 GAST, M.: 'A technical comparison of TTLS and PEAP', The O'Reilly Network,
http://www.oreillynet.com/lpt/a/2827, October 2002
2 3GPP TS 33.102 v4.3.0: '3rd Generation Partnership Project', Technical Specification Group Services and System Aspects, 3G Security, Security Architecture,
Release 4
3 HOWARD, P. and GOSSET, P. (Eds): 'ASPeCT: Advanced security for personal
communications technologies, D20 – Project final report and results of trials,
http://www.esat.kuleuven.ac.be/cosic/aspect/, December 1998
4 SHAMAN (Security for Heterogeneous Access in Mobile Applications and
Networks): IST-2000-25350 SHAMAN, http://www.ist-shaman.org
5 KNOSPE, H. and SCHWIDERSKI-GROSCHE, S.: 'Future mobile networks:
ad-hoc access based on online payment with smartcards', 13th IEEE International
Symposium on *Personal, Indoor and Mobile Radio Communications* (PIMRC
2002), September 2002, pp. 197–200
6 KNOSPE, H. and SCHWIDERSKI-GROSCHE, S.: 'Online payment for access
to heterogeneous mobile networks', IST Mobile & Wireless Telecommunications
Summit 2002, Thessaloniki, Greece, June 2002, pp. 748–52
7 MENEZES, A., VAN OORSCHOT, P. C. and VANSTONE, S. A.: 'Handbook
of applied cryptography' (CRC Press, Boca Ration 1996), http://www.cacr.math.
uwaterloo.ca/hac/
8 DIFFIE, W. and HELLMAN, M.: 'New directions in cryptography', *IEEE
Transactions on Information Theory*, 1976, **IT-22** (6), pp. 644–654
9 RIVEST, R., SHAMIR, A. and ADLEMAN, L.: 'A method for obtaining digital
signatures and public key cryptosystems', *Communications of the ACM*, 1978,
21, (2), pp. 120–26
10 WALKER, M. and WRIGHT, T.: 'Security', in HILLEBRAND, F. (Ed.):
'GSM and UMTS: The creation of global mobile communication', (John Wiley,
New York), 2001

11 BORISOV, N., GOLDBERG, I. and WAGNER, D.: 'Intercepting mobile communications: the insecurity of 802.11', in Proceedings of MOBICOM 2001, http://citeseer.nj.nec.com/article/borisov01intercepting.html

12 IEEE Standard 802.1X-2001: 'Port-based network access control'

13 BLUNK, L. and VOLLBRECHT, J.: 'PPP Extensible Authentication Protocol (EAP)'. Internet RFC 2284, March 1998

14 DIERKS, T. and ALLEN, C.: 'The TLS protocol version 1.0,' Internet RFC 2246, January 1999

15 KENT, S. and ATKINSON, R.: 'IP encapsulating payload (ESP)'. Internet RFC 2406, November 1998

16 HARKINS, D. and CARREL, D.: 'The internet key exchange (IKE)', IETF RFC 2409, November 1998

17 ZIMMERMANN, P.: 'PGP™ user's guide, volume II: special topics', ftp://ftp.pgpi.org/pub/pgp/2.x/doc/pgpdoc2.txt

18 DANKERS, J., GAREFALAKIS, T., SCHAFFELHOFER, R. and WRIGHT, T.: 'PKI in mobile systems', this issue (Chapter 2)

19 IP Security Protocol Working Group (IPSEC): Charter, http://www.ietf.org/html. charters/ipsec-charter.html

20 KAUFMAN, C. (Ed.): 'Internet Key Exchange (IKEv2) Protocol', IETF Internet Draft, draft-ietf-ipsec-ikev2-04.txt, January 2003

21 SIMPSON, W.: 'The Point-to-Point Protocol (PPP)', STD51, RFC 1661, July 1994

22 ABOBA, B. and SIMON, D.: 'PPP EAP TLS Authentication Protocol', Internet RFC 2716, October 1999

23 FUNK, P.: 'EAP Tunneled TLS Authentication Protocol (EAP-TTLS)', Internet Draft, draft_ietf-pppext-eap-ttls-02.txt, November 2002

24 ANDERSON, H., JOSEFSSON, S., ZORN, G., SIMON, D. and PALEKAR, A.: 'Protected Extensible Authentication Protocol (PEAP)', Internet draft, draft-josefsson-pppext-eap-tls-eap-05.txt, September 2002

25 ASOKAN, N., NIEMI, V. and NYBERG, K. 'Man-in-the-Middle in tunnelled authentication', Cryptology ePrint Archive: Report 2002/163, http://eprint.iacr. org/2002/163/, November 2002

Chapter 9

Security in personal area networks

Christian Gehrmann and Kaisa Nyberg

This chapter discusses security for personal area networks (PANs). An overview of different PAN security issues and solutions is given. We define a PAN reference and trust model. A PAN security architecture based on the model is described. Especially we provide new solutions to the PAN device security initialisation problem using manual authentication techniques. We show that PAN key management can be substantially simplified using trust delegation or a personal 'Public Key Infrastructure (PKI)'. Internal PAN communication security as well as secure configuration and access control is discussed.

9.1 Introduction

The next generation of mobile communications is expected to be different from current systems. We foresee changes both for the type of accesses to the networks and the terminals used to access the networks. We expect future multi-function mobile terminals to consist of several different configurable components that may be worn about the body and are connected through local wireless communication. Hence, we address security problems for distributed dynamically configurable terminals. A distributed terminal consists of several components within physical proximity to each other and to the user or users. Components are interconnected with local communication links such as short-range wireless connections, for example, Bluetooth. This type of personal local network used to be called a Personal Area Network (PAN). We treat the security problems related to the configuration and communication in a PAN. This chapter is based on the work by the SHAMAN project [1].

9.1.1 PANs and PAN security

In this chapter, we focus on personal networks consisting of a limited number of components within the proximity of a person. Only components owned and controlled by one user and the components directly communicating with these components are considered. Using this limitation, we can define a PAN reference model of reasonable complexity that is applicable for the distributed terminal scenarios we would like to cover.

We have as a goal to provide security architecture applicable to our PAN reference model. The basis for the architecture is a trust model that describes the basic security relations between different PAN components. We have decided to use a component centric trust model. We view the surrounding component in relation to a PAN reference component. Once the basic trust relations are defined, we can work with solutions of how to set up the security associations between the different components. A security association in combination with appropriate security protocols can be used to secure the local communication interface between components in the PAN. The level of security needed for a communication service offered by one component to other components is determined by security policies. Furthermore, the access to particular services should be restricted and should be a part of the component security policy. In summary, this chapter covers the following security topics:

- PAN reference and trust model;
- PAN device security initialisation and the personal PKI concept;
- manual authentication;
- internal PAN communication security;
- security configuration and access control.

Next, we illustrate how these different areas fit together when providing secure configuration and communication for a typical PAN scenario, a PAN business meeting example.

9.1.2   Business meeting example

We consider a business meeting scenario where two persons, an employee and a guest, meet in a room equipped with a video projector. The two persons in the room are both carrying one laptop each. The laptops contain presentation information that the users would like to present to each other using the video projector. Furthermore, after the presentation, the guest would like to send over his presentation to the employee. We assume that the video projector and the laptops support common short-range wireless interfaces that they use for the communication.

Hence, we have a PAN scenario with three different components:

- a video projector,
- a guest laptop,
- an employee laptop.

The situation is illustrated in Figure 9.1. In this example, we have only three components and no complex trust relations can be expected. We can choose to consider the whole network from any of the three components' points of view. If we consider

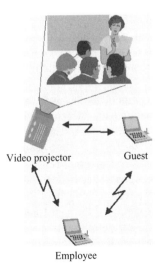

Video projector   Guest

Employee

Figure 9.1   Business meeting

the view of the video projector, it can be connected to either the employee laptop or the guest laptop. It is reasonable to assume that, since the video projector and employee laptop both belong to the same organisation, the employee laptop would trust the video projector more than the guest laptop. In Section 9.2 we will introduce a trust model that reflects this type of differential trust using three trust classes.

Since the presentation material might be sensitive information, the employee and the guest would like to have the local wireless communication protected from eavesdropping. This can be provided by a proper security protocol and a shared secret that is a part of a security association between the two components. It is reasonable to require that a security association between the video projector and employee laptop should be possible to create with no or almost no user interaction. On the other hand, it is also reasonable to assume that a security association between the guest laptop and the video projector (or the one between the two laptops) need some user interaction. Furthermore, some user actions might be required in order for the video projector to grant the guest laptop access to the projector services. These types of security aspects and requirements are included in the security architecture described in this chapter.

9.2   PAN reference and trust model

9.2.1   PAN reference model

In order to provide a PAN security architecture we first need to define a PAN. We have defined a PAN reference model that we think is easy to understand and work with, and that is applicable to practical use cases.

Inspired by the work of the IEEE 802.15 [2], we define a PAN as follows.
'A PAN is a collection of fixed, portable, or moving components within or entering a Personal Area, which form a Network through local interfaces. A Personal Area is a sphere around a person (stationary or in motion) with a typical radius of about 10 meters.'

The definition includes components that are carried, worn or located near the user's body. In particular, components, which can be networked, include (but are not limited to) computers, personal digital assistants (PDAs)/handheld personal computers (HPCs), printers, microphones, speakers, headsets, bar code readers, sensors, displays, pagers, mobile phones and smartcards.

We make a distinction between a 'Personal Network' and a 'Personal Area Network'. The first might include remote devices of the same user, whereas the latter consists of devices in the proximity of the user (possibly including devices of others within reach). In addition, our definition of a Personal Area Network does not restrict the ownership and/or usership of the participating components to a single person. Rather, components belonging to different owners/users will be considered in the context of a PAN. Below, we give the rest of the basic PAN reference model definitions. The terminology we introduce is used in the rest of the chapter.

Component. A PAN consists of components. Each component is an independent computing unit. That is, it must have processing capabilities as well as digital memory. A component might have a user interface, but it need not. A component must have at least one local interface that it can use to connect directly to at least one other component. A direct connection can be wired or wireless, but it excludes intermediate hops, and implies physical proximity between the connected components. Note that a direct connection between every pair of components is not required, but each component must be able to connect to at least one other component. A component can be both stationary and mobile.

Service. A service is a communication or computing service offered by a component either locally (i.e. through a user interface of some sort) or remotely to other components. A service need not be security related. Each component keeps a list of services it offers as well as rules/policies for access and service discovery and/or advertisement.

Application. An application is a process running on a component. An application can be a service offered within a component or to other components. An application might try to connect to other components and utilise the services they offer.

User. The user of a component is the person who physically controls and operates the component in accordance with the policies configured in the component. In most cases, this requires the user to authenticate to the component (e.g. by a PIN), although certain components and/or policies may not have this requirement. Local services are only available to the user. The user of a component might change, but at any given time each component has a single user.

Owner. Each component has a single owner. By specifying an appropriate policy, the owner of a component might allow users to temporarily use his or her device, but each component has only *one* owner. In case the owner is a person, he or she is likely to be the sole user of the component, but that is not required.

Local interface. Each component has at least one local communication interface suitable for direct connection to other PAN components. This interface can be a cable or wireless interface. We consider both fixed and wireless PAN interfaces. However, wireless interfaces are most interesting and the security architecture is built with the assumption that wireless interfaces are used to interconnect any components. This assumption sets the basic communication security requirements for communication over the local interface.

Global network interface. Apart from one or several local interfaces a component may also have a global network interface. We consider a network interface as a communication *service* different from a local interface.

Security policy. Each component has different security policies. We distinguish between two different types of security policies: local and remote. The local security policy determines which resources on the component that a user is allowed to manage and if authorisation is demanded or not. It also describes how configuration and executables should be installed. The remote security policy determines the requirements on access to the component services and the communication between the service and the entity in the PAN that utilise the service. This includes authentication and encryption requirements as well as access rules.

9.2.2   PAN trust model

The concept of trust is often thought of as relations between humans. When we speak about trust between humans, we often mean that a person is trusted, if he or she can be relied upon. Ideally, we would like our components (PAN devices) to act regarding trust exactly as we do ourselves. However, it is very hard (if at all possible) to implement a human trust model in a communication system. But, in order to provide secure PAN communications and services, we need a model for the trust relationship between different components. If we cannot distinguish trusted from non-trusted components, it is not possible to protect our systems from hostile components. On the other hand, if no component is trusted, we cannot offer any useful communication services. Since we find it hard to directly apply a human model, for implementation and usability reasons, we need to find a reasonable model that can be *accepted* by humans and that at the same time is easy to implement. Furthermore, it must be easy for humans to understand the model. These facts are the background to the trust model we have chosen for PANs. A new model is introduced, since no suitable model was found to exist.

The trust model we seek should be natural and self-explanatory. The chosen model is very simple. A simple model is easy to implement and understand, but might not be sufficient to describe more complex and advanced PAN relationships. However, as we will show, the basic simple model can rather easily be extended to cover more complex relationships. It is clear that trust is always a question of relations. Trust on a person or device is always in relation to another person or a device. In the PAN setting, we can choose any component with local connectivity and describe the trust relationship between this component and the rest of the components within communication range, that is, all components in the same PAN. In this possible large set of devices each device must be able to make decision on to what extent it trusts all other devices in the PAN. This is the same situation as when a group of humans gather at a meeting. Each person has a certain relation to the rest of the members in the group. The complete picture of all trust relations between the persons in the group can then be obtained by asking each of them about their trust relation to the other people in the group. Even if this complete picture might be nice to have, it will not change the relations or views of the individual persons. We apply the same thing to the PAN model. What is interesting is the trust relation between *each* of the devices in the PAN and the rest of the devices. The complete trust situation in the PAN can then be obtained by going over all devices in the PAN and finding out the different relations.

The facts discussed above have inspired us to use a trust model where trust relations are described in relation to one particular component. We call this component the PAN reference component. We view all other components in relation to this single component. This allows us to describe any trust relation between any given component and all other components in a PAN.

Next, we introduce some assumptions and observations.

- When we say that a device is trusted, we mean that it has been designated as trusted.
- A user or an organisation has a limited number of devices (a rather small set) that is a reasonable candidate set for representing highly trusted components. These components should then be under control by the very same user or organisation and have protection from illegal usage by other persons.
- Obviously, only components belonging to the same user or organisation should consider the other devices belonging to the same user or organisation as highly trusted.
- There are as many sets of highly trusted devices of the type described above, as there are PAN users or PAN organisations.
- Components not under the control of the same user or organisation might for some reason be considered (or actually designated) as trusted to some extent.

These assumptions and observations allow us to define a trust model that is easy to understand. As we have explained, we view all trust relations in relation to a reference component. In relation to this reference component, we came across three broad levels of trustedness:

- not trusted at all,
- trusted to some extent,
- highly trusted – there is no reason why the component cannot be trusted completely.

We use these three classes as our basic trust classes. The terms we will assign to them are:

- untrusted components,
- second party components,
- first party components.

The details of how we classify components according to these three sets are given below.

9.2.2.1 Untrusted components

Untrusted components are by definition all PAN components that the reference component has no security relations with. For example, any new component that a user buys is an untrusted component from the perspective of all the other components belonging to the same user. In fact, all components that are not designated to any of the other two trust classes by the user/organisation in control of the component belong by default to this trust class.

9.2.2.2 Second party components

A second party component is not fully trusted. It might be trusted for some actions but not for others. The user or organisation that controls the component assigns a component to this trust class. This must be done through default configuration or a direct action. It must also be possible to add or delete components from this trust class. It must also be possible for the reference component to identify another component in order to know if it belongs to this trust class or not. Hence, we require that the reference component share a security association with all components that are assigned to this trust class. This means that second party component identities can be verified, that is, authenticated. We require that the communication with trusted components belonging to the second party class can be protected. Thus, it must be possible to protect confidentiality and/or integrity of the communication between the reference components and a second party component.

The finegrain level of trust given to a particular second party component (i.e. is which actions it is trusted for and which not) is determined by the remote security policy of the reference component. The person or organisation that is in control of the component can set this policy. In the simplest case, all second party components are given the same default second party component remote security policy rights.

9.2.2.3 First party components

A first party component has the same owner as the reference component. The user or organisation that controls the component assigns a component to the first party trust class. It must also be possible to add or delete components from this trust class. By definition, a first party component is considered highly trusted and will get access to most services provided by a component. This allows all first party components to uniformly share resources among themselves. Similar to the situation for the second party class, it must be possible to determine if a component belongs to this class or not. Hence, the reference component must be able to identify another component in order to know if it belongs to this trust class or not. We require that the reference component shares a security association with all components that can be used to authenticate the component, and to confidentiality and/or integrity protect the communication between the reference components and a second party component.

9.3   Security initialisation and the personal PKI concept

In this section, we discuss how we can initially create security associations and set the security policies of the PAN components. We refer to this as security initialisation. Here, we slightly extend the 'resurrecting duckling' model used by Stajano and Anderson [3]. They assume that at manufacture a component does not have an owner or users. Instead, the component is made first party by an initialisation or (referring to Reference 3) at an imprinting phase. At the imprinting phase the necessary keys are created or transferred to the component. These keys can be used to identify and authenticate the component to all other components of the same owner. It is the owner that is responsible for the primary imprinting of his/her devices. We extend this principle of initialisation to not only be true for the time when the user buys the component, but also whenever a new security association needs to be established and consequently a new security policy must also be defined.

We start by discussing different aspects and solutions to the security initialisation problem in the PAN context. We consider assignment of first party trust relations particular important. This can be done using both symmetric and asymmetric cryptographic techniques. For the symmetric case we discuss trust delegation as one user-friendly way for first party trust assignment. In the public key initialisation case a natural option is to use a special personalisation device or certification device. Public keys can be used to create security associations and to assign components to the first party trust class. These principles we call the personal PKI concept. We explain our ideas behind the personal PKI concept and how it can be used for secure initialisation and trust class configuration.

9.3.1   Component security initialisation

The administrator of a component is able to change security policies, add/remove components from a trust class, etc. In most cases the administrator will be a human and often (but not always) this will also be the owner of the component. Hence, we require that security initialisation shall be possible to perform by an ordinary user of a component. A completely different approach is that some manufacturers do the initialisation during production and equip their devices with initialisation keys or device identity certificates (and trusted root public keys). In this case, the whole initialisation procedure takes place during production. This is quite different from the case when the user after production would like to initialise his or her devices. Here, we concentrate on the latter case. This means that we would like the procedure to be as user-friendly as possible. That is, the number interactions and the complexity of the interactions should be minimised. We have investigated low complexity and manually assisted security initialisation procedures.

In principle, a security initialisation procedure contains the following three steps:

1. establish an initial secure channel;
2. configure cryptographic parameters for autonomous secure PAN connection establishment;
3. configure/negotiate (default) policies for remote configuration and resource access.

9.3.1.1   Establish an initial secure channel

If it shall be possible to create security associations in a secure way, an initial secure channel must be established between the involved PAN components. With a secure channel we mean that there must be a way for the components to communicate such that they can be sure that they exchanged information with the components they intend to and that the communication is not intercepted or modified by any hostile device or person.

A direct wired connection (e.g. serial, USB, Firewire, Ethernet) between two components may provide sufficient protection in most cases and is a good initial secure channel candidate. The principle of direct physical connection was proposed for the 'imprinting' in Reference 3.

In a PAN environment, wireless connections are expected to be more common and to find and connect via cables is not an especially user-friendly procedure. Indeed, many PAN devices need not even have a wired interface; hence, wireless interfaces must also be considered. Since wireless interfaces are generally not physically secure, they are not secure channels. This has caused us to investigate alternatives for the wireless only cases. In a PAN, the components are always close to each other. Furthermore, in most cases there is at least one human that controls the components. This means that we can use the human directly as the 'secure channel'. A human can be asked to read, check and/or enter values into the components. Indeed, this is all we need for the initial secure channel. Different from the physical connection case though, we cannot ask a human to transfer any large amount of information (which would simply be too tedious for the user to do). This has implications on how we can use the initial secure channel when we create security associations. Hence, in the wireless case, we need convenient and secure procedures that combine the human secure channel with the security association establishment procedure. We have chosen to call these Manual Authentication (MANA) protocols. Several different MANA protocols are presented in Section 9.4.

Another way to establish initial secure communication is to optically read information from a paper or slip. Then, we assume that the person who would like to create a security relation between his/her components will uses his/her optical component to read some code and/or network address information. The code can, for example, be a one-way hash of public key or a public key certificate. The information is transmitted from the optical component (in a secure way like over a cable or secure wireless connection) to the component for which the security association is to be established (if different from the optical component). The code can form the basis for the creation of the wanted security association. If a mutual security association is wanted, both components have to read information from each other optically.

9.3.1.2   Configure cryptographic parameters

The security functions of a PAN, such as communication security (see Section 9.5) and access control (see Section 9.6), require certain cryptographic parameters configured on the PAN components. Depending on component capabilities and/or preferences, these parameters may include secret keys for symmetric cryptosystems or public and private key pairs for asymmetric cryptosystems. The initial secure channel can be used to securely transfer key information between the devices. This information in its turn can then be used to perform an authenticated key establishment. We will call shared secret or shared trusted public keys between components a security association. When a security association is in place, we have the means for authentication of the other component. Furthermore, we are able to protect integrity and/or confidentiality of the information exchange over an insecure channel between the components.

A security association can be between just two components. This might, for example, be the initialising and initialised components or it can be a group association. In both cases both symmetric and asymmetric cryptosystems can be used. Hence, in principle, we have the following four different initialisation cases:

- security association between two components using symmetric keys;
- security association between two components based on asymmetric keys;
- security association between a group of components using symmetric keys;
- security association between a group of component using asymmetric keys.

Obviously, the first case is a special case of the third case and the second a special case of the fourth case. However, we consider these cases particularly important and treat them separately. We discuss the different cases in the subsequent sections.

9.3.1.2.1   One association, symmetric keys

In the symmetric case, a two party security association is normally based on a shared secret between the two components. This might be a secret key that can be used directly for authentication, integrity and confidentiality protection or it is a 'master key' that can be used to derive encryption keys and/or integrity keys. The component must know which key to use when communicating with other components. Hence, as part of the initialisation procedure, there is also a need to associate the key with an identity (temporary or long-lived) of the other component. The key and identity configuration must be done in a secure manner. Hence, both the exchange of the key and the corresponding identity mapping must be done in a secure way. There are several options for utilising the initial secure channel we discussed previously. A practical option is to use the manual authentication protocol discussed in Section 9.4.

Once the components have established the shared secret and exchanged identity information, any standard way (see Section 9.5) of secure communications set up can be used to secure further communication between the two devices. It is good security practise though to periodically update the shared secret. As part of the key exchange the component must also decide to what extent the other component in the association is trusted. Following our trust model, it is up to the user at the initialisation occasion to assign a trust class to the association, that is, first or second party component.

9.3.1.2.2 One association, asymmetric keys

In the asymmetric case, a two party security association is based on trusted public keys. This means that the first component holds the public key or a secure one-way hash of the public key of the second component and that it has marked this key or hash as a trusted key and vice versa. The trust class can be the first or second party class. The user does the trust class assignment during initialisation. The trust assignment must be done in a secure way. Here, the initial secure channel is used. Depending on the type of channel, either the public key or one-way hash of the public key is directly transferred or it is transferred while being protected by a manual authentication protocol as described in Section 9.4. Once marked as trusted, the public key(s) can be used directly for authenticated key exchange. Then, the agreed symmetric keys can be used for authentication, integrity and confidentiality protection as described in Section 9.5.

9.3.1.2.3 Group associations, symmetric keys

During the initialisation, we might not only want to create a security association between the initialising and initialised components, but also between several components. This can be done using a protocol between the involved components. What we want in this case is a common shared secret, a group key, known to all components in the group. We have several options for how such a key can be securely distributed to all the components. One possibility is the following principle.

1. One component acts as a master component.
2. The master component uses any of the methods described in Section 9.3.1.2.1 with every other component in the trust group to establish a shared secret with each of them.
3. Then, any of the components can generate a group key, which is then distributed encrypted and possibly integrity-protected to the rest of the components in the trust group.

If this principle should work, all the involved components must be within communication range. This might not always be the case. But, still we would like that it is possible to distribute a group security association. Allowing trust delegation can do this. Next, we describe one trust delegation procedure based on symmetric keys.

Each component supporting the trust delegation method must have an internal trusted group keys database. The database contains a list with at least the following two entries for each record:

- a group key index,
- a secret key corresponding to the index.

Each device might be pre-configured by the manufacturer with at least one key index chosen at random and a corresponding group key. Alternatively, it is shipped with no group key at all. When the user would like to connect two components (which have not previously been in contact with each other), called here the first and second components, the following procedure applies.

1. The first component sends the list of key indices from its database to the second component.
2. The second component receives the list of indices and checks the list against the internal list of trusted group keys. If the component finds a match between any of the received indices and the internally stored list of indices, the device chooses an arbitrary index among the matches and returns this index to the first device, and jumps to step 5. If no match is found step 3 is executed.
3. The first component requests a security initialisation. If the second device refuses this request, the procedure is aborted. If the second device accepts the request for security initialisation the next step applies.
4. The two devices perform a symmetric key exchange according to any of the principles described in Section 9.3.1.2.1. As part of the user interaction at the key exchange, the user is asked if the bonding will be performed with a highly trusted component. If the bonding is to be performed with a highly trusted component, step 6 applies. Otherwise, the procedure ends here.
5. The two components perform an authenticated key exchange. The authentication is based on the group key corresponding to the agreed index. The authenticated key exchange can, for example, be an authenticated Diffie–Hellman key exchange.
6. The two components switch to an encrypted connection using the recently agreed secret key. The second component then sends a list of trusted group key indices (if any) and the corresponding group keys to the first component. If the component holds no group keys from the start, the device first generates a group key and the corresponding index, both uniformly and randomly from some suitable spaces.
7. The first component receives the list of group keys and indices and adds all the indices and keys to its trusted key database. Next, it sends its own list of trusted keys indices and the corresponding keys to the second component.
8. The first component receives the list of group keys and indices and adds all the indices and keys to its trusted key database. This ends the procedure.

The procedure above can only be used to delegate trust among 'highly trusted' components. If any component that takes place in the trust delegate is compromised, the principle is a great security risk. But, for first party trust delegation this might be a quite useful way to propagate a group security association. The advantage compared to pure pairwise initialisation is that the number of manual interactions is considerable reduced.

9.3.1.2.4   Group associations, asymmetric keys

It is also possible to create group associations using asymmetric key techniques. PAN components could, for example, be configured with a certificate issued by a commercial CA and instructed to 'trust' the root public key of that CA. However, online connection to a central server and certificate-based systems require access to the CA for enrolment, verification and revocation checking. As the availability of global access at all times cannot be guaranteed (and should not be necessary to operate PAN components) in a PAN environment, the use of external trusted third parties as a basis for PAN is maybe not the best option. We have investigated an alternative approach, where the user himself or herself issues certificates using a component dedicated to this task, a certification device. Not only must certificates be issued in order to accomplish group associations, but also a common trusted root key must be distributed. We call this concept of creation of group associations based on root keys and certificates in a PAN a personal PKI. The personal PKI concept is treated in more detail in Section 9.3.2. Similar to the symmetric group association case, the personal PKI concept implies automatic trust delegation. Hence, it is most suitable to distribute first party trust relationships since arbitrary trust delegation compromises security.

9.3.1.3   Configure policies for remote resource access

The service and policy configuration/negotiation step brings the most significant difference for initialisation of the three different trust classes. In a first party initialisation, the initialised component trusts the initialising component entirely (and possibly vice versa), and thus it allows all its services and access policies to be configured via the initialising component. Alternatively, each component might have a default policy for all first party initialisations. The latter is especially attractive from the user perspective. It is possible to also have a default policy for second party relationships. Then, at the initialisation phase, these policies are automatically configured in the components. However, for the second party case, the owner might want to individually configure the resource access policy. It is then important that this can be done in an intuitive and easy way. The same applies to the untrusted class. If the owner wants untrusted components to have extended access rights, he or she would obviously be allowed to configure this. But, since there is no possibility to distinguish different untrusted components from each other, such extended rights will apply to all components in the untrusted class. When it comes to the enforcement and description of the access rules we have several different options. We discuss some approaches in Section 9.6.

9.3.2   The personal PKI concept

In a 'conventional' PKI model, a Certification Authority (CA) issues a public key certificate. The CA is responsible for checking that the public key in an issued certificate corresponds to a private key that the holder (with the ID given in the certificate) of the certificate possesses. This is necessary in order to maintain the security of a global or very large PKI. The drawbacks of a central CA include the following.

- It must issue all certificates used by the communication units, and all units must share trusted public root keys; this can be a tedious process that the user of a communication unit would like to avoid.
- It is very costly to maintain a well-controlled highly secure certification process that can handle thousands of users.

- A user who wants to manage his/her own local environment, such as a PAN, will gain few benefits within the PAN from employing a centralised CA.
- The user might not want, for privacy reasons, to delegate the CA operation to a centralised entity outside his personal environment.

Nevertheless, as discussed in Section 9.3.1.2.4, a PKI would be a convenient solution for creating group security associations in a PAN. We seek a solution that is adapted to the local PAN environment and that minimises the necessary user interaction. Furthermore, we would like to maintain a reasonable security level. The personal PKI concept provides this.

A personal PKI is different from large-scale or global CA functions. An ordinary user for home or small office deployment uses the personal PKIs. As with any other PKIs, we would like all units in a communication network to share common root public keys and use certificates issued by a trusted CA corresponding to the public root key. In order to use PKI technology in such an environment we need to reconsider the CA policies. One of the personal components must act as a 'personal CA' or personal certification device. Such a component is able to issue certificates to all other personal components. Hence, since all the personal components can be equipped with certificates issued by the same CA, that is, the personal CA, they will all share a common root public key. Consequently, the public keys in the certificates can be used to exchange session keys or authenticate personal components in a PAN.

We assume that one of the PAN components is defined as the personal certification component. Preferably, the component should have a display and a keypad. Examples of possible personal certification components are mobile phones, PDAs or PCs. A personal certification component might be pre-configured (at the manufacturer) with a private/public key pair or might be able to generate such a key pair. In general, the personal PKI key pair should be securely generated within the device, or securely generated and transferred to the device at manufacture, and (in both cases) the private key securely stored when on the device. The personal certification component is used to initialise other PAN components according to the principles discussed in Section 9.3.1. At the initialisation phase the necessary keys and certificates are securely transferred to the component that can be used to identify the component to all other first party components of the same owner. Here, any of the initial secure channels we discussed can be used. The case when the only available secure channel is a human channel is considered especially important. How the certification can be performed using the MANA protocol is discussed in Section 9.4. PAN components should be able to verify certificates issued by the personal certification component, and check certificate validity and revocation status when appropriate.

Once a mobile device has been imprinted and provided with a public key certificate by the personal certification device, there is a need for ongoing management of key pairs and certificates. There are three main issues that need to be resolved within the PAN.

- *Certificate and key pair update*, that is, methods to be used when a device wishes to use a new key pair or when the certificate for a current key pair has expired.

- *Key status management*, that is, disseminating information regarding revoked public keys across the PAN.
- *Trust management*, that is, managing the relationship between the mobile device and the personal CA, including CA (root) key update and the possible replacement of personal CA devices (especially in the event of lost or stolen personal CA devices).

Details, solutions and recommendations are treated in Chapter 3 [4].

9.4   Manual authentication protocol

An important step of the initialisation is to perform the initial authentication of the device and to provide the device with the necessary information needed to prove the identity to other PAN devices (first or second party). In Section 9.3, the initialisation procedure was described. In this section, we shall present a number of protocols that are recommended for use when the PAN device is identified for the first time and equipped with a cryptographic key. The type of the key can vary a lot depending on its subsequent use in security mechanisms. If the key is related to a public key cryptosystem, then the device is also issued a certificate, which can be verified by other parties.

The wireless link is the basic communication channel between the devices. It is inherently insecure, that is, passive eavesdropping, channel hijacking, as well as active impersonation and tampering of data is possible. The procedure of initial key management, where the initial secret cryptographic parameters are set in the device, is the most sensitive part of the communication. If tampering or eavesdropping the imprinting step is possible, then the security of all future communication based on imprinting is ruined. Therefore, initial key management requires some auxiliary secure channel to be used. In the PAN context, the secure channel can be based on one of the following technologies:

- fixed connection such as cable, USB interface, bar-code reader, smartcard reader;
- human interfaces, such as key pads, displays and voice;
- second wireless, for example, infrared;
- other proximity-based technology, for example, low-power channel.

All the technologies listed above require human involvement. For better security, as well as for user convenience, it is desirable to minimise the human involvement, and to make it as easy and robust as possible. The main goal of this section is to present special authentication protocols for this purpose. In these protocols, the interface between the PAN device and the external world is typically one of the types listed above, and, therefore, suitable only for restricted communication. The protocols are called Manual Authentication (MANA) protocols to indicate that a constrained communication interface is used. It does not imply that the interface is limited to ones operated by human hands, even though this is the example interface used in the description of the protocols.

The goal of initial key management is to equip a set of devices with a cryptographic key. The set can consist of one, two or a larger number of PAN devices. If the goal is to provide the devices with a shared key of a symmetric system, then at least two devices are involved in the initial key management procedure. If some public key based key agreement protocol such as the Diffie–Hellman protocol is used, the passive eavesdropping attacks can be eliminated. The task of authentication remains and can be performed using a very narrowband physical channel as will be shown in Section 9.4.1. The cryptographic authentication technique is called MANA and is based on unconditionally secure message authentication codes. MANA-based key agreement protocols are discussed in Section 9.4.2. MANA protocols can also be used to authenticate public keys using very lightweight techniques.

9.4.1   *Protocols for manual authentication*

Human operated authentication methods were previously mentioned by Stajano and Anderson in [3], but the authors were rather sceptical about their usability. They proposed using the Diffie–Hellman key agreement protocol, and compute a hash value of the results in both devices. The hash values are then displayed to the users, who should compare them. If the displayed values are the same the user can be convinced that a secure link between the two devices has been established. Such hash values must protect against an active attacker, a man-in-the-middle that tries to establish its own Diffie–Hellman secrets with the legitimate parties. Therefore, the hash values must be collision resistant, and a length of 160 bits would be recommended. Stajano and Anderson concluded that it would be too cumbersome for the users to compare 160-bit (or 40 hexadecimal digit) hash values. Also, it would be too easy for the user to just let the test pass even if it fails.

Initial authentication protocols based on hash codes were also more recently developed by Balfanz *et al.* [5]. They separate the pre-authentication stage from the key agreement protocol. In this respect, their protocols use a very similar approach as the MANA certificates developed by the SHAMAN project, see Section 9.4.1.2. The essential difference is that because deterministic hash codes are used, the protocols presented in Reference 5 require authenticated channels capable of transmitting 160-bit hash codes. Hence, they are not suitable for use in applications where a very constrained (user operated) communication channel is used to transmit the initial authentication information.

The purpose of this section is to present protocols for manual verification where, instead of an authenticated channel based on deterministic hash values, short secret authentication codes are used. The short authentication codes are keyed using short randomly generated keys. The security is based on the fact that for each check a new key is generated randomly. Explicit security estimates can be derived for specific constructions as will be shown in Section 9.4.1.5.

Three protocols with different properties are presented. All three protocols are suitable for manual authentication of any pre-established secret or public data items. The first two, MANA I and II, were first presented in Reference 6. MANA I and II are closely similar; they differ only in the types of interfaces the parties are using. For

MANA I, the first device has an output interface suitable for short strings of alphabetic and/or numeric symbols and a simple input interface. The second device has an input interface suitable for short strings of alphabetic and/or numeric symbols and a simple output interface. For MANA II, both devices have output interfaces capable of outputting short strings of alphabetic and/or numeric symbols. Additionally, both devices have simple input interfaces for the user to enter acknowledgements.

The third protocol, MANA III, is a slight modification of a protocol presented in Reference 7. It assumes that both devices have input interfaces for entering short strings of alphabetic and/or numeric symbols. Additionally, both devices have a simple output interface.

The exact requirements of the capabilities of the interfaces become clear from the protocol descriptions. Typically, the output interface is a display, and the input interface is a keypad. Further, it is assumed that the two devices share an algorithm for computing a message authentication code (MAC). For an example of a suitable MAC algorithm, see Section 9.4.1.6. The description of the protocols is based on the assumption that one user handles both devices. If there are different users for different devices then the users need to communicate with each other over a secure channel.

9.4.1.1 MANA I

The steps of MANA I are as follows (see also Figure 9.2).

1. The first device outputs a signal to acknowledge that it has data D ready for verification. The user notifies the device the first device that the protocol can start.
2. The first device generates a random key K, where K is suitable for use with a MAC function shared by the two devices. Using this key K, the first device

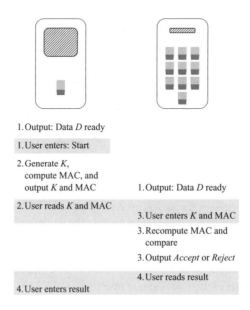

1. Output: Data D ready
1. User enters: Start
2. Generate K, compute MAC, and output K and MAC
2. User reads K and MAC

1. Output: Data D ready
3. User enters K and MAC
3. Recompute MAC and compare
3. Output *Accept* or *Reject*
4. User reads result
4. User enters result

Figure 9.2   Manual authentication protocol MANA I

computes a MAC as a function of the data D.The MAC and the key K are then output by the output interface of the first device. The user now reads the MAC and key K from the output of the first device.

3.   The user enters the output of the first device to the second device using the input interface. The second device has its version of the data D. The second device uses the key K to compute the MAC value as a function of D. If the two MAC values agree then the second device outputs a success signal to the user. Otherwise it gives a failure signal.

4.   The user enters the result to the first device. In case of success, the devices accept the data D.

9.4.1.2   MANA certificates

The MANA I protocol has the property that no authentication information is transmitted over the insecure channel. Therefore, it does not make any difference in MANA I if the manual authentication values K and MAC are transferred from one device to another already before the latter has received the actual data D. Naturally, such an approach is applicable only to situations where one device generates the data D to be authenticated. But if this is the case, MANA I offers an authentication means for data to be received some later time. Such an authentication is called the MANA certificate.

The MANA certificate comprises of K and $MAC(K, D)$. An example of data items in D could be a public key of the device, its identity, the domain of service, etc. In Section 9.4.2, an example is given of how MANA certificates can be used to establish a shared secret key between two devices.

9.4.1.3   MANA II

The steps of the second variant of the manual authentication protocol, MANA II, are as follows (see Figure 9.3).

1.   Both devices output a signal so as to acknowledge that they have received data D and that they are now ready for the verification. The user receives the signal from both devices and enters a signal into one of the devices (called the first device in the sequel) to notify that the verification can start.

2.   The first device generates a random key K, where K is suitable for use with a MAC function shared by the two devices. Using this key K, the first device computes a MAC as a function of the data D.The device outputs the MAC, and transmits the key K to the second device over the insecure wireless link.

3.   The second device uses the key K to compute the MAC value as a function of its stored version of data items D, and outputs the key K and the computed MAC value.

4.   The user compares the K and MAC values in both devices. If they agree, then the user enters a signal of acceptance in both devices, and the devices can accept data D as the basis of their subsequent operation. If the MAC values are different then data D must not be accepted.

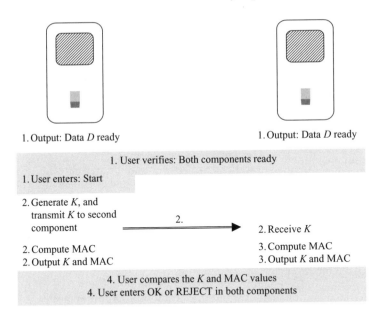

1. Output: Data *D* ready   1. Output: Data *D* ready

1. User verifies: Both components ready

1. User enters: Start

2. Generate *K*, and
   transmit *K* to second
   component   2.   2. Receive *K*

2. Compute MAC   3. Compute MAC
2. Output *K* and MAC   3. Output *K* and MAC

4. User compares the *K* and MAC values
4. User enters OK or REJECT in both components

Figure 9.3   Manual authentication protocol MANA II

9.4.1.4   MANA III

The MANA III protocol makes use of a special technique of commitments. The steps of the protocol are as follows (see also Figure 9.4).

1. Both devices output a signal so as to acknowledge that they have received data *D* and they are now ready for the verification. The user receives the signal from both devices and generates a short random key *K*, where *K* is suitable for use with a MAC function shared by the two devices. The user enters the key in both devices, and enters a signal in one of the devices (called the first device in the sequel) to notify that the protocol can start.
2. The first device generates a random number R_1. Using the key *K*, the first device computes a MAC_1 as a function of the data *D* and the random number R_1. The first device transmits the MAC_1 to the second device over the insecure wireless link.
3. The second device generates a random number R_2. Using the key *K*, the first device computes a MAC_2 as a function of its stored version of data items *D* and the random number R_2. The second device transmits the MAC_2 to the first device over the insecure wireless link.
4. The first device sends its random number R_1 to the second device.
5. The second device verifies that the received MAC_1 corresponds to the MAC value computed using the stored values of *K* and *D* and the received value R_1. If verification is successful the device outputs *accepted* (OK) and sends its random number R_2 to the first device.

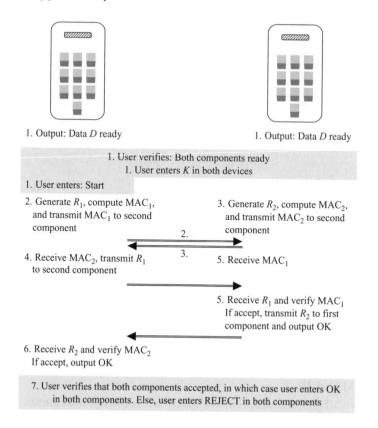

Figure 9.4   Manual authentication protocol MANA III

6. The first device verifies that the received MAC_2 corresponds to the MAC value computed using the stored values of K and D and the received value of R_2. If verification is successful the device outputs *accepted* (OK).

7. The user verifies that both devices accepted the verification, and enters OK in both devices.

9.4.1.5   Security of the MANA protocols I and II

All data to be transferred are assumed to be public, even if in some cases part of data D may be secret. The security goal of the MANA protocols is to protect the integrity of the data, not the confidentiality. The necessary integrity protection is performed using a checking procedure based on the MAC.

A MAC function is a mapping f from a data space Δ and a key space Γ to a tag space Σ, such that

$$f : \Delta \times \Gamma \rightarrow \Sigma, (D, K) \mapsto C$$

A MAC is used to protect integrity of some data. In manual authentication, short MAC values are used. Therefore, in manual authentication the security is unconditionally based on the security of the MAC function rather than computational security, which is the case if hash functions with long hash codes are used. The unconditional security of MAC codes is based on results developed by message authentication theory [8, Section 4.5]. Two main types of attacks are normally considered:

- impersonation attack,
- substitution attack.

In an impersonation attack, the attacker tries to convince a receiver that some data are sent from a legitimate sender without observing any prior data exchange between the sender and the receiver. In a substitution attack, on the other hand, the attacker first observes some data D and then replaces it with some other data $\hat{D} \neq D$. The probabilities for the attacker to succeed in an impersonation attack and in a substitution attack are denoted as P_I and P_S, respectively, and they can be expressed as

$$P_I \hat{=} \max_{C \in \Sigma} P(C \text{ is valid})$$

$$P_S \hat{=} \max_{\substack{C, \hat{C} \in \Sigma \\ C \neq \hat{C}}} P(\hat{C} \text{ is valid} \mid C \text{ is observed})$$

The security of the MANA protocols depends on the probability for an attacker to replace the observed data D with some other data $\hat{D} \neq D$. The attacker succeeds if $\hat{D}$ is accepted by the device as valid data. Since we assume that both devices are physically close to each other and we do not accept any data unless both devices actually signals that they are ready, the impersonation attack does not apply to the MANA scenario. Furthermore, the normal MAC situation is that both the data and the MAC function output, the check value, are transmitted and can be observed by the attacker. This is not the case for the MANA protocols I and II. Here, only the data are sent over a public channel and the attacker does not know the output of the MAC. This simplifies the security analysis and the expression for a successful substitution attack. Hence, the probability of a successful substitution attack for MANA I and II can be expressed as

$$P_S = \max_{\substack{D, \hat{D} \in \Delta \\ D \neq \hat{D}}} P(f(D, K) = f(\hat{D}, K) \mid D \text{ is observed})$$

Note that even if the key is sent over the open channel in MANA II the user compares the two key values and the attacker has no chance to choose, replace or modify the key. In MANA I, the key is transferred over the secure channel and we have the same situation. Thus, given that the key is chosen uniformly at random from the key space, K, the probability above can be expressed as

$$P_S = \max_{D, \hat{D} \in \Delta, D \neq \hat{D}} \frac{|\{K \in \Gamma : f(D, K) = f(\hat{D}, K)\}|}{|\Gamma|}$$

where $|\Gamma|$ denotes the cardinality of the set Γ. It follows from this equation that in order to provide high security, the collision probability of the MAC function must be low. This can be achieved by using MAC functions obtained from error correcting codes. This will be shown in Section 9.4.1.6.

9.4.1.6 MAC construction examples

Next, we discuss possible MAC constructions and the probability of successful attack for different constructions. Considering the expression for a successful attack in Section 9.4.1.5, a straightforward approach is to use constructions from coding theory. The relation between error correcting codes and message authentication codes has been treated in Reference 9.

Before we can go into concrete examples, we need to recall a couple of basic definitions from coding theory. For simplicity, we will treat codes over a finite field F_q. Denote a q-ary code over F_q by V. Suppose the code words have length n. The code is a mapping from messages to code words. Each message has its corresponding unique codeword. Then, the code V consists of all vectors $\mathbf{v} \in V = \{\mathbf{v}^{(D)} : D \in \Delta\}, \mathbf{v}^{(D)} = v_1(D), v_2(D), \ldots, v_n(D)$, where $v_i(D) \in F_q$. We need two more definitions.

Definition. If x and y are two q-ary tuples of length n, then we say that their Hamming distance is

$$d_H(x, y) \hat{=} |\{i \in \{1, \ldots, n\} : x_i \neq y_i\}|$$

Definition. The minimum distance of a code V is

$$d_H(V) \hat{=} \min_{x, y \in V, x \neq y} d_H(x, y)$$

Now, we will show how we can create a MAC for the MANA I and II protocols based on a code. The construction is very simple and the mapping from the message and key space is obtained as

$$f(D, K) = v_K(D)$$

where $K \in \Gamma = \{1, \ldots, n\}$. Hence, we obtain a MAC with a key space size equal to n and with message space size equal to the coding space size. What about the probability of successful substitution attack for this construction? From the expression for P_S in Section 9.4.1.5 we directly get:

$$P_S = \max_{D,\hat{D}\in\Delta,D\neq\hat{D}} \frac{|\{K \in \Gamma : f(D,K) = f(\hat{D},K)\}|}{|\Gamma|}$$

$$= \max_{D,\hat{D}\in\Delta,D\neq\hat{D}} \frac{|\{K \in \Gamma : v_K(D) = v_K(\hat{D})\}|}{|\Gamma|}$$

$$= \max_{D,\hat{D}\in\Delta,D\neq\hat{D}} \frac{n - d_H(v^{(D)}, v^{(\hat{D})})}{n} = 1 - \frac{d_H(V)}{n}$$

Now that we have an exact expression for the probability of a successful substitution attack, we can continue by looking into some concrete constructions. We look into rather long codes with very high minimum distance. This property holds for the very well known Reed–Solomon codes [10] (RS codes). An RS code can be constructed over an arbitrary finite field, F_q. The calculation of a codeword is very simple and is done by polynomial evaluation over the finite field (generalised one). Express the data (message) D to be encoded as q-tuple of length t over F_q, $D = D_0, D_1, \ldots, D_{t-1}, D_i \in F_q$. Then, the RS encoding polynomial is given by

$$p^{(D)}(x) = D_0 + D_1x + D_2x^2 + \cdots + D_{t-1}x^{t-1}$$

The MAC mapping function is directly given by evaluating the polynomial in an arbitrary point $K \in F_q$,

$$f(D,K) = v_K(D) = p^{(D)}(K) = D_0 + D_1K + D_2K^2 + \cdots + D_{t-1}K^{t-1}$$

The RS code has (for the generalised one) the following properties:

$$n = q = |\Gamma|$$
$$|\Delta| = q^t = n^t$$
$$d_H(V) = n - t + 1$$

This implies that $P_S = (t - 1)/n$ for a MAC obtained from the RS code. The probability increases with the size of the message space Δ. Hence, a good approach is to *first* apply a good one-way hash function like SHA-1 to the data and *then* use the output from the one-way hash as input to the Reed–Solomon code. This implies that we keep a low probability without considerably increasing the key length or the length of the output of the MAC. By using this approach, a message length of around 128 bits (truncated SHA-1) gives sufficient security. In Table 9.1, we show a couple of construction examples and the corresponding probabilities of successful attacks.

As one can see from Table 9.1, a code with four hexadecimal digits of key and MAC size gives a forgery probability of around 2^{-12} or less. If the MAC size is increased to five hexadecimal digits, the probability decreases to around 2^{-17} or less.

Table 9.1   *RS-code   MAC   construction examples   with   probability   of successful substitution attack,* P_S

$\log_2 |\Delta|$	$\log_2(n)$	P_S
128	16	$2^{-13}\text{--}2^{-16}$
256	16	$2^{-12}\text{--}2^{-16}$
128	20	$2^{-17}\text{--}2^{-20}$
256	20	$2^{-16}\text{--}2^{-20}$

9.4.1.7   Security of MANA III

Because of using random number commitments, the security of the MANA III protocol is based on different principles than the ones for MANA I and II. Without going into details, estimates of suitable parameter lengths are given. The size of key K of about 16 bits is recommended for this case also, but the MAC should be longer. The size of the output of the MAC function to be used for the MANA III protocol should be 128–160 bits. Similarly, the random strings R_1 and R_2 should be about the same size, 128–160 bits. Usual timing procedures should also be implemented for detection of possible interruptions of the protocol.

9.4.2   Using MANA for exchange of symmetric keys

Requirements for secure exchange of shared secret keys between two PAN components were discussed in Section 9.3.1.2.1. In practice, a communication channel that is capable of transmitting the required key material securely, that is, preserving confidentiality and authenticity of the keys, may not always exist. Even if the devices are in close proximity, a wireless communication channel is preferred over the hardwired one due to the flexibility of the wireless channel. Wireless communication is prone to eavesdropping, and, therefore, insecure. The Diffie–Hellman key exchange method and other public key based key agreement protocols can be used to thwart passive eavesdropping attacks. But still the active, man-in-the-middle type attacks remain possible. Such active attacks can only be removed by application of proper measures of authenticating the key exchange messages sent by the communicating devices. If the devices belong to the same PKI, then the corroboration of authenticity is typically based on public key certificates. But if no such authentication infrastructure exists, other methods must be used.

In proximity scenarios authentication of two devices can be performed using various physical means. Balfanz *et al.* [5] were mainly interested in methods capable of handling a cryptographic hash code, which is typically about 160 bits, in an authenticated manner. Secure handling of such long strings of authentication data requires standard hardware interfaces, which sets another hurdle. The most flexible solutions are offered by methods that exploit the human capabilities of handling different man–machine interfaces the user would have to handle anyway. But then the string of 160 bits would be too long for the user, for example, to enter to a device using a key pad. Hence, the length of authentication data must be reduced. On the other hand, users are able to keep such authentication data confidential. It may be difficult if these data are used multiple times, such as the usual PIN codes that need to be memorised by people. But if the data are used only once for a short event of authentication negotiation, then the risk of accidental disclosure is minimal.

The MANA protocols presented in Section 9.4.1 are cryptographic authentication protocols that reduce the capacity requirement of the secure communication channel to the minimum. In most cases sufficient security is achieved using strings of 12 decimal digits, which corresponds to a MAC length of five hexadecimal digits (or 20 bits), see Table 9.1. While these constraints are adopted to accommodate the needs of a human user, they can significantly facilitate implementation of other types of physical security interfaces such RFID channels and bar code readers.

In this section, two variants of Diffie–Hellman based key agreement are described. In the first protocol the devices are in symmetrical position and mutually authenticated based on the same authentication method. The structure of the second protocol is very similar to a TLS-based network access protocol [11], where the network entity (server) is authenticated based on its public key certificate and the client requesting access is authenticated based on a secret passkey. Such examples are also discussed in Section 9.5.2.3. The only difference here is that the network certificate and the client passkey are used only once (or a small number of times), and that the certificate and the passkey are logically bound together. The network certificate is the MANA certificate as defined in Section 9.4.1.2.

9.4.2.1 Authenticated Diffie–Hellman key agreement

The protocol described here is the basic anonymous Diffie–Hellman protocol, which is subsequently authenticated using the MANA protocol. The Diffie–Hellman protocol is described in terms of a group G, and an element g in G, which has a sufficiently large order. The steps of the protocol are as follows.

1. The first device generates randomly and privately an integer x, computes g^x and sends this value to the second device.
2. The second device generates randomly and privately an integer y, computes g^y and sends this value to the first device.
3. The devices execute the MANA protocol for data $D = (g^x, g^y, text)$ where *text* is any additional data, for example, each other's identities. The exact content of the *text* field must be defined before the protocol is used.
4. If the result of the MANA protocol is successful the devices can compute the shared Diffie–Hellman key as $K = g^{xy}$. The devices can then derive cryptographic keys of required length and format from the shared secret Diffie–Hellman key K using some predetermined key derivation function.

The roles of the first device and the second device as defined by the MANA protocol need not be the same as in the anonymous Diffie–Hellman protocol in steps 1 and 2. In the protocol to be described in Section 9.4.2.2, the roles of the devices are fixed.

9.4.2.2   Diffie–Hellman key agreement and MANA certificates

The protocol to be described here is based on a Diffie–Hellman protocol, where one of the Diffie–Hellman public keys is authenticated using MANA certificates. The second entity is authenticated using an encrypted passkey. Such a protocol is suitable for situations where the first device is a network access point or a small device with no output interface of its own. In such situations, the keys and the MANA certificates are generated by a proxy device, which then inputs the necessary data to the first device using some management channel.

Similarly, as in Section 9.4.2.1, the Diffie–Hellman protocol is described in terms of a general group G, and an element g in G, which has a sufficiently large order. The roles of the two devices are essentially different and they are called the first device and the second device. The protocol has two stages, Stages A and B. In Stage A, the first device generates its Diffie–Hellman secret key, computes the public key and produces a MANA certificate. The MANA certificate is transferred to the second device. In Stage B, the second device receives the public key of the first device, verifies it, generates its own secret and public Diffie–Hellman keys. Further, both devices compute the shared Diffie–Hellman secret, from where a symmetric key for passkey encryption is derived. Finally, the first device verifies the encrypted passkey it receives from the second device. As the result, the Diffie–Hellman secret shared by the two devices has been authenticated.

The analogy with a PKI situation is apparent. Stage A corresponds to the PKI registration phase, where also the PKI certificate is generated. The difference lies in the fact that the MANA certificate is a one-time certificate given to one specific other device. The MANA certificate is authenticated by physical means, while in PKI the authenticity of the certificate is reduced to the authenticity of the root key used for signing the certificate.

Stage A

1. The first device generates randomly and privately an integer x and computes g^x. Then, the first device creates the MANA certificate on data D that comprises g^x and possibly some other data. The MANA certificate (K, MAC) is transferred and stored to the second device by the user channel. The first device stores x and g^x, and possibly the other data items comprising D.

Stage B (is initialised by either entity)

2. The first device sends g^x and possibly some other data to the second device. The second device verifies the authenticity of g^x based on the stored MANA certificate.
3. The second device generates randomly and privately an integer y, computes g^y. The second device computes the Diffie–Hellman shared secret as $S = (g^x)^y$ and

uses S to encrypt the passkey K.The second device sends the encrypted passkey $e_S(K)$ and its Diffie–Hellman public key g^y to the first device.

4. The first device computes its copy of the shared secret as $S = (g^y)^x$. Then it decrypts $e_S(K)$, and verifies that K is the correct passkey. If K is correct, then the first device can accept S as authenticated.

The devices can then derive cryptographic keys of required length and format from the shared secret Diffie–Hellman key S. In the description of the protocol the value K from the MANA certificate is used as passkey in steps 3 and 4 to authenticate the second device to the first device. This is just one option. Instead, any other secret value, for example, the MAC or some special purpose passkey value, agreed between the parties at Stage A, could be used.

This protocol is very practical for ad hoc network access. Then, in Stage A the role of the first device is taken by a service booth, which issues the MANA certificate. If the accessing device (the second device) is also to be authenticated, then the necessary details of the second device are stored in a network server that can be accessed by the network access points. The protocol also supports anonymous access. In this case, the network (the first device) uses the same Diffie–Hellman key with all devices, and, therefore, no second device specific information needs to be stored by the network. In anonymous access the network is authenticated, and the accessing device can use the resulting secret key to protect its communication to the network, for example, to provide anonymity over the wireless link.

9.5 Communication security in PANs

9.5.1 Background and requirements

It may be foreseen that in the future personal devices are capable of running a generic transport protocol for transporting communication between the devices. Then, the natural approach would be to perform authentication of the devices above or at the transport layer, and to secure the communication data also at the transport layer. However, this is still not always the case, and may not be in the future, either. The new applications such as ubiquitous computing make use of PAN networks consisting of simple devices that may have only link layer capabilities. Securing such applications is possible only using security services available at the link layer.

In any case, encryption and integrity protection is necessary at the link layer if higher layer addresses and other signaling communication must be protected over the wireless link.

In particular, an essential question of PAN communications security is how the device's identity and location privacy should be protected. The device identity, or information derived from it, is often communicated in packet headers, which must remain unencrypted in the corresponding layer. A particular challenge is presented by the fact that the identity the PAN devices often consists of several pieces of information, such as IEEE device addresses, IP headers, channel access codes, etc., which are coupled together, and then used repeatedly not only over the whole duration of the communication event, but also between the different communication layers.

Very straightforward attacks can be launched on the physical radio layer against communication channels over the wireless link. Brute force jamming can cause Denial of Service (DoS) for the entire link. Also, targeted jamming might be possible if the individual channel access codes are available. Physical attacks cannot, however, be launched without being detected, and they do not release the denied service for use by the attacker. Also, the frequency hopping techniques and Code Division Multiple Access (CDMA) channel coding methods make targeted attacks on the physical layer very difficult to launch. Much more serious threats are created by an attacker who operates on the wireless link layer and tampers the link layer signalling messages. Such an attacker can cause much more serious losses in quality of service and in availability of the access service. Moreover, the legitimate users may never detect such attacks.

Although it is ideal to implement all security services at the link layer, in some cases, link layer security will not be sufficient to secure PAN communications. A typical scenario would be the case, where different wireless technologies are used within one PAN and some PAN devices can only communicate using one wireless technology. PAN gateways (which can use both technologies) are then used to pass traffic between PAN devices. The same scenario also applies to inter-PAN communications.

In a typical PAN set-up, where there is only one hop (all using same wireless technology) between each device, the link layer security will be equivalent to an end-to-end security solution. However, when two PAN devices communicate through gateways the link layer security cannot be treated as end-to-end security. Instead, each link will usually be secured using different link layer security associations. In other words, if there is any routing (beyond the first hop) in the PAN then the link layer security mechanisms implemented will not be sufficient and network layer security services should be deployed.

The other limitation of link layer security is the lack of reliable implementations of security services in current wireless technologies. For example, cryptographic data integrity protection is not provided by the current wireless technologies such as Bluetooth and IEEE 802.11. As it requires considerable time and investment to make these existing wireless technologies provide reliable data integrity protection, a higher layer based solution is usually used to provide the same service. IPsec with ESP (using integrity protection) can be used to provide data integrity for PAN devices even when they are directly (first hop) connected to each other. The bandwidth overheads of IPsec may be traded off with the requirements for integrity.

In addition, some security mechanisms, such as source and destination address (for the link layer) authentication, are not generally provided at the link layer by current wireless technologies. Not having these mechanisms allows attackers to spoof traffic inside a PAN to either launch targeted DoS or replay attacks if other security mechanisms are not deployed.

Generally, implementations of reliable link layer security mechanisms are limited by the performance impacts of these features on the wireless technologies. Complexity, that is, cost of implementation is another important factor for wireless technology developers to include reliable security services at the link layer. Considering the wide usage of the public Internet, network layer protocols and associated security services are expected to be used by the PAN devices. For example, Bluetooth SIG is developing a specific PAN profile to specify IP networking over Bluetooth [12].

The main link layer technologies, Bluetooth and IEEE 802.11, will be discussed below. IP-based communication can be secured in a network layer, transport layer or application layer. A brief overview of the IP layer security (IPsec) and transport layer security (TLS) is given. A large variety of application-dependent security services are also available for securing specific applications in a PAN. But in this article the discussion is restricted to application-independent network, transport and link layer security.

From a PAN point of view, the important question is how the security is initialised. Some security technologies discussed below can be bootstrapped from a sufficiently long shared secret known to the communicating peers. But only Bluetooth provides a method, the Bluetooth pairing procedure, using which such a shared secret can be established between two devices. All other security systems rely on some physical key management channel using which sufficiently long secret keys can be configured to the devices. In Section 9.4, cryptographic protocols, the MANA protocols, were presented, using which the requirement of full key transport can be reduced to secure the transport of a short passkey. The IP-based security technologies IPSec and TLS also support authenticated public keys to be used as the basis for the initial security context. The MANA protocols presented in Section 9.4 can also be used to provide authenticity of public keys in PANs without external key management facility.

9.5.2 Link layer security

Personal area networking is often most convenient over wireless links. Two types of physical means have been developed for establishing wireless short-range connectivity and allowing ad hoc networking. They are infrared light waves and radio frequency. Infrared comes in the form of diffuse infrared or directed infrared. Both forms are suitable and widely used for ad hoc networking and distributed terminals. Diffuse infrared allows many-to-many connections, does not require direct line of sight and can be uni- or bi-directional. It is based on visible light. Financial trading floors are an example of diffuse infrared. Direct infrared is point-to-point, typically one-to-one communication, and requires line of sight and is a secure form of data transmission and reception. Infrared Data Association (IrDA) is an example of directed infrared. The security of infrared is based on its visibility within a room, and is claimed to be as secure as using a cable. No infrared-specific link layer security systems have been developed.

Two major wireless technologies based on radio frequency are Bluetooth and IEEE 802.11 wireless connections. Radio frequency is not secure in that it is inherently broadcast, it can penetrate walls and is subject to uncontrolled interference. Therefore, cable replacement based on radio frequency is not possible without additional measures of protection. Therefore, dedicated security systems have been developed for Bluetooth and IEEE 802.11.

9.5.2.1   Bluetooth

The Bluetooth system has been developed by Bluetooth Special Interest Group (Bluetooth SIG) as a cable replacement for short-range connectivity. In Bluetooth, effort has been taken to develop and standardise adequate security mechanisms and procedures for protecting the wireless radio link. This set of mechanisms is defined in the Bluetooth Baseband specification [13] and is referred to as Bluetooth Baseband security. It is based on strong cryptographic algorithms and well-established security principles. Still, more work is required to integrate Bluetooth Baseband security into various applications that may have very different link layer security requirements. Bluetooth Baseband security is implemented in the Bluetooth module and is common to all Bluetooth units.

The basic idea in the security concept is that trust between *devices* is created at a *pairing* procedure. A pairing is performed between two Bluetooth units. The purpose of a pairing is to create a common shared secret between two units. The common shared secret is called a *link key*. Two types of link keys are specified: *unit keys* and *combination keys*. The link key can also be established at a higher communication layer of the device and then imported to the Bluetooth unit using the host controller interface (HCI).

A Bluetooth unit with restricted memory resources might use a unit key. A unit uses the same unit key for *all* its connections. During the pairing procedure the unit key is transferred (encrypted) to the other unit. No pairing is possible between two units that *both* would like to use a unit key. In Reference 14 (see also Section 3.1 of Reference 6), unit key drawbacks are discussed. The unit key option is not recommended by the Bluetooth SIG for general use [15].

A combination key is a key that is unique to a particular *pair* of devices. The combination key is only used to protect the communication between these two devices. The combination key is calculated during the pairing procedure. Since a link key is used to protect the wireless link between two Bluetooth devices, each unit needs to store the link key it is supposed to use when communicating with a unit with which it has already established a link key. Hence, each unit needs to keep a link key database. The database contains the device address and the corresponding link key. The Bluetooth pairing procedure has been criticised, for example, in References 14 and 16 due to the fact that it is not cryptographically secure when short PIN values are used. Enhancements to Bluetooth pairing using MANA protocols were proposed in Reference 6.

The link key is used to authenticate other units. During the authentication process one unit, the verifier, sends a random value to the other unit, the claimant. The claimant has to process the random value together with the link key, to obtain a correct response value. The response value is sent back to the verifier who compares the received value with an expected value pre-calculated by the verifier. The authentication works only one way. If the units want mutual authentication, two consecutive authentication processes must be performed. As a side result, the authentication process generates a bit string, the Authentication Ciphering Offset (ACO). The ACO is used for ciphering key generation. The ciphering key is calculated as a cryptographic hash of the link key, a random value and the ACO. For broadcast encryption within a piconet the ACO parameter is replaced by the device address of the piconet master.

9.5.2.2   Wireless LAN and extensible authentication protocol

The original IEEE 802.11WLAN [17] security mechanism is known as Wired Equivalent Privacy (WEP). To access a WEP-protected network, a user must know the identity of the serving network, SSID, and the shared WEP key. The network identity is usually broadcasted by the access point, so that the user can select it from a list. The shared WEP key is the same key that is used for all stations. WEP link layer encryption is used to encrypt the data. WEP security has many serious problems, such as the lack of integrity and replay protection [18]. The encryption algorithm is the RC4 stream cipher with a short initialisation value. The mechanism is weak and allows attackers to recover the keys. Frequent re-keying is necessary, as noted by Fluhrer *et al.* [19]. For further discussion on WEP flaws see References 16, 20–21. Therefore, VPN solutions are recommended for use in WEP-protected WLANs.

One of the main problems in plain WEP is the lack of user authentication. The same key is used for all users, and anyone who knows it can get in. It is also difficult to change the key, since it is usually manually configured in the access point. The IEEE 802.1X authentication and key management framework has been developed to enable authentication of individual devices and to distribute WEP keys. The 802.1X authentication framework makes use of the Extensible Authentication Protocol (EAP), which is a standard interface for any authentication protocol. Using EAP, devices can be authenticated based on, for instance, passwords or public keys. The EAP interface has become widely used also in other types of networks, and will be discussed in more detail in Section 9.5.2.3.

In 2002, project 802.11i started to develop new and adequate security mechanisms for WLAN networks. A 'snapshot' of the future 802.11i was published in late 2002 in the specification of Wi-Fi Protected Access (WPA) endorsed by the Wi-Fi Alliance. This specification was also known as Safe Secure Network (SSN). It is not an IEEE standard. Some vendors have developed similar, but proprietary and incompatible products. WPA also uses the 802.1X authentication and key management functionality, but replaces the WEP encryption with Temporal Key Integrity Protocol (TKIP). TKIP changes the way RC4 keys are used, and also adds message integrity and replay protection.

In addition to 802.1X, WPA also supports a Pre-Shared Key (PSK) mode, which allows manually entered keys or passwords. This mode is intended for use in standalone PANs without access to external key management facility.

The 'full 802.11i', currently called Robust Security Network (RSN), is still under development and is not expected to be available until 2004 [23]. The main change is that the RC4-based TKIP will be phased out, and is replaced by the Counter with CBC MAC (CCM) Protocol, which uses the AES block cipher standard to encrypt and integrity protect the traffic. The standard may also include the Wireless Robust Authentication Protocol (WRAP), which is a similar protocol also using AES. RSN will also add several other enhancements. Previous WLAN security solutions have only supported the Basic Service Set (BSS) mode, that is, a network controlled by a single access point. RSN will also provide security for two other types of network configurations, Independent Basic Service Set (IBSS) and Extended Service Set (ESS). IBSS is a standalone network such as a PAN, without network access point and ESS is formed by a set of multiple access points with the MAC layer handoffs and access point roaming.

The new RSN will also provide a means for initial authentication and key agreement for WLAN-based PANs. RSN specifies how 802.1X and EAP authentication is used in IBSS. Each device in IBSS can take on the Supplicant, Authenticator and Authentication Server roles as defined by the 802.1X authentication. In this manner, using a suitable EAP method two devices in IBSS can negotiate the security association and establish a pairwise master key (PMK), from which they can derive keys for securing the subsequent communication. For more details on 802.1X and EAP, see Section 9.5.2.3.

9.5.2.3   IEEE 802.1X and EAP

The IEEE 802.1X standard [24] specifies an authentication mechanism intended for wired LANs, using the EAP. EAP specifies a 'transport protocol' where a client and a back-end authentication server exchange messages through the access point (or Ethernet switch). The access point does not have to know the details of various EAP authentication methods. When the user has successfully authenticated (e.g. by proving knowledge of a password or private key), the back-end server sends an 'accept' message to the access point. The access point and back-end authentication server usually communicate using the RADIUS protocol. 802.1X is an authentication framework, not a full authentication protocol, which must be taken into account when it is implemented as pointed out in Reference 25.

Some wireless-specific extensions have been added to 802.1X to support key management. EAP authentication methods intended for wireless LANs usually also provide mutual authentication (the server is also authenticated to the client) and key derivation. In key derivation, the parties negotiate a Master Session Key (MSK). The MSK is sent from the back-end authentication server to the access point, and is used to derive keys for protecting rest of the communication.

As its name implies, EAP is extensible: new authentication methods can be implemented without changing the base EAP protocol, or modifying the access points. Typically, EAP methods re-use existing authentication methods and related user databases and credentials. For example, in the EAP-TLS the client and the server authenticate using public keys and X.509 certificates. Another common example is MS-CHAPv2, which provides mutual authentication (the server must also know the password), and derives session keys. It can re-use existing Windows user databases. Other password-based methods include MD5-Challenge, MS-CHAPv1, SRP and SKE. These methods were designed for remote client-server authentication, where special measures must be taken to secure the usage of the password over the remote connection. The resulting protocols are computationally heavy. In PAN scenarios where proximity-based security methods can be used, it is more favourable to use proximity-based authentication and key agreement protocols like MANA and in such manner reduce the computation and memory requirements.

Authentication protocols using Token cards or one-time passwords can also be transported within the EAP. Typically, these methods do not provide mutual authentication or derive session keys. Recently, 'tunnelling' EAP methods, such as PEAP or EAP-TTLS, have been proposed. They combine weak legacy client authentication with public key based server authentication and can also provide session key derivation.

9.5.3 IP-based communication

9.5.3.1 IPsec

IPsec was developed by the IPsec working group of the IETF. The purpose of the IP security architecture [26, 27] is to provide various security services for the IP layer applicable to both IPv4 and IPv6. These services include data origin authentication, connectionless integrity, confidentiality (encryption), (windowed) replay protection and, to a certain degree, traffic flow confidentiality (by random padding of the transmitted data). IPsec uses two protocols to provide the above-mentioned services: Authentication Header (AH) and Encapsulation Security Payload (ESP). AH provides integrity protection whereas ESP supports encryption and optionally (but highly recommended) also integrity protection. Both protocols can be run in tunnel and in transport modes. Figure 9.5 shows (with TCP being the transport protocol) how original IP packets are modified in tunnel and transport modes.

In the transport mode and using AH, the whole IP packet is authenticated, including the IP header. The usage of ESP in transport mode provides for data confidentiality of the IP data, but offers only limited authentication of the IP header. In the tunnel mode and using AH, the complete newly created IP packet, including the old IP packet, is authenticated, but AH provides no data confidentiality. ESP in the tunnel mode provides for authentication and data confidentiality of the original IP packet.

In order to provide its functionality, an IPsec implementation must have access to a Security Association Database (SADB) and to a Security Policy Database (SPD). The SADB provides the IPsec engine with the necessary security association parameters

Original IP packet	IP header	TCP header	Payload data		
Transport mode protected packet	IP header	IPsec header	TCP header	Payload data	
Tunnel mode protected packet	IP header	IPsec header	IP header	TCP header	Payload data

Figure 9.5 Format of protected IP packets in different IPsec modes

(key, lifetime, sequence number, etc.) whereby the database is indexed by the triple <protocol type, SPI, destination address>. The SPD provides information that traffic should receive protection and how the protection should look like. The SPD is organised as a set of rules having a condition part and an action part indicating whether to apply IPsec processing, drop or permit (i.e. forward packet to further IP-related processing).

IPsec allows manually and automatically configured security association. The protocol designed to be used for the IPsec security association establishment is the Internet Key Exchange protocol (IKE). IKE is a complex protocol with many options and many authentication modes, and is complex to implement and use. IKE is divided into two phases. IKE phase I provides authentication and establishes a bi-directional ISAKMP security association. The first phase has two authentication modes: Main and Aggressive Modes. These two modes offer four authentication methods each: Pre-Shared Secret, Public Key Encryption, Digital Signature and Revised Public Key Encryption. The Main Mode requires six messages and promises to provide user identity confidentiality (user identity of the initiator) and prevention against DoS attacks. Aggressive Mode requires only three messages at the expense of the mentioned properties. The second phase of IKE (Quick Mode) is then responsible for establishing an IPsec security association by negotiating the algorithms and required parameters including the derivation of the key material.

A new and improved version of IKE, called IKEv2, is under development by the IETF [28]. Among other improvements IKEv2 will also provide support for legacy authentication methods. This means that the client can be authenticated using any EAP method. Hence, more flexible solutions can be expected in the near future to be available for the establishment of IPsec security association in a PAN also.

The MANA authenticated Diffie–Hellman protocol described in Section 9.4.2.1 can be used to provide a shared secret between peer devices in a PAN. This shared secret can be used as a pre-shared key for the ISP security association negotiation protocol. IKE and the current version of IKEv2 provide a security association negotiation protocol that is based on pre-shared secrets.

9.5.3.2 Transport layer security protocol

Transport Layer Security protocol (TLS) is specified in RFC 2246 [26] and is currently widely used to provide confidentiality, integrity and authentication of communication between web browser and server. TLS consists of four different protocols: TLS handshake, TLS change cipher spec, TLS alert and TLS record protocols.

Using the TLS handshake protocol the client and the server authenticate each other, agree upon the cryptographic algorithms (cipher suite) and the keys to be used. After the handshake protocol the record protocol protects communication between client and server.

The critical issue in TLS is that client and server can authenticate each other, after that everything goes on automatically. Peers can authenticate each other based on public keys using some of the public key based authentication methods specified by TLS.

At least two approaches use the TLS protocol in the initial authentication of peer devices in a PAN. First, it is possible to use EAP-TLS in the 802.1X authentication framework for authentication and establishment of the pairwise master link key (WLAN PMK or Bluetooth link key) as described in the context of wireless LANs in Section 9.5.2.2. The second approach is to use the full TLS protocol at the transport layer. The TLS handshake authentication mechanisms are based on public key certificate and digital signature verifications. Therefore, in both cases, the PAN devices must possess relatively high capabilities in terms of computation power and memory. One other possibility is to use certificates issued by a personal certification device according to the personal PKI concept, as described in Section 9.3.2.

9.5.3.3   IP networking over Bluetooth

The Bluetooth PAN profile defines how to do IP networking over Bluetooth [12]. In a PAN profile it is possible to use either Bluetooth security procedures or network layer security solutions, or both. A PAN User (PANU) is able to communicate with other devices in the Group Ad Hoc Network (GN) or Network Access Point (NAP) services using IP on a PAN. Both the NAP/GN and a PANU may require a certain level of security as part of the PAN service establishment.

When a PANU enters the NAP/GN coverage area and detects the presence of NAP/GN a part of the connection establishment procedure can be that either the NAP/GN or PANU request Bluetooth security procedures. Three security modes are defined as part of the Generic Access Profile. They can be applied for PAN profile as follows.

- *Security mode 1.*   In this mode, neither the mobile node nor the NAP requires security procedures invoked, so this case needs no further consideration.
- *Security mode 2.*   This mode does not mandate use of any security procedures in the link layer before the link set-up is completed but can be invoked by the service layer. Thus, it is possible to set up a connection and retrieve information about the discovered device if it is a NAP, more about the system that it is connected to. Using Security mode 2 it is possible for the devices to run an authentication and key agreement procedure at a higher layer, and in this manner to establish a common Bluetooth link key.
- *Security mode 3.*   In this mode, security is enforced by the link layer and requires the security procedures to be executed before link set-up is completed. If no link key exists pairing is initiated using the common passkey. The passkey must be known and distributed to both units in advance, or entered to the devices manually using an appropriate user interface. In closed environments and systems with a limited number of access units this approach may be appropriate. In systems with multiple access points, which may be changed without notice of the mobile node, it is not feasible to use this approach, but more automated access security solutions must be used.

9.6   Security configuration and access control

Next, we treat the security policy configuration and access control problems in a PAN. There are (at least) four different issues to consider.

- There must be rules to decide what functions a particular user (i.e. potentially other than the owner) may access on a given device.
- Various processes running on a device may need to be controlled in their access to resources/functionality local to the device.
- There must be rules (the remote security policy) for which service those external PAN components are allowed to access on a PAN component.
- The access rules must somehow be enforced, that is, we need control mechanisms.

The first two issues apply to any computing device in control by a human. Hence, this section focuses on the last two problems. Regarding access rules, the remote security policies must be configured so that the desired services become available to the PAN users. We follow our three-level trust model and suggest policy configuration based on this trust model. In addition, we discuss different options for describing authorisation statements and access control.

A component that offers services to other components has to make access control decisions on which of its services are available to what components under what conditions. Such decisions are made based on policies, which can be expressed in some form of 'authorisation statement(s)'. In a PAN we can think of authorisation statements either as tickets presented by the component requesting service or as a local configuration on the component using a policy database. In a more complex authorisation model, authorisations are allowed to be delegated using, for example, public key certificates like in the SPKI system [29]. In SPKI, certificates are used as 'tickets' to express authorisation statements and delegate authorisations. These principles fit quite nicely into the dynamic PAN environment and preferably combined with a public key based initialisation as described in Section 9.3. However, certificate delegation is quite complex to handle and often hard to express for a user in a way that is easy to understand. Hence, we will here instead discuss the alternative approach using a policy database. It is worth noticing though that usage of delegated authorisations in PANs is an interesting area that needs more investigations.

The authorisation information in a security policy database must be handled in a systematic way. The simplest form of policy database has just two policies: one for first party components and one for second party components. The first party component policy description is a list of all services that should be made available to all first party components. Similarly, the second party component policy is a list of all services that should be made available to all second party components. It is also useful to include in the policy whether encryption and/or integrity protection is required for the available services. Preferably though, all first and second party services are confidentiality and integrity protected. In addition to these policy settings there must be an identifier list for all first and second party components. Recall that we also require that it shall be possible for the reference component (in this case the resource under access control provider) to securely identify first and second party components and to know which category (first or second) it belongs to. In the simplest form, this is obtained by just securely storing the identities of all first, respectively, second party components. In addition, it is useful to store information on how they are authenticated and the cryptographic capabilities of the different first and second party components.

For the second party members we might have a more finegrain authorisation hierarchy. Then, we need to store more information. Typically, the following records need to be stored in the database:

- resource descriptor,
- subject identities or group of identities together with authentication information,
- validity.

The resource descriptor is a description of the service under access control. A serving component maps the resource descriptor unambiguously to a local service/object. Ideally, subject(s) should also understand what resource they are authorised to use, although strictly speaking this is not a requirement. The subject identity is the identity of the service requestor. If secure access control shall be possible, the serving component must be able to securely identify the requesting component. A condition that must be satisfied at the time the access control decision is made is for how long the access shall be granted, that is, the validity. Conditions can be set, for example, on dates, times of day, length of session, required encryption strength, mutual authentication, user confirmation, etc.

We also need methods for managing authorisation statements (i.e. policy entries) and for registering and locating resources. Default identifiers and secret/private keys are configured during component initialisation as described in Section 9.3. Later ID assignments and policy management may be performed using a new initialisation procedure, using a remote configuration (e.g. from a personal certification device) or through manual configuration over a user interface.

9.7 Conclusions

In this chapter, we have presented a PAN security architecture. We have covered the security problems we think are the most important to address in a distributed terminal environment. Different from general ad hoc networking studies, we have limited our scope to a PAN utilising a natural PAN reference model. The reference model was complemented with a PAN trust model. We suggested the usage of a trust model with only three basic trust levels: first, second and third party components. In this model, trust is always measured in relation to a reference component. First party components are 'highly trusted' and the idea of introducing this class is to simplify security configuration for the user and allow automatic trust delegation. Hence, resources can automatically become available in a controlled manner to all devices within this trust class. Less trusted components are sorted into the second trust class. Second party components have typically more limited access right, but still secure connection can be created between the reference and second party component. In our model, all components without any security relation to the reference components belong to the third party (untrusted) class.

We have extended the Stajano and Anderson 'resurrecting duckling' work and suggested different approaches and solution to the PAN security initialisation problem. The configuration of cryptographic parameters is the most challenging problem. We have discussed several different options both based on symmetric and asymmetric key techniques. In the case of group security associations, we need practical methods for trust delegation. In the symmetric key case this can be achieved by the usage of group key propagation. In the asymmetric key case, we suggested the usage of a personal PKI.

MANA protocols provide a useful tool for establishing initial security associations in an ad hoc manner between different devices. The practical applications are not limited to the Diffie–Hellman protocols. In a similar manner, public keys with respect to other public key systems such as RSA signature and encryption mechanisms or elliptic curve based systems can be authenticated (certified) using MANA protocols.

Temporary ownership is often practical to indicate using manual actions of the user. A typical example of such a scenario would be a meeting involving a number of different devices without any previous trust relationship. In such a scenario, a shared secret is distributed to the participating devices at the beginning of the meeting. Similarly, ad hoc connections to computing appliances, such as printers, projectors, etc., are often practical to set up by establishing a shared secret with the appliance. The shared secret can be used directly as a key for a security mechanism to protect the wireless link, or a simple key exchange protocol can be executed for derivation of suitable keys. It was also discussed how MANA certificates can be used to secure network access without any *a priori* security relationship with the serving network.

In Section 9.5, several building blocks of PAN communication security were discussed. The communication security mechanisms, authentication, confidentiality and data integrity can be provided at the link layer, network layer or transport layer. The main focus was on security methods that could be taken into use without relying on key management service provided by some external third party. The initial authentication and key management problem must be solved in all cases. Key management should become an integral part of the configuration of the PAN. No standard solutions exist yet. The MANA protocols presented in Section 9.4.1 have been developed to facilitate initial key mangement, and to make it secure and feasible even if the initial security information is communicated to the devices using human operated interfaces.

Finally, in Section 9.6 we discussed PAN security policy configuration and access control. We suggested following the three level trust model we have introduced and apply uniform access rules to all first part components. The more finegrain access rules needed between the reference and second and possibly third party components can be treated using a security policy database. Access control enforcement is preferably based on authentication of components and the policy in the database.

Acknowledgements

The authors gratefully acknowledge the SHAMAN project during which most of the results presented in this chapter were developed. We thank all participants of the SHAMAN WP2 for fruitful discussion and collaboration. We also thank Pasi Eronen for a useful survey of the recent developments in WLAN security.

References

1 SHAMAN (Security for Heterogeneous Access in Mobile Applications and Networks): IST-2000-25350 SHAMAN, Deliverable D13 A2, http://www. ist-shaman.org

2 IEEE Std 802.15.1: 'IEEE Standard for Information technology – Telecommunications and information exchange between systems – Local and metropolitan area networks – Specific requirements Part 15.1: Wireless Medium Access Control (MAC) and Physical Layer (PHY) Specifications for Wireless Personal Area Networks (WPANs)'

3 STAJANO, F. and ANDERSON, R.: 'The resurrecting duckling: security issues for ad-hoc wireless networks', 1999 AT&T Software Symposium, available at CHRISTIANSON, B., CRISPO, B., and ROE, M. (Eds), 'Security protocols', 7th International Workshop Proceedings, LNCS, vol. 1796 (Springer-Verlag 1999)

4 MITCHELL, C. and SCHAFFELHOFER, R.: 'The personal PKI', this book (Chapter 3)

5 BALFANZ, D., SMETTERS, D. K., STEWART, P. and CHI WONG, H.: 'Talking to strangers: authentication in ad-hoc wireless networks', in *Network and Distributed System Security Symposium* Conference Proceedings: 2002 (NDSS 2002), http://www.isoc.org/isoc/conferences/ndss/02/proceedings/papers/ balfan.pdf

6 GEHRMANN, C. and NYBERG, K.: 'Enhancements to Bluetooth baseband security', *in* Proceedings of *Nordsec 2001*, 1–2, November 2001, Technical University of Denmark, Lyngby, Denmark

7 LARSSON, J.-O.: 'Higher layer key exchange techniques for Bluetooth security', Opengroup Conference, Amsterdam, 24 October, 2001

8 STINSON, D.: 'Cryptography – theory and practise' (CRC Press, Boca Raton, 2002, 2nd edn.)

9 KABATIANSKII, G., SMEETS, B. and JOHANSSON, T.: 'On the cardinality of systematic A-codes via error correcting codes', *IEEE Transaction on Information Theory*, 1996, **IT-42**, pp. 566–78

10 REED, I. S. and SOLOMON, G.: 'Polynomial codes over certain finite fields', *Journal of the Society of Industrial and Applied Mathematics*, 1960, **8**, pp. 300–4

11 BLUNK, L. and VOLLBRECHT, J.: 'PPP Extensible Authentication Protocol (EAP)', RFC 2284, 1998

12 Bluetooth SIG: 'PAN Profile', v 0.95a, June 2001, http://www.bluetooth.com/ dev/specifications.asp

13　Bluetooth SIG: 'Specification of the Bluetooth system', Volume 1, v 1.1, February 2001

14　JAKOBSSON, M. and WETZEL, S.: 'Security weaknesses in Bluetooth', *in* Proceedings of the RSA Conference 2001, San Francisco, USA, 8–12 April 2001, Springer Lecture Notes in Computer Science, Vol. 2020, ISBN 3-540-41898-9, http://www.rsasecurity.com/rsalabs/staff/bios/mjakobsson/bluetooth/bluetooth.pdf

15　Bluetooth SIG: 'Bluetooth Security White Paper', v 1.0, April 2002, http://www.bluetooth.com/upload/24Security_Paper.PDF

16　KARYGIANNIS, T. and OWENS, L.: 'Wireless network security, 802.11, Bluetooth and handheld devices', National Institute of Standards and Technology (NIST), Special Publication 800–48, November 2002

17　IEEE: 'IEEE Std. 802.11, 1999 Edition, Part 11: Wireless LAN Medium Access Control (MAC) and Physical Layer (PHY) Specifications'

18　WALKER, J.: 'Unsafe at any key size – an analysis of the WEP encapsulation', October 2000, http://grouper.ieee.org/groups/802/11/Documents/DocumentHolder/0-362.zip

19　FLUHRER, S., MANTIN, I. and SHAMIR, A.: 'Weaknesses in the key scheduling algorithm of RC4', August 2001, http://www.cs.umd.edu/~waa/class-pubs/rc4_ksaproc.ps

20　NEWSHAM, T.: 'Cracking WEP Keys', Black Hat Briefing, http://www.lava.net/~newsham/wlan/ and http://www.lava.net/~newsham/wlan/WEP_password_cracker.ppt

21　IOANNIDIS, S. J. and RUBIN, A.: 'Using the Fluhrer, Mantin, and Shamir attack to break WEP', AT&T Labs, August 2001, http://www.cs.rice.edu/~astubble/wep/

22　WAGNER, D. *et al.*: 'Intercepting mobile communications: the insecurity of IEEE 802.11', MOBICOM 2001, January 2001

23　IEEE Std 802.11i/D3.02: 'Draft Supplement to ISO/IEC 8802–11/1999(I) ANSI/IEEE Std 802.11', 1999 edition, Specification for Robust Security, April 2003

24　IEEE Std 802.1X-2001: 'Standards for local and metropolitan area networks: port based access control', June 2001

25　MISHRA, A. and ARBOUGH, W.: 'An initial security analysis of the IEEE 802.1X Protocol', University of Maryland, February 2002, http://www.cs.umd.edu/~waa/1x.pdf

26　DIERKS, T. *et al*: 'The TLS Protocol Version 1.0', IETF RFC 2246, January 1999

27　DORASWAMY, N. and HARKINS, D.: 'IPsec the new security standard for the Internet, intranets, and virtual private networks' (Prentice Hall PTR, 1999)

28　KAUFMAN, C. (Ed.): 'Internet Key Exchange (IKEv2) Protocol', version 0.7, April 2003, work in progress, http://www.ietf.org/internet-drafts/draft-ietf-ipsec-ikev2–07.txt

29　ELLISON, C., FRANTZ, B., LAMPSON, B., RIVEST, R., THOMAS, B. and YLÖNEN, T.: 'SPKI Certificate Theory', *IETF RFC 2693*, September 1999

Chapter 10

Towards the security of routing in ad hoc networks[1]

Po-Wah Yau and Vaia Sdralia

Mobile ad hoc networking is a technology designed to ease the burden of network management through the use of distributed solutions. Whereas in the past the design of such networks was aimed at satisfying military scenarios, recent growing awareness of the scope for commercial use has accelerated research into high-performance, self-configuring ad hoc networks. However, the performance advantages that ad hoc networks offer are counterbalanced by security vulnerabilities, which are not present in conventional networking. This chapter discusses these vulnerabilities, focusing on the network layer, and presents a threat model classifying the types of threats to ad hoc networks. A variety of different security requirements can be extracted from the threat model. The latter part of the chapter discusses security mechanisms which have been proposed to satisfy these requirements, and identifies areas for future research.

10.1 Introduction

Mobile ad hoc networks are a step towards achieving autonomous networking. An ad hoc network is a temporary or permanent collection of nodes that can communicate with each other. Typically these networks will be completely self-configured and self-maintained. Control within an ad hoc network is assumed to be totally distributed, and the reliance on a central entity to control network functions is removed.

Mobile ad hoc networks can be seen as part of the move towards ubiquitous computing. The ease of information distribution that they imply, comes with many issues and responsibilities when using and securing that information. This chapter describes the wide variety of threats to mobile ad hoc networks, as well as some of the mechanisms and protocols that have been proposed to address these threats. The functionality of mobile ad hoc networks can be secured at different layers, although in this chapter the emphasis will be focused on the network layer. While many threats exist to layers both above and below the network layer, they are outside the scope of this chapter. In the network layer, the routing protocol will need to be secured so that functionality in the layers above, such as the Transport Control Protocol (TCP) and other network applications, can have some assurance that routing is reliable.

After a brief history of mobile ad hoc networks, Section 10.2 presents a description of their salient properties, along with how these properties relate to security for ad hoc networks. Section 10.3 then describes the general routing operations possible in ad hoc networks. The ad hoc threat model in Section 10.4 introduces the threats posed to the routing fabric, revealing how nodes can behave and what threats they present. Section 10.5 then discusses what security features are required to address the identified threats. This is followed in Section 10.6 by a discussion of protocol design and test issues. Section 10.7 surveys the mechanisms and protocols that have been proposed to secure different aspects of mobile ad hoc networks. Finally, security issues which have not been addressed by current research are highlighted.

10.2   What are mobile ad hoc networks?

Mobile ad hoc networks have existed since the 1950s, with the inception of the DARPA packet radio network [1]. Since then, like many other technologies, their use has been restricted to military scenarios, but over the past couple of years the realisation of the potential for commercial use has grown. One particular widely discussed scenario where ad hoc networks offer major advantages is in emergency situations, such as responses to an earthquake, where network infrastructure may be damaged or even not exist. Other future scenarios have been depicted in films, for example, in 'Minority Report',[2] where the lead character walks into a shopping centre and devices in shops offer clothes and items tailored to his specific user profile. This film scene also highlights another important factor – security.

The Internet Engineering Task Force (IETF) has set up a working group called MANET,[3] with the objective of selecting the most suitable routing protocols for ad hoc networks. In parallel with this, research into the security of mobile ad hoc networks has recently mushroomed, resulting in the publication of many papers.

This section describes the main characteristic features of a mobile ad hoc network, and underlines how each property affects the provision of security. Specifically, ad hoc networks are multi-hop, where information travels from the originator to the destination node often via intermediate nodes. The link between two nodes is considered as one hop. Each characteristic will affect the provision of security in different ways. Some present similar problems to conventional wired networks, while others introduce new issues particular to ad hoc networks.

10.2.1   Infrastructure

A salient characteristic of ad hoc networks is their lack of a predefined infrastructure. The implication of this property is that any network operations should be totally distributed, since no central control entity can be assumed. Hence, any security functionality will ideally be distributed, although traditionally many such functions, such as access control, are maintained by a central server. Also, security mechanisms involving online trusted third parties may no longer be viable in ad hoc networks. As nodes may be mobile, arbitrarily joining and leaving the network, security will have to be scalable to cope with the resulting dynamic topology.

10.2.2   Dynamic topology

The lack of pre-defined infrastructure means that nodes are free to move in any direction at any speed. Communication and, indeed, security can no longer rely on the continuous availability of any one link. Also, the use of wireless technology means that asymmetric links may exist, where communication can flow one way but not the other. Therefore, any security measures must be tolerant of the fact that a route may become broken at any time, in at least one direction if not both.

10.2.3   Mobility

It is intrinsically difficult to secure mobile ad hoc nodes. They could be very small, highly mobile devices with limited processing power, or part of a large network, such as those which exist in cars. Any security mechanisms will have to be able to operate at a reasonable speed using the limited processors that may exist.

10.2.4   Energy constraints

Another consequence of node mobility is the potential lack of available power. Any operations related to security should be resource efficient. Power levels may be an overriding factor in deciding the performance of any ad hoc operation. A commonly discussed threat is 'sleep deprivation torture' [2, p. 4]; this denial of service attack exists in many different forms, as discussed below. Furthermore, any security measures will have to allow for nodes which cannot participate in distributed operations because they are trying to conserve power or recharge (see Section 10.4).

10.2.5   Bandwidth constraints

The use of wireless communications technology will typically mean that bandwidth will be limited compared to that achievable using wired technology. Therefore, the amount of information sent by security mechanisms should be minimised. However, the level of security and assurance required may directly affect the bandwidth used, leading to a trade-off between performance and security.

10.2.6   Interaction with other networks

Only recently has research started to take into account internetworking between ad hoc networks and other network technologies such as wired networks, leading to hetero-geneous networks. The implications of this are extremely important for addressing, routing and security. Most routing protocols assume that some ad hoc nodes may also be gateway nodes, that is, nodes with a network interface connected to another network. These nodes are vulnerable to attacks that disable the connection between the two networks. It is prudent to ensure that gateway nodes are not at risk to attacks such as sleep deprivation by, for example, ensuring a permanent power supply. The effect of denial of service attacks could be mitigated by ensuring that there are always other gateway nodes available.

There is also an issue surrounding the interaction between security mechanisms in wired networks and any security mechanisms within ad hoc networks. As discussed earlier, energy constraints in ad hoc networks necessitate the use of security mechanisms that minimise computational overheads. Any security services provided in wired networks may thus have to be modified for interaction with ad hoc networks. An example is provided by IPsec [3], where the cryptographic algorithms chosen to secure communications between a wired host and a wireless ad hoc host would have to suit the resources available to both hosts, whilst being sufficiently strong to suit the context of the data being sent.

10.2.7   Interactions between multiple domains

All existing ad hoc network routing protocols assume that all nodes will be fully cooperative in forwarding data packets. This may become an unrealistic assumption in networks where multiple domains exist. There are thus many issues surrounding the provision of security in multi-domain networks. Trust relationships will need to be defined; for example, can nodes from one domain trust nodes from some domains more than others to route data packets, or to help provide distributed security? If there is no way to manage trust amongst the multiple domains, a powerful policing scheme may be needed to monitor, detect and react to security threats. Policies will then need to specify the appropriate reactions.

10.2.8   Security policy and management

One fundamental issue for security of ad hoc networks is how they will be managed, and where existing security policies will need to be updated to incorporate the new risks and threats that exist. Existing uses of wireless networks have shown less than positive results. In currently deployed wireless networks, an unacceptable number of risks are being taken [4]. An example is provided by the use of IEEE 802.11 in infrastructure mode [5], where nodes communicate with a wired network through an access point (also a realistic scenario for ad hoc networks). Although security features have been provided within IEEE 802.11, they are not being used by many network managers and operators.

Techniques such as 'war-driving', in which an attacker armed with the appropriate equipment travels down a stretch of road detecting access points for wireless networks, have revealed an alarming number of unsecured access points [6]. As for routing, security should ideally be as automatic as possible to simplify management. Nevertheless, the level of security is often chosen as a trade-off with performance.

One possible approach to security management is that, depending on the context of use, basic security (however that may be defined) could operate autonomously, but extra security mechanisms can be used at the users' discretion, so as to balance security and performance.

10.3 Routing protocols

Routing is a major area of research in ad hoc networks, as the characteristics of ad hoc networks pose many new challenges by comparison with traditional wired area networks. Existing protocols are likely to be too resource intensive to be suitable for ad hoc use, so solutions using a wide variety of methods have been proposed. This section describes the types of protocols which have been proposed to MANET, before presenting the threats which exist, and what security mechanisms will be required to prevent threat realisation.

There are two main types of ad hoc network routing protocols, namely *proactive* and *reactive* protocols. Reactive protocols discover routes only when necessary, whereas proactive protocols seek to find routes ahead of time. Within these categories, different implementations use a variety of techniques to find and maintain routes. *Hybrid* routing protocols also exist, which use both proactive and reactive techniques to control a hierarchical architecture. Other hierarchical protocols have been proposed that operate with existing MANET routing protocols. Finally, *geographical location assisted* routing has also been proposed, where location information is used to assist in network routing. Most routing protocols are table-driven, where information is processed and stored in routing tables, but other novel methods have been proposed.

10.3.1 Reactive routing

Reactive protocols are typically divided into a *route discovery cycle* and *route maintenance procedures*. A node initiates route discovery when it needs to send a data packet to a destination for which no route is known. This typically involves broadcasting some form of route request message, where an intermediate node, or the destination node itself, can provide the originator node with a reply containing a route to the destination. For example, in the Ad hoc On-demand Distance Vector (AODV) protocol [7], any node receiving a route request creates a route pointing back towards the node that originated it. The same applies when a node receives a reply to the request, that is, it creates a route pointing to the destination node. Thus, in AODV, the route discovery operation actually calculates both the forward (downstream) and reverse (upstream) paths.

The Dynamic Source Routing (DSR) protocol [8], uses 'source routes' where each node receiving the route request message adds its address to the 'source route'. When an intermediate node with knowledge of a route to the requested destination, or the destination itself, receives the route request, a route reply is sent which contains the accumulated source route. This explicit source route is the path that any messages sent by the originator node for that destination will follow.

Another family of reactive routing protocols uses geographical information to assist in routing. An example is Location Aided Routing (LAR) [9]. Using the Global Positioning System (GPS), LAR restricts its route requests to a certain geographical angle range, using prior knowledge of where the destination node previously was. The main advantage is the increased efficiency when compared to network-wide route request floods in other protocols. Route maintenance is required in reactive routing, as there are no periodic route update messages. Instead, when a link break is detected between two nodes, one or both nodes are responsible for propagating error information about the broken link to all affected parties.

10.3.2   Proactive routing

Proactive protocols use periodic topology updates to disseminate route information throughout the whole network, but try to minimise the information being sent in order to save bandwidth. Various techniques are used to achieve this. The Optimised Link State Routing (OLSR) protocol [10] requires each node to select 'multi-point relay nodes' from amongst its neighbours. Multi-point relay nodes are responsible for forwarding data packets for nodes which select them, and thus exchange route information between themselves. The optimisation is achieved through the decrease in routing update messages by comparison with a classic link-state routing protocol, in which every node has to periodically exchange route update information.

Topology Broadcast Reverse Path Forwarding (TBRFP) [11] requires nodes to compute and report a whole topology tree using periodic topology update messages. Changes to this tree are then reported in smaller but more frequent update messages. Bandwidth is saved since only differential updates are exchanged.

A number of hierarchical protocols have also been designed which function on top of another routing protocol. Fisheye State Routing (FSR) [12] attempts to introduce optimisations by organising an implicit hierarchy. Multiple scopes can be managed, where a node advertises route updates for inner scope nodes more often than route updates for nodes outside of the scope. Routing update messages are thus minimised, as they do not include every single possible destination (unlike classic link-state routing protocols). Landmark Routing (LANMAR) [13] is another hierarchical routing protocol that utilises proactive routing within each hierarchy. It assumes that the members of a subnet have similar mobility patterns. Each subnet nominates a 'landmark' node. Information about routing packets to other subnets is only exchanged between landmark nodes.

10.3.3   Hybrid routing

To achieve scalability, hybrid protocols combine both reactive and proactive techniques to form hierarchical routing protocols. The Zone Routing Protocol (ZRP) [14, 15] is a hybrid protocol that actually combines three sub-protocols. The Interzone Routing Protocol (IERP) is a proactive table-driven protocol used by nodes within a zone to provide up-to-date routing information. When a route is requested for a node outside of the zone, nodes on the border of the zone use the Intrazone Routing Protocol (IARP) and Broadcast Resolution Protocol (BRP), an on-demand method of requesting a route by asking other border nodes of other zones. ZRP allows a certain degree of modularity, as IERP and IARP can be, respectively, replaced by different proactive and reactive routing protocols.

10.4 A threat model

This section gives a threat model for ad hoc networks. Section 10.4.1 defines the security services that are covered in the threat model and distinguishes between two broad classes of threat, while Sections 10.4.2 and 10.4.3 elaborate on how ad hoc nodes can misbehave.

10.4.1 Ad hoc network routing threats

The main threats to an ad hoc network routing protocol are as follows. The following list also provides the basis for a generic list of security requirements.

- *Confidentiality.* The primary confidentiality threat in the context of routing protocols is to the privacy of the routing information itself, which leads to a secondary privacy threat to associated information, such as the network topology, geographical location, etc.
- *Integrity.* Network integrity is where every node has correct and up-to-date routing information. Thus, threats to integrity either introduce incorrect routing information or alter existing information.
- *Availability.* This is defined as access to routing information at all times upon demand. If a route exists to a mobile node, then any node should be able to get that route when they require it. Also, a routing operation should not take an excessive amount of time to perform. Related to this, a node should be able to carry out normal operations without excessive interference caused by the routing protocol or security.
- *Authorisation.* An unauthorised node is one that is not allowed to have access to routing information, and is not authorised to participate in the ad hoc routing protocol. There is no assumption that there is an explicit and formal authorisation protocol, simply an abstract notion of authorisation. As discussed below, formal identity authentication is a key security requirement, needed to provide access control services within the ad hoc network, and clearly authentication is a prerequisite to effective authorisation control.
- *Dependability and reliability.* One commonly envisaged application for ad hoc networks is in emergency situations, when the use of wired infrastructure is infeasible. In such a case routing must be reliable, and emergency procedures may be required. For example, if a routing table becomes full due to memory constraints, a reactive protocol should still be able to find an emergency route to a given destination.

The threat model used here distinguishes between external and internal attacks – see also Reference 16, p. 25. External attacks are performed by unauthorised nodes or entities and are likely to be more easily detected than threats from internal nodes. Internal attacks are posed by internal nodes, that is, they are performed by authorised nodes within the ad hoc network, and are thus likely to be more difficult to detect as they arise from trusted sources.

In the text that follows, 'correct' data packets and 'correct' procedures are simply those that adhere strictly to the routing protocol in use. By contrast, 'incorrect' data packets and 'incorrect' procedures are those that are in any way different to the format and behaviour as stated in the protocol. 'False' data packets are data packets that are of the correct protocol format, but contain false information.

10.4.2 External threats

In the presence of an authentication protocol to protect the upper layers, external threats are directed at the physical and data link layers. Physical layer security is intrinsically difficult to provide due to the mobile nature of ad hoc nodes.

In order to secure the ad hoc network, the hardware itself will have to be secured. Many nodes could be unsupervised, as is the case with wireless sensors, for example, as used to obtain environmental information in areas which are difficult to access. Data stored in ad hoc nodes may be very important, because operations are now distributed. Each node may store significant information about the network, for example, routing information, as well as authentication and identification metadata about itself or other nodes. Thus, hardware may have to be made tamper resistant [2, 17], so that data are kept confidential and integrity is maintained, that is, the data are protected from unauthorised reading and modifications. Tamper detection may also be a desirable feature in some contexts, where attempts to break into an ad hoc device can alert the user or other nodes. It is prudent to note that security against sleep deprivation attacks cannot be achieved in the physical layer, even though power constraints are a physical layer attribute. Since power levels affect all ad hoc network operations, securing such networks is uniquely difficult.

We divide external threats into two major categories: *passive eavesdropping*, where the adversary simply listens to transmitted signals, and *active interference*, where the opponent sends signals or data designed to disrupt the network in some way.

10.4.2.1 Passive eavesdropping

This can allow unauthorised principals to receive messages, including routing updates. An unauthorised node could gather data enabling it to infer the network topology, along with other information such as the identities of the more heavily used nodes. Hence, techniques may be needed to hide such information. Eavesdropping is also a threat to node location privacy. Note that passive eavesdropping also allows unauthorised nodes to discover that a network actually exists within a geographical location by just detecting that there is a signal present. Traffic engineering techniques have been developed to combat this.

Passive eavesdropping is traditionally addressed in the Medium Access Control (MAC) layer, where link layer encryption is the most popular defence. The scope of security provisions in the MAC layer will depend on the network technology in use. Current 'popular' MAC protocols include IEEE 802.11, Bluetooth and HIPERLAN[4] [5, 18]. To ensure that there is some defence against passive eavesdropping, routing updates may have to be encrypted in the network layer, because of the lack of security in the link layer.

10.4.2.2 Active interference

Active interference is any attack by an external node that results in a deviation from correct use of the routing protocol. These attacks can be divided into those which cause denial of service and those which affect network integrity.

10.4.2.2.1 Denial of service

The major threat from active interference is a denial of service attack caused by blocking the wireless communication channel or distorting communications. The effects of such attacks depend on their duration and the routing protocol in use.

With regard to the routing of data packets, reactive routing protocols may see a denial of service attack as a link break. Route maintenance operations will then cause most protocols to report the link as broken so that participating nodes can find an alternative route. By contrast, proactive routing protocols do not react immediately to non-delivery of data packets. If the route is believed to be broken, it will eventually be timed out and deleted.

Route request messages are broadcast, so the effect of a denial of service attack varies. An attack concentrated on one node will prevent it from discovering new routes, as it cannot transmit its route request. The success of a similar attack on an intermediate node depends on the topology surrounding the attacked node. For example, an attack on an intermediate node in an area densely populated by network nodes should have minimal effects, as route request messages are broadcast to every node. Therefore, the route request should be able to find its way to a node which contains the route being asked for.

Route replies are generally unicast back to the originator node from either the destination or an intermediate node that has knowledge of the route requested. A denial of service attack can thus be concentrated on the path of the route reply message. Reactive protocols employ an exponential back-off technique to re-initiate route discovery, while proactive protocols typically mitigate the effect of denial of service attacks through the frequent periodic transmission of routing updates. Thus, some loss of routing information is tolerable. A denial of service attack has to be permanent to prevent a route from ever being used in a proactive protocol – for example, by damaging a node involved in the route so that it cannot transmit.

Denial of service attacks can disrupt reactive route maintenance by preventing route error messages from getting to the nodes affected. These nodes will continue to try to use broken routes, where the node upstream of the break will become a bottleneck for packets still trying to use the link.

Probably the most serious type of denial of service attack is a sleep deprivation torture attack [2, p. 4], where node energy is deliberately wasted. With limited power and resources, prevention of such attacks is of utmost importance. Security against such attacks has already been extensively studied and developed by the military for packet radio networks. Spread spectrum technology is designed to be resistant to noise, interference, jamming and unauthorised intrusion [19, 20].

10.4.2.2.2   Network integrity

Threats to integrity include an external attacker attempting to replay old messages in order to reintroduce out-of-date information, or to change the order of messages. Out-of-date routing information could lead to further attacks, as nodes try to use old but invalid routes, or delete current valid routes. If the routing protocol utilises neighbour sensing by monitoring received data packets, replaying old packets may falsely lead nodes into believing that an 'old' link with a neighbour has become active and usable again.

Network integrity can be protected using well-established security techniques. The inclusion of unique sequence numbers in every data packet enables nodes to cope with out-of-sequence delivery, and to detect and ignore duplicate messages. This is a preventative measure but, as revealed later, there are other threats to network integrity where additional security mechanisms are needed.

10.4.3   Internal threats

The threats posed by internal nodes are very serious, as such nodes will have the necessary information to participate in distributed operations. Internal nodes can misbehave in a variety of different ways; we identify four main categories of internal node misbehaviour:

- failed nodes,
- badly failed nodes,
- selfish nodes and
- malicious nodes.

Failed nodes are simply those unable to perform an operation. Badly failed nodes still perform route operations but perform them incorrectly. Selfish nodes try to exploit the routing protocol to their own advantage. Finally, malicious nodes deliberately disrupt the operation of the routing protocol. Note that two misbehaving nodes within the same category may exhibit different degrees of incorrect node behaviour. For example, some nodes will be more selfish than others. Also, a node may demonstrate behaviours from more than one category – indeed, this may even be the typical case.

10.4.3.1   Failed nodes

Failed nodes are simply those unable to perform an operation; this could be for many reasons, including power failure and environmental events. The main ad hoc routing issues are failing to update data structures, or failing to send or forward data packets, including routing messages. This is important as the data packets may contain important information pertaining to security, such as authentication data or routing information. A failure to forward route error messages will mean that originator nodes will not learn of broken links and continue to try to use them, creating bottlenecks. The threat of having failed nodes is most serious if failed nodes are needed as part of an emergency route, or form part of a secure route.

10.4.3.2 Badly failed nodes

Badly failed nodes exhibit features of failed nodes such as not sending or forwarding data packets or route messages. In addition, they can also send false routing messages, which are still correctly formatted, but which contain false information and are a threat to the integrity of the network. For example, false route requests for a node that does not exist may circulate in the ad hoc network using valuable bandwidth, as no node can provide a suitable reply. Unnecessary route requests for routes which badly failed nodes already have, might also be sent. False route replies in response to a true route request may result in false routes being set up and propagated through the network. False route error messages will cause working links to be marked as broken, potentially initiating a route maintenance procedure.

Protocols that rely on neighbour sensing operations are also vulnerable, as false messages may cause nodes to 'sense' extra neighbours. This is especially true in protocols such as LANMAR, which do not rely on a specific neighbour sensing message. However, if a routing control message containing an unknown source address is received directly, then that address is used as a new neighbour [13, p. 2].

Protocols such as AODV include within the route error messages a list of affected nodes to which the route errors should be unicast [7, p. 211]. If this list is large, then the threat not only affects network integrity, but is also a denial of service attack, as resources and bandwidth are being used up by the large volume of route error messages sent, and the unnecessary route requests and replies used to find alternative routes.

The effect of badly failed nodes on proactive routing depends on how long a node sends false update messages. The false information will still be a threat to network integrity. Resulting route table calculations using false topology updates could create false routes to destinations that may or may not exist, and could, for example, lead to loops being formed. However, as route tables are periodically updated using fresh information, any false route information only remains as long as the badly failed node is active.

10.4.3.3 Selfish nodes

Selfish nodes exploit the routing protocol to their own advantage, for example, to enhance performance or save resources. Selfish nodes are typified by their unwillingness to cooperate as the protocol requires whenever there is a personal cost involved, and will exhibit the same behaviours as failed nodes, depending on what operations they decide not to perform. Packet dropping is the main attack by selfish nodes; most routing protocols have no mechanisms to detect whether data packets have been forwarded, with DSR being an exception. Another possible pattern of behaviour is partial dropping, which could be difficult to prevent and detect. It is important to emphasise that, in this model, selfish nodes do not perform any action to compromise network integrity by actively introducing incorrect information.

Some neighbour sensing protocols have an implicit defence by deleting selfish nodes from their neighbour lists, due to the lack of communication. In most neighbour sensing protocols, this means that selfish nodes will no longer be included in any route calculations. This has both advantages and disadvantages for selfish nodes. Since they are not part of any routes they will save resources by not having to participate in processing routing messages. However, this also means that selfish nodes will not be able to receive data packets, as there are no routes pointing to them for other nodes to use. Nevertheless, there is no mechanism to prevent selfish nodes from sending information, akin to a covert spy working in another country who is able to send information back to base but unable to receive information because the base does not know where the spy is. So it may be advantageous for a selfish node to just maintain active links with selected neighbours, allowing those neighbours to know where to send data addressed to the selfish node.

10.4.3.4 Malicious nodes

Malicious nodes aim to deliberately disrupt the correct operation of the routing protocol, denying network services if possible. Hence, they may display any of the behaviours shown by the other types of failed nodes. The impact of a malicious node's actions is greatly increased if it is the only link between groups of neighbouring nodes. We next indicate some of the main types of attack which could be launched by a malicious node.

10.4.3.4.1 Denial of service attacks

The simplest threat is a denial of service attack, for example, the 'sleep deprivation torture' attack [2, p. 4]. There are a variety of ways for a malicious node to achieve the objective of a denial of service attack. For example, in proactive protocols a malicious node could advertise topology updates with many false routes and addresses so that the route table calculation will take more time and resources. This latter threat is also an attack on network integrity; indeed many denial of service attacks exploit the routing protocol to introduce incorrect routing information.

10.4.3.4.2 Attacks on network integrity

The more densely populated the area in which a malicious node attacks, the more nodes will be affected. Protocols such as OLSR use a pure flooding mechanism, so that false information will be relayed to every node [10, p. 4]. With the hierarchical FSR protocol, participating nodes far away from the malicious node will be less sensitive to injected false information, especially in a multi-scope implementation [12, p. 2]. Hence, the preferred environment for a malicious node in a FSR controlled network is to have few scopes of a large size, to make sure false information is periodically broadcast as often and to as many nodes as possible.

10.4.3.4.3 *Attacking neighbour sensing protocols*

Malicious nodes can either force nodes to incorrectly add neighbours when they do not exist, or cause nodes to ignore valid neighbour nodes. The method will depend on the neighbour sensing protocol, but most require the receipt of some form of message. As with a badly failed node, a malicious node can send a neighbour sensing message with a false source address to cause the same effects.

For a malicious node to cause another node to ignore its neighbours, it could perform an active denial of service attack similar to that available to an external node; however, this could also be easily detected. Thus, this attack will be more successful for the malicious node if it could exploit some other operation, such as a blacklist. If a bi-directional MAC protocol is in use, DSR uses a blacklist for neighbours it believes it has asymmetrical links with. A malicious node could thus try to block transmission in just one direction to cause the node to be blacklisted. Blacklisted entries either expire or are deleted when bi-directional communication has been confirmed; thus, conversely, a malicious node could try to force a node to incorrectly delete neighbours from its blacklist by masquerading as a blacklisted node to forward a route request, whose source header contains details of the blacklisted node (its IP address, etc.). Similar attacks apply to AODV.

10.4.3.4.4 *Misdirecting traffic*

As previously mentioned, a malicious node can usually masquerade by just using a false source address in the data packets it sends, as described in FSR [12]. In FSR, nodes examine the IP Header source address and use it as a neighbour address. If the malicious node uses a false address which belongs to another node, then it can affect network integrity by getting all nodes in the network to point their routes to the malicious node, instead of the true owner of the source address. A malicious node can do this in a reactive protocol by replying to route requests before the original owner, and the same effect can be achieved in a proactive protocol by just advertising false routes in the hope they get accepted before the true routes. The malicious node will then receive any information intended for the owner of the address. This attack has been named the 'black hole' attack [21, p. 4].

Another reason for masquerading in this way is when the malicious node targets another node and causes excess traffic to be routed to it, resulting in a targeted sleep deprivation attack. A malicious node could send false route requests on behalf of this node, so that other nodes will then direct route replies to the node. Malicious nodes can advertise routes with attractive route metrics and high sequence numbers so that the likelihood of the false route being accepted is increased. However, the further away a malicious node is, the less successful this attack will be in getting the false routes accepted before true routes.

However, as identified and addressed by Hu *et al.* [22, p. 2], another attack exists with a similar effect. In the 'Wormhole' attack, a malicious node tunnels packets. In a reactive routing network, tunnelling route requests close to the destination node will result in the tunnelled route being replied to, and all other route requests being ignored. Thus, the malicious node has injected itself into the route. In a proactive routing network, the same technique can be used to tunnel neighbour sensing messages, in order to attack neighbour sensing protocols as described above.

10.4.3.4.5   Exploiting route maintenance

Malicious nodes can simply propagate false route error messages so that valid working links are marked as broken. Resources will be used in attempts to repair the links or find alternative routes. An alternative attack involves a malicious node coercing another node into sending route error messages by blocking an operational link (e.g. by blocking acknowledgments in DSR [8, p. 147]). This attack can also be performed by an external attacker.

10.4.3.4.6   Attacking sequence numbers and duplicate mechanisms

Unique sequence numbers prevent replay attacks using old data packets. However, this mechanism can also be exploited to cause a denial of service. A malicious node could flood the network with messages with false source addresses containing high sequence numbers. Any true messages sent will be discarded as duplicates or out of sequence messages.

This attack is possible because most protocols require nodes to maintain their own sequence number counter, and do not take into account the sequence numbers of received messages. Note that this issue refers to message identifier sequence numbers and not the sequence numbers used to guarantee route freshness, as used in AODV and OLSR.

10.4.3.4.7   Attacks on protocol specific optimisations

There are many protocol specific attacks. As an example we describe an attack on DSR salvaging, which is used to find alternative routes when a link break is detected [8, p. 151].

Using the attacks described earlier, the malicious node injects into the network as many routes with as many different next hops as possible, all of which do not exist and all point to the same target. The malicious node then sends a data packet addressed for that non-existent target. This is a denial of service attack, as an intermediate node will now attempt to send route error messages for each next-hop from which it tries to gain an acknowledgement. The intermediate node then tries to salvage the data packet by finding an alternative route. The alternative route is likely to be another false route and this carries on until the data packet has been the subject of *MAX_SALVAGE_TIMES* salvage attempts. The malicious node can just send another data packet for the same target to repeat the attack.

10.5   Security requirements and their scope

This section gives an overview of the security services that may be required for mobile ad hoc networks. These requirements have been developed from the threat model and the ISO 7498-2 security architecture standard [23]. An overview is also provided of how specific security services may address the threats identified above.

10.5.1   Security requirements

The following is a list of security services that may be required in ad hoc networks.

- *Confidentiality.* Data confidentiality is the service that prevents the unauthorised reading of data and routing packets by external users. Network confidentiality is the service preventing an external attacker from detecting that a mobile ad hoc network is present at a certain location. Traffic flow confidentiality may also be required to prevent external attackers inferring information from data traffic volumes.
- *Integrity.* Data integrity is the service whereby nodes can detect modification, insertion, deletion or replay of any packets received. This can occur on an end-to-end basis, between the originator and destination node, or at a peer-to-peer level between intermediate nodes. Network integrity provides the service whereby a node can be sure that it is receiving correct and up-to-date routing information.
- *Availability.* If a node transmits a data or routing packet, the intended next-hop will receive the data packet. This is peer-to-peer availability. End-to-end availability is where an originator node has assurance that, if it generates and sends a packet, the destination node will receive the packet within a reasonable time. The availability service also includes preventing denial of service attacks, including sleep deprivation torture.
- *Authentication.* Both entity and origin authentication can be used to prevent masquerade attacks. An external node is typically prevented from impersonating an internal trusted node because it lacks the necessary authentication credentials. Authentication will also be required to prevent internal malicious nodes from masquerading as another internal node.
- *Access control.* Access control ensures that only authorised nodes can participate in routing. This service must exist in conjunction with authentication to allow nodes to deny access to unauthorised nodes.
- *Non-repudiation.* Proof of origin is required so a node cannot deny sending a data packet. The sending node can also require that proof of delivery is provided. Again, these services can be applied on a end-to-end or peer-to-peer basis.
- *Dependability and reliability.* If an originator node needs to discover a route to send a data packet to a destination node, this service guarantees that the originator node will obtain such a route in a reasonable time as long as the destination node is in operation and receiving data packets. Moreover, if a route breaks, the originator node will be notified of the break within a reasonable time.
- *Accountability.* This service ensures that any actions affecting security can be selectively logged and protected, allowing for appropriate reactions to attacks. As explained above, the misbehaviours demonstrated by different types of nodes will need to be detected, if not prevented. Event logging will also help provide non-repudiation, for example, preventing a node from repudiating involvement

in a security violation. This service will especially help to detect attacks on the ad hoc network by internal nodes.

10.5.2   Design of security protocols

Table 10.1 indicates how the above security services can address the identified threats – 'Y' indicates that a service can help protect against a threat. Preventing the realisation of threats should always be the primary goal, that is, protocols should be designed so that it is impossible or too expensive to attack them. Ideally, only authorised nodes can participate in the network, and an internal node should have no other option but to follow the protocol.

Unfortunately, it may be impractical to prevent some attacks, and thus some means of detection of misbehaviour is needed. The learning latency is a measure of how long it takes for nodes to identify misbehaving nodes, whether this is performed locally or in a distributed manner between several nodes. Ideally, this latency should be as low as possible, so that normal routing operations can continue with little interference from misbehaviour. In practice, there will be a trade-off between the time for which an attack is allowed to occur, and the time in which an attack is confirmed. This is important because any attack will result in bandwidth and resources being wasted, both of which are important commodities in ad hoc networks.

Ad hoc networks are an example of the distributed computation model, and many proposed security solutions use a distributed scheme involving inter-node collaboration [24–26]. However, such distributed solutions have associated hazards, as they must take into account the multitude of scenarios that may occur in ad hoc networks. These can be as trivial as just two stationary nodes communicating in a room, to many thousands of nodes moving at random and at high speeds. Compromising one of the required principles underlying the distributed computation should not enable a simple denial of service attack.

Another important design factor is the use of packets containing security information. The more such security packets are needed, the greater will be the opportunity for an attack. Similar issues apply regarding the distance security packets have to travel – security packets that have to travel end to end may take too long to reach their destination, or may not even reach it at all. Thus, the distance that security packets should travel must be minimised, as shown in Reference 27, p. 2, which suggests that localised solutions are the most efficient. Security packets will also have an adverse effect on network performance, which is always an important consideration.

The design requirements for any security protocols or mechanisms must be carefully defined, explicitly stating which threats they will address. The same threat can arise from different node misbehaviours, and thus a single threat may necessitate a variety of countermeasures. For example, both failed and selfish nodes may refuse to forward data packets. While many solutions attempt to mitigate the effect of selfish nodes by excluding them [25, 26, 28], they do not always take into account failed nodes that have to conserve power and enter into sleep modes. There is thus a risk that failed nodes will be excluded, when it may be reasonable to allow them to rejoin

Table 10.1   Threats versus services

Source of threat	Description of attack	Confidentiality	Integrity	Availability	Authentication	Non-repudiation	Dependability	Accountability
External	Passive eavesdropping	Y						
	Actively blocking transmission			Y			Y	
	Sleep deprivation			Y			Y	
	Insertion, deletion, modification and replay of packets		Y					
	Misdirecting traffic (wormhole attack)		Y	Y				
	Masquerade as an internal node				Y			
Failed internal	Not forwarding packets			Y		Y	Y	Y
Badly failed internal	Sending false route information		Y				Y	Y
	Sleep deprivation			Y			Y	Y
Selfish internal	Not forwarding packets			Y		Y	Y	Y
Malicious internal	Sleep deprivation			Y		Y	Y	Y
	Sending false route information		Y			Y		Y
	Actively blocking transmission			Y				Y
	Misdirecting traffic		Y	Y		Y		Y
	Masquerade as internal node				Y	Y		Y
	Insertion, deletion, modification and replay of packets		Y					Y

the network. Of course, the task of distinguishing between failed and selfish nodes is a non-trivial problem.

10.6   Protocol, mechanism and simulation design issues

A motivation for the threat model described in Section 10.4 is to produce more realistic scenarios and simulations to properly test security mechanisms designed to cope with all the likely types of node behaviour, producing solutions which will work in a practical system. In the past, simulations of ad hoc networks have often been too 'clean' or 'vanilla', where the system runs too smoothly and uniformly. Network topology has a major impact on the effect of the realisation of threats. Thus, the following properties affect the possible threats identified in the threat model, and should be considered when designing secure protocols and simulation studies to test such protocols.

- *Size.* It is a prerequisite that the network size should be changeable. Indeed, size should be made a dynamic variable which changes as nodes enter and leave the ad hoc network. Simulations should ideally also cover the case where large numbers of nodes simultaneously enter or leave the network.
- *Density of nodes.* Denial of service threats will be more difficult to achieve in crowded than sparsely populated areas. This is because the more nodes there are, the more possible alternative routes exist. However, threats to network integrity may be more serious in dense areas, as any false information will propagate faster.
- *Position of nodes.* This describes the problem of nodes which drift in and out of transmission range of the ad hoc network at the perimeter, and also any partitioning where only a few nodes link two groups of nodes. Here the perimeter of the network may be the actual physical perimeter, or the logical perimeter of a collection of network nodes which are cooperating with each other. Denial of service threats are more serious here, as boundary nodes will have fewer routes to the rest of the ad hoc network. In both scenarios, there are fewer routes to attack to perform a successful denial of service attack.
- *Grouping.* Allied to the position of nodes should be some means of simulating the grouping of nodes. The property of locality of reference refers to the fact that nodes will communicate with a common group of nodes, often those closest to it. In any case, scalability will probably see ad hoc routing using some form of hierarchical model, as with the Internet. Grouping may be temporary, with a high number of nodes leaving and joining (e.g. a hotspot location), or quite permanent (e.g. a Personal Area Network).
- *Mobility.* It must be emphasised that not all ad hoc nodes will be mobile. Therefore, protocol simulations should include stationary nodes. Indeed, research may reveal that ad hoc networks cannot function without some semi-central entities. Thus the number of stationary nodes should be an automatically dynamic variable. In addition, when simulating mobile nodes, care must be taken to make their movement random [29]. Their speed should not be constant and their movement made of different trajectories. Further, groups of nodes will also move together,

for example, nodes in a car. Since some nodes may move very quickly, protocols will have to cope with nodes entering and leaving at a high rate. For example, nodes belonging to people and shops in a street may want to network together and share services, as there will be some link persistence. They may, however, wish to ignore nodes in a car that drives past, as the relationship will not be long-lasting enough for any practical benefit to result.

- *Relationships.* Most current simulations include a node randomly sending another node a data packet or route request. Although this may be too complex to implement, simulations should ideally include relationships of differing length. Short-term relationships of seconds and long-term relationships of many minutes should both be covered. The motivation for this arises from recent reputation research [24, 25]. The differing lengths of relationships can dictate how good or bad reputations will be considered and managed.

10.7 Some current proposals

This section focuses on existing proposals addressing the security requirements outlined above. The discussion is divided into four main categories: secure routing, coping with selfish behaviour, key management and authentication. Note that an analysis of the advantages and disadvantages of each approach is outside the scope of this paper. This section does not present an exhaustive list of proposals – the rapid recent growth of the area makes this impractical; however, efforts have been made to describe a representative sample of schemes (although inclusion of a scheme does not imply that it is better than the schemes not mentioned). Terms are used as defined in their original papers – where terms have been used that differ from those used in this chapter, they are explicitly defined.

10.7.1 Secure routing

There are many ways in which routing in ad hoc networks can be secured. Extra security layers can sit above and below the network layer, or the routing protocol can be extended to include security mechanisms.

- Security-Aware ad hoc Routing (SAR) [30] extends AODV by using a security metric in addition to a distance metric, so that routes with multiple security levels can be discovered.
- The Secure Efficient Ad hoc Distance vector (SEAD) routing protocol [31], and its sister Ariadne [32], both use one way hash chains to authenticate routing information. In addition, Ariadne incorporates the advantages of TELSA, a broadcast authentication mechanism where loose clock synchronisation is used as the basis of key management.
- Secure AODV (SAODV) [33] uses hash chains to detect the modification of mutable data, in particular the hop count field in AODV routing packets. A hashed Message Authentication Code is then used to protect the rest of the data.

- The Secure Routing Protocol (SRP) [34], uses the same premise as SAODV, but has been adapted for DSR using symmetric cryptographic techniques.

We now consider the last two of these schemes in a little more detail.

10.7.1.1   Secure AODV

SAODV [33] is an extension of the AODV protocol, satisfying the following requirements:

1. import authorisation,
2. origin authentication and
3. integrity.

Import authorisation is to detect whether incoming route information is true. Origin authentication verifies the source of route information, and integrity is used to detect if route information has been modified accidentally or maliciously.

In terms of securing ad hoc networks, the scheme distinguishes between data packets and routing packets. Data packets can be secured in the upper layers of the network stack, although this is not possible with routing packets. The information in a routing packet can be classified as mutable and non-mutable. Mutable information is dynamic information which is changed at each hop, for example, the hop count, whereas non-mutable information is that which has to stay fixed. SAODV suggests ways of securing both.

10.7.1.1.1   The protocol extension

SAODV makes one major assumption. A public key management system is in place and operational. All participating nodes are able to obtain public keys and are also able to verify the authenticity of each public key. In order to use SAODV, each and every AODV routing packet must make use of SAODV signature extensions [33, p. 9].

10.7.1.1.2   Hash chains

The only mutable information of concern in AODV routing packets is the hop count. This leads to a possible attack where a malicious node reduces the hop count before forwarding route request messages.[5] This increases the malicious node's chances of being included in the resulting route between the originator of the route request and the requested destination node. SAODV recursively uses hash functions to detect if the previous hop has modified the hop count. As the hop count has no relevance in route error messages, there is no need to apply the mechanism to such packets.

For each route request or reply, the originator node has to generate a random seed. This random seed is passed through a hash function x times, where x is equivalent to the Time-To-Live (TTL) value of the packet. The result *Top_Hash* will now be used as a comparison by other nodes receiving the route request or reply. The original seed, referred to as *hash*, the hash algorithm used, and the TTL value are sent as part of the appropriate SAODV extension for the route discovery packet. Each node receives the route discovery packet and increments the hop count, before checking that:

$$Top\_hash = h^{\text{TTL–hop count}}(hash)$$

where h is the hash function in use.

The node can trust that the hop count has not been modified if the check succeeds. Before forwarding the route discovery packet, the node will have to apply the hash function once more to the received *hash* value. This method works because any attempt to change the hop count will produce a different result to *Top_hash*.

10.7.1.1.3 *Digital signatures*

Digital signatures are used in order to prevent malicious or accidental modification of the SAODV extension, making the use of hash chains secure. Therefore, all information other than the '*hop count*' and '*hash*' value are signed, so that modifications can be detected. The signature encompasses the entire AODV route discovery message as well as the SAODV extension, so as to provide data integrity.

Digital signatures also provide origin authentication. Hence, digital signatures prevent masquerade attacks when a malicious node tries to send false route requests and replies. One problem with applying this to AODV is that intermediate nodes can reply to route requests if they have a route to the destination node. Thus, an intermediate node needs a way to sign a route on behalf of the destination. SAODV includes the following procedure to allow this.

When generating a route request, an originator node a will add a SAODV signature extension containing a's signature on the route request. Intermediate nodes receiving this route request must verify the signature before creating a reverse path to node a. Once verified, the intermediate node can store the reverse route. In order to enable the intermediate node to reply on behalf of node a, node a will need to add a SAODV double signature extension, which also includes a's private key for signing. Thus intermediate nodes can use a's signature to sign a route reply.

For any route replies on behalf of node a, the intermediate node will also need to include the old lifetime of the route so everyone can use it to check node a's original signature. On top of node a's signature, the intermediate node will need to sign the new route reply itself, producing a SAODV double digital extension for a route reply, akin to the way one is produced for a route request.

To conclude, an intermediate node will only reply to a route request if it has the requested destination's signing key and the old lifetime value of the route. If either value is absent, the intermediate node will rebroadcast the route request as if it did not have the route. Destination nodes replying to route requests themselves only need to use the SAODV single signature extension.

10.7.1.1.4 *Route error messages*

In a note about route error messages, the authors of the SAODV scheme simply suggest that route error messages are only signed and verified between neighbour nodes. For example, node a signs a route error message and its neighbour node b verifies the signature. Node b then removes node a's signature extension and produces its own, before passing the route error message to its neighbours, who will perform the same process.

10.7.1.2   Secure routing protocol

The SRP [34] is a reactive routing protocol which uses source routes. Its main aim is to detect malicious modification of routing packets, so that a node originating a route request can detect whether received route replies have come and from legitimate nodes and contain truthful information.

10.7.1.2.1   Assumptions

SRP assumes that some predefined security association exists between the originator node and the destination node that will receive the route request. The authors suggest that public key cryptography could be used if a viable infrastructure exists, but state that a shared secret key can also be used. Other assumptions are that only bi-directional links are used, all devices operate in promiscuous mode and that each device has only one network interface.

10.7.1.2.2   Neighbour lookup protocol

SRP uses the Neighbour Lookup Protocol (NLP) to sense neighbours. Each node maintains a 'Neighbour Table', which stores the IP and associated MAC addresses extracted from overheard communication. Each entry will timeout if no new messages are received from the corresponding neighbour. Multiple IP addresses are not permitted. NLP is used to detect and discard packets if the following occurs:

- a neighbour has used a different IP address from the one stored in the Neighbour Table;
- two different neighbours have used the same IP address (indicating a possible IP spoofing attack); and
- two neighbours have used the same MAC address (indicating a possible data-link layer spoofing attack).

When a routing packet is received, the last address of the source route is checked to ensure it matches the neighbour's address from which the packet was received, in addition to the NLP checks.

10.7.1.2.3   Securing route discovery

Three fields are used within SRP messages to provide security.

- Each node maintains a *query sequence number* for the lifetime of a security association with a destination node. This is incremented for every route request issued for that destination node. Thus, the destination node only processes a SRP route request if the message contains a higher *query sequence number* than the destination has stored itself.
- Each route request also includes a unique *query identifier*, produced by hashing a random seed. Any routing packets received with a query identifier that has been used before will be dropped, preventing replay attacks. The *query sequence*

number and the *query identifier* will be included in any resulting route replies to the originator of the route request.

- Finally, each route request must contain a Message Authentication Code that encapsulates:
  - the IP header,
  - the route request and
  - the shared key between the originator and destination nodes.

The mutable fields of the IP header and the source route are not included in the Message Authentication Code. When the destination node receives the route request, it can re-compute the Message Authentication Code to check that the route request has not been modified, and that it has come from a trusted node with which the destination has a security association. The destination node will then compute a Message Authentication Code on the route reply, which now contains the full source route, so that the originator node can perform the same check upon receiving the route reply.

Another important feature of the route request is the value of N_RREP, which is the expected number of route replies the originator wants before it will accept the route as valid.

10.7.1.2.4 *Preventing denial of service attacks*

A method is proposed 'to guarantee the responsiveness of the routing protocol'. When a destination node is receiving route requests from multiple originator nodes, higher priority is assigned to those nodes that are generating requests at a lower rate. Unprocessed route requests are eventually discarded. Within each class, a round robin scheme is adopted. This optimisation is designed to prevent excess flooding of route requests by malicious nodes.

10.7.1.2.5 *Allowing intermediate nodes to reply*

A destination node can allow other intermediate nodes to reply on its behalf, through the use of a shared group key (which implies a group security association). Each route request is now concatenated with a 'Intermediate Node Reply Token' (INRT), which is essentially the same as the Message Authentication Code used previously, using the group key rather than the shared key. Each group member maintains the latest *query identifier* used by its peers. Any route replies by intermediate nodes are exactly the same as before, but with the new INRT. The route from the replying intermediate node to the destination is also appended to the source route. Destination nodes also generate route replies as before, but append an additional Message Authentication Code encapsulating the INRT and route reply.

10.7.1.2.6 *Route maintenance*

Route error messages are generated by intermediate nodes which fail to deliver a packet (although it is not clear how a node detects a delivery failure). The route error is routed back along the prefix of the source route before the link break. Each node which receives the route error message verifies that the last address of the source route matches the address of the neighbour from which the route error was received, forwarding the packet if no mismatch occurs. The end node will be the originator node, who will verify that the source route contained in the route error belongs to a stored source route. Note that securing route error messages has been delegated to a separate protocol called the Secure Message Transmission protocol, details of which are yet to be published.

10.7.2 Coping with selfish behaviour

Current routing protocols make an assumption that all nodes in an ad hoc network will cooperate to provide each other with routing information and to forward packets for each other. However, this may not always be the case, especially when nodes are selfish and want to save power. The Collaborative Reputation Enforcement (CORE) scheme [25], CONFIDANT [24], and the 'nuglets' model [28] are examples of mechanisms that try to detect selfish node behaviour and provide ways to deal with them. Both CORE and CONFIDANT are based on quantifying reputation, where reputation values are distributively monitored to detect misbehaviour. Misbehaving nodes are then excluded from accessing network services, such as routing. The 'nuglets' mechanism was produced as part of the 'terminodes' project,[6] and is a currency model. Each node has to earn 'nuglets' by doing work for other nodes, for it is these 'nuglets' that are required to be able to send data packets for oneself.

10.7.2.1 Collaborative reputation enforcement

Michiardi and Molva [25] propose a reputation mechanism that enforces node cooperation when nodes in an ad hoc network do not have any prior trust relationship. They define an ad hoc network as a community where each member must contribute to the running of the community to remain a trusted entity. Any member not contributing will find their good reputation deprecating until there are gradually excluded because of their bad reputation.

A member can be a requestor, who requires a service from one or more providers. Providers are required to broadcast an explicit denial of service message when a member with a negative reputation tries to request a service. Thus, neighbouring nodes can use these denial of service messages to detect any inconsistencies. Allied to this are peer nodes, who are present to monitor community misbehaviour such as a member denying a service to a reputable member.

They are three types of reputations – subjective, indirect and functional. Each community member will combine these reputations to form a global reputation value for another community member. Each calculation is normalised so that a reputation ranges from -1 (bad) to $+1$ (good). Thus, 0 represents a neutral view and is used when there are not enough observations to gain an accurate reputation.

Subjective reputation. This is a locally calculated reputation value, where node a calculates the reputation of a neighbour node b, at a given time for a particular function. In the calculation, more emphasis is given to past reputation values to eliminate any current sporadic behaviour.

Indirect reputation. Here, reputation values are accepted by node a from node c about node b. Only positive reputation values are used, to eliminate the threat of a malicious node transmitting negative reputations to cause denial of service.

Conversely, this does not prevent a malicious node from boosting the reputation of nodes which have bad reputations. However, the mechanism does give some protection, in that subjective reputation is given more weight than indirect reputation.

Functional reputation. Functional reputations are subjective and indirect reputation values relating to a certain function, such as packet forwarding, route discovery, etc. Each function is given a weight. For example, data packet forwarding may be deemed more important than route request forwarding, so greater weight will be given to the reputation calculations regarding packet forwarding.

10.7.2.1.1 *Validation mechanism*
Reputation values are based on observations. If the observed behaviour matches the expected behaviour then the *k*th observation will be positive. If not, the *k*th value is negative. To be able to perform this validation reliably is fundamental to the functionality of the CORE scheme; the authors have suggested using the 'Watchdog' mechanism. The expected result is stored in a buffer until the matching observed result occurs. While the expected result is still present in the buffer, the reputation relating to the observed function is gradually decreased.

10.7.2.1.2 *The CORE protocol*
The CORE scheme is generic so it can be applied in many different scenarios. In the context of ad hoc networks and routing, it can sit above the Network Layer to help achieve cooperative routing. Each node in an ad hoc network is either a requestor or a provider of a routing function. For example, a requestor node *a* could ask its neighbour nodes for a route to destination node *d*, where node *a*'s neighbours *b* and *c* are provider nodes.

Each node will maintain a reputation table. This table consists of the reputations of other nodes with each entry consisting of a unique ID, recent subjective observations, recent indirect observations and the composite reputation for a given function. A reputation table has to be maintained for each function that is monitored.

10.7.2.1.3 *Reputation table updates*
There are three ways in which a reputation table is updated.

- The first case is a local calculation. A requestor monitors what its provider is doing. Suppose node *a* is a requestor and its neighbour node *b* is the provider. If node *a* requests a service from node *b*, but node *b* refuses to perform the service, node *a* will decrease its perceived reputation of node *b*. This is a local calculation of node *b*'s subjective reputation.
- In the second case, there is a global distribution of reputation within a reply phase. The reply phase contains a list of entities which successfully cooperated in providing the function. Recall that only positive values are given in indirect reputations.
- The final case is when reputations are gradually decreased to a null value when there is no interaction with the observed node.

10.7.2.2 CONFIDANT

CONFIDANT detects malicious nodes using observations or reports regarding attacks, by employing 'neighbourhood watch', and allowing nodes to route around misbehaved nodes and to isolate them from the ad hoc network [24]. The components of a node are:

- a monitor for observations,
- reputation records for first-hand and trusted second-hand observations about routing and forwarding behaviour of other nodes,
- trust records to control trust given to received warnings and
- a path manager that adapts the node's behaviour according to reputation and takes action against malicious nodes.

CONFIDANT (Figure 10.1) operates as follows. Every node monitors its neighbours and changes its reputation records if a suspicious event has occurred more often than a predefined threshold, which is high enough to distinguish malicious behaviour from coincidences such as collisions. The rating of the node is then re-calculated, and if it is categorised as malicious the path manager is called, which deletes all routes containing the malicious node. Additionally, an ALARM message is sent by the trust manager, which contains information such as the type of protocol violation, the number of occurrences observed, whether the message was self-originated by the

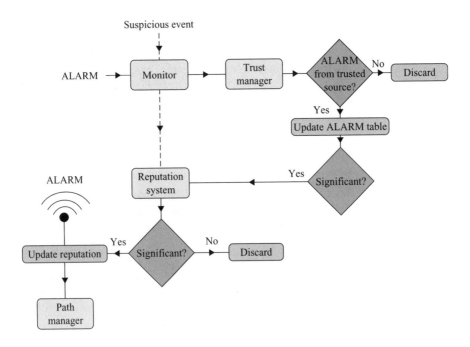

Figure 10.1   The CONFIDANT scheme

sender, the address of the reporting node, the address of the observed node and the destination address.

On receipt of an ALARM the monitor component of the node passes it on to the trust manager, where the source of the message is evaluated. If the source is at least partially trusted, the table containing the ALARMs is updated. If the source of the ALARM is fully trusted, or several partially trusted nodes have made consistent reports and their respective assigned trust adds up to the value of one entirely trusted node or more and the node reported in the ALARM is malicious, the information is sent to the reputation system where it is again evaluated for significance, number of occurrences and accumulated reputation of the node.

10.7.2.2.1   CONFIDANT components

The monitor. In an ad hoc environment, the nodes that can detect non-compliant behaviour are those in the neighbourhood of the malicious node as well as the source and destination. Specifically, the nodes of the 'neighbourhood watch' can detect deviations from correct behaviour by the next node on the source route by either listening to its transmission (passive acknowledgement) or by observing route protocol behaviour. By keeping a copy of a packet while listening to the transmission of the next node, any content change can also be detected. The monitor registers these deviations from normal behaviour and calls the reputation system as soon as malicious behaviour occurs.

The trust manager. The trust manager deals with outgoing and incoming ALARM messages sent by other nodes as a warning of malicious behaviour. Outgoing ALARMS are generated by the node itself after having experienced, observed or received a report of malicious behaviour. The recipients of these messages are administered in a 'friends' list. There is not yet a trusted method for the creation of such a list. The authors of the scheme are currently using for this purpose the method described in Reference 35, where users are authenticated by 'imprinting'. However, this method does not ensure the initial inclusion of a malicious or selfish node in the 'friends' list of its neighbours.

Incoming ALARMs originate from nodes that are either friends or not. Therefore, the source of an ALARM must be checked for trustworthiness before deciding on a response. This is achieved by filtering the incoming ALARMs according to the trust level of the reporting node. The trust level of each node is calculated by employing the trust management mechanism used in Pretty Good Privacy (PGP) for key validation and certification. Specifically, the trust manager consists of the following components:

- an alarm table containing information about received ALARMs;
- a trust table managing the trust levels of the nodes to determine the trustworthiness of an ALARM;
- a friends list containing all friends to which a node may potentially send an ALARM.

In the case of routing and forwarding, trust is important when making a decision about:

- providing or accepting routing information;
- accepting a node as part of a route; and
- taking part in a route originated by another node.

The path manager. The path manager is responsible for the following functions:

- path re-ranking according to the security metric;
- deletion of paths containing malicious nodes;
- actions on receiving a request for a route from a malicious node, for example, to ignore it; and
- actions on receiving a request for a route containing a malicious node in the source route.

10.7.2.2.2   The reputation system

The reputation system maintains a table consisting of local rating lists and/or black lists at each node; the table is potentially exchanged with 'friends'. However, since there is no reliable method for creating the 'friends' list, a malicious node can boost the reputation of nodes that have bad reputations. Additionally, when a node requests a route it can ask for a malicious node to be avoided, informing the rest of the nodes. On receipt of such information the node checks whether the sender is included in its black list before forwarding the message.

Some of the disadvantages of this mechanism are that differentiating between suspected and confirmed malicious nodes is difficult, as well as problems with list overflow. The authors propose for this purpose the use of timeouts and frequent revocations of well-behaved nodes within a specified period or time. When sufficient evidence of malicious behaviour has occurred a number of times exceeding a threshold (to rule out coincidences), the rating for the node is changed according to a rate function that assigns different weights to the type of behaviour detected. This scheme, which assumes a trust relationship between the nodes, gives greatest weight to a node's own experience, lesser weight for observations by nodes in the neighbourhood, and an even smaller weight for reported experience, since nodes trust their own experiences and observations more than those of other nodes. If the rating of a node falls out of a tolerable range, the path manager is then called.

Clearly, the reputation system is built on negative experience rather than positive impressions. The latter is an issue that is yet to be addressed.

10.7.2.3   Nuglets – a virtual currency

Nuglets are a virtual currency model proposed in Reference 28. Each node contains a tamper resistant security module which contains a nuglet counter. When a node wants to originate a packet, it must first estimate the number of intermediate nodes n that the packet will have to traverse. The node only can proceed with packet transmission if its nuglet counter is greater than n. The nuglet counter is then decreased by n after the packet has been sent.

In principle, if the memory buffer is large enough, a node that runs out of nuglets can store packets it wants to originate until it receives packets from other nodes to forward.

10.7.2.3.1 The mechanism

The tamper-resistant security module prevents user modifications. It contains:

- the nuglet counter,
- a private key,
- a public key certificate authenticating the private key, signed by the manufacturer,
- the public key of the manufacturer, and
- the public keys of all trusted manufacturers.

10.7.2.3.2 Security association

When two nodes become neighbours, their security modules form a security association. This association includes a shared secret session key, a sequence number (which is incremented every time a packet is sent between the two nodes) and a pending nuglet counter. The pending nuglet counter is the number of nuglets that node B collects to give to node A.

The authors assume that the security module runs the routing algorithm. Whenever, a node A wants to originate a packet, it gives the packet to its security module X. The security module then estimates the number of intermediate nodes n. Only if the A's nuglet counter is greater than n can the node proceed with packet transmission. The nuglet counter is then decreased by n. The security module X, will lookup the identifier of Y, the security module in the next hop node B. X will then produce a security header consisting of:

- X's identifier,
- Y's identifier,
- a sequence number,
- the result of a hash function on (a session key, X's ID, Y's ID, the sequence number, the packet for transmission), and
- the identifier of Y's owner, node B.

The security header is passed to node A to send with the packet to next hop B containing Y. Y now processes the security header by first recomputing the hash function to verify the authenticity. Once verified, the sequence number can be extracted to check for a replay attack or if old information has been received. Y can update its own sequence number if all is well.

If node B receives a packet from node A that was not originated by A, node B needs to increment its pending nuglet counter for node A. This is finally transferred to A using a nuglet synchronisation protocol (details of which are not specified) and B resets its pending nuglet counter for node A to zero, ready for the next packet from A. Thus, the scheme takes account of the fact that this mechanism does not guarantee that a node will receive nuglets it is owed.

10.7.2.3.3   *Identified problems*

The authors have identified several problems with their mechanism. A node with multiple interfaces and multiple security modules, could conceivably bounce packets back and forth between its own security modules to increase nuglet counters. The mechanism also only supports unicast traffic, and does not take into account packets of different sizes.

The scheme has been designed only for data packets and not route control packets. The problem with including route control packets is the question of who should pay for them, since routing information is often beneficial for all. Finally, different devices use different amounts of transmission power, so there are fairness issues.

10.7.3   *Key management*

Key management is difficult in ad hoc networks as it is assumed that there is no central authority for issuing and revoking keys. One possible solution to some of the key management problems is to use threshold cryptography [35].

- Zhou and Haas [16] introduce a system model where secure communication takes place between server nodes sharing a system private key (as discussed in more detail in the text that follows).
- Yang *et al.* [26] also propose using threshold cryptography as a basis for secure routing.
- Hubaux *et al.* [17] have suggested using PGP-style public key certificate chains to authenticate a public key in a self-organised way. Each user is responsible as its own certificate authority. Thus, a user authenticates another user's public key if a certificate chain exists between the two users' public key certificates.

10.7.3.1   The Zhou–Haas scheme

A key management scheme for mobile ad hoc networks based on threshold cryptography was introduced by Zhou and Haas [16]. The system possesses a public/private key pair. All nodes have the public key, and are able to verify any information signed by the private key. It is assumed that each node can query other client's public keys, and also send update packets to request a change to its own private key.

The key property of the system is that the system private key is shared amongst special nodes called server nodes. This is used to guarantee confidentiality of a private key, so that the compromise of one server node does not compromise the private key. Each server node owns their own public/private key pair, and knows the public keys of the other servers. The system uses a (n, k) threshold scheme [35], where n is the total number of server nodes, and k is the minimum number of server nodes needed to compute a complete private key (i.e. k server nodes have to be compromised in order to compute the private key, and $k - 1$ server nodes cannot compute the private key). In addition, n has to be greater than $3m + 1$ so that Byzantine behaviour can be detected [36].

The private key is divided into n parts, where each server node receives one part. Each server node can therefore generate a partial signature, and send it securely to a combining server. The combining server node can construct the complete signature from the k or more partial signatures it receives.

There are many possible problems. If k partial signatures are needed, then an easy denial of service attack would be to just allow the combining server to receive $k - 1$ partial signatures. Indeed, some of the n shares needed may have been lost to transmission errors, or a badly failed server node could have sent a false partial signature. The false partial signature can be verified using the badly failed node's public key. In the case of missing partial signatures, any k shares can be used, so in a (n, k) system the system will work as long as there are no more than $n - k$ compromised server nodes.

Proactive security is needed to prevent a malicious node from gaining the private key by collecting partial signatures over a period of time. Thus, share refreshing is needed, where each server generates a set of sub-shares using its own part of the private key. A server then gives each share to its assigned owner via a secure channel. Each server can then combine the received sub-shares and combine them with its old share to compute the new share of the private key. This proactive scheme must tolerate missing or erroneous sub-shares.

10.7.4 Authentication

Most of the solutions discussed above assume that authentication techniques will be used. Stajano and Anderson [2] have given a general model for authentication in mobile ad hoc networks, which they call the 'Resurrecting Duckling' security model. We now briefly discuss their scheme.

The 'Resurrecting Duckling' security model attempts to address the problem of authentication in ad hoc networks. The main property that the 'Resurrecting Duck' model provides is secure transient association between two devices, that is, a master and a slave. In this mechanism, 'secure' refers to the fact that the slave only obeys the master. The transient property dictates that the slave can have a new master if its ownership changes or the slave becomes inoperable.

10.7.4.1 Imprinting

This is the process of establishing a shared secret between the master and the slave, and also exchanging other information such as access-control lists and security polices (as shown in Figure 10.2). Due to the wide range of devices, powerful devices will have to securely communicate with weaker devices, so public key cryptography is seen by the authors as too resource intensive. What they suggest instead is some form of physical contact where the shared secret can be communicated.

The slave's interactions are now governed by the security policy that the master has imprinted. In general, slaves can communicate with each other, although the nature of such communications is determined by the security policy imprinted by the master. Peer-to-peer communication is handled through the use of temporary master–slave relationships between slaves. The authors have suggested a multi-level integrity system, although not in any great detail, where different parts of a security policy are ranked in order of importance. Low integrity decisions can be re-written

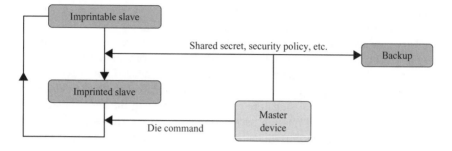

Figure 10.2   The 'Resurrecting Duckling' imprinting process

by a designated 'godparent', that is, a master who has the ability to upload security policies, but high integrity decisions affecting sensitive actions such as re-imprinting are still only controlled by the master.

10.7.4.2   Integrity

Another aspect of the model is node integrity, defined as ensuring that the node has not been maliciously altered so that it sends incorrect information (i.e. it has been compromised). One proposal is for the slaves to perform digital signatures asymmetrically by using the master's public key or symmetrically through use of some pre-established shared secret key. This highlights requirements for tamper resistance to prevent the private key being extracted. In order to prevent masquerading, the authors suggest the use of tamper-evidence where a bearer certificate is contained in the slave within sealed enclosures. The bearer certificate is a mechanism whereby proof of possession of a valid bearer certificate is all that is required to establish trust, and an attacker will have to be able to produce a valid bearer certificate and also produce the seal.

Such seals will have to be resistant to non-intrusive attacks, such as power manipulation. When using seals, the time before a broken seal is noticed and the likelihood of a successful attack on the type of seal will determine if this mechanism is feasible.

10.8   Scope for further research

This section discusses research issues that are yet to be completely solved (or at least not in the public domain). However, since none of the current proposals have as yet been widely used, there is almost certainly scope for further work in the more 'popular' areas.

As mentioned earlier, geographical location assisted routing has being proposed, where location information is used to assist in network routing. However, major issues surround the privacy of personal location and context information. Unauthorised entities must be prevented from acquiring context information from ad hoc users. Specifically, user context information, as required to support context-aware services, for example, calendar data, user location data, social context, personal preferences and type/status of the communications network infrastructure, is potentially very privacy sensitive. Unrestricted use of such data will conflict with end-user privacy requirements, and thus new cryptographic techniques are needed to allow users to protect their contextual information, but still allow calls to be appropriately billed, and services to be delivered to mobile devices. Additionally, there is a need to understand how privacy protection for end users can be balanced against the legitimate requirements of law enforcement agencies. Developing cryptographic techniques (such as extended escrow capabilities) may help to strike this balance.

The identity of an entity can be regarded as the combination of information related to that entity. In the case of a person, a single piece of information, unambiguously bound to an individual, can be sufficient to identify that person. Examples include a fingerprint, a social security number or a credit card number. In other cases, information needs to be combined in order to allow proper identification; for example, a name in itself does not necessarily identify a person, whereas a name combined with an address usually does. From a privacy point of view, the collection of personal user data is a threat, especially if the individual to whom the data belongs can be identified. Identity management allows a personal user to act in a certain role without revealing their true identity, or without disclosing any 'personal' information, for example, by using pseudonyms or even multiple identities. For some services or applications, the use of a pseudonym rather than the 'true' identity might be required or preferred. This includes corporations, devices, etc. In any case, identity management can help the user to decide what 'role' they are going to use for specific tasks. Single and multiple identity ownership will certainly affect security mechanisms for routing protocols, as solutions rely on using identities as a way of rerouting or restricting information.

Connected to identity management is the possible use of identity-based cryptosystems in ad hoc networks [27, p. 21]. The idea of identity-based cryptography dates back to Shamir [37]. In Shamir's model, a Trusted Authority calculates a user private key as a function of his/her identity and a system master secret. Public keys can be derived purely from identities (in fact from any string at all), and no certificates are needed. In 2001, two new identity-based encryption schemes were published almost simultaneously [38, 39] and the subject has undergone rapid development since then. Identity-based cryptography may itself be an attractive means of key management in certain types of ad hoc networks.

When new types of network, for example, ad hoc networks, are integrated into existing networks, problems may arise. Whereas pre-established services exist for fixed networks, ad hoc networks fit better to an on-demand paradigm. This applies to all services including routing, and creates new challenges for security, since there are no central entities that all devices trust. Routing itself is a service that has to be secured against misuse and intrusion. Currently, many security protocols are known and standardised, for example, IPsec or SSL/TLS [40], and some layer 2 protocols, such as IEEE 802.11 and Bluetooth, include security features. However, without some pre-established security associations, especially over wireless and dynamic connections, building low-cost secure channels and handling a public key infrastructure is still a challenge. It is also unclear how security mechanisms such as those in IPsec can interwork with mobile IP [41] and firewalls.

Networks of low-power devices may be obliged to share access to physically secure subsystems storing secret and private keys, implementing complex cryptographic functions. For this to be done in a satisfactory way, delegation of responsibility will need to managed securely with appropriate protocols. Novel cryptographic solutions are likely to be required to enable low-power devices to securely delegate the computation of a security sensitive function (e.g. a digital signature) to another device in the same Personal Area Network (PAN) (see Chapter 9 for more information on PAN security). This issue becomes even more problematic if delegation of functions needs to take place in devices that are not fully trusted, for example, if a device wishes another device to help with a cryptographic computation without revealing its secret or private key.

There is a clear trend towards integrating broadcast and mobile wireless technologies. Secure communication requires that the content and context is authenticated, encrypted and integrity-checked. The group members and the service providers want to be sure that their communication cannot be overheard, and that members leaving the group cannot continue to read messages. This brings a range of issues, many of which relate to Digital Rights Management (DRM), given that broadcast channels are largely used for distributing proprietary content (see Chapter 15 for more information on DRM). Supporting the secure distribution of content to a future mixture of mobile and broadcast enabled devices will require careful design and optimisation of techniques for distribution of keys to multiple recipients of proprietary digital content.

Finally, another area in which further research is needed is middleware security. An example is provided by mobile agents acting as a policing authority within an ad hoc network. There are several issues surrounding such a scheme. For example, the very presence of a mobile agent may force a node into behaving correctly, what reaction should a mobile agent perform if misbehaviour is detected, and what happens when a mobile agent cannot move because its path is blocked by a failed node (see Chapter 12 for more information on mobile agent security)?

10.9   Conclusions

The properties of ad hoc networks affect the provision of security in many different ways. Because of node mobility and the dynamic nature of ad hoc networks, link breaks are likely to be common rather than rare. This means that the network layer, in which routing takes place, is a vitally important focus for mobile ad hoc network security measures.

External and internal threats exist to network routing, where internal threats will be more difficult to address. One major difficulty is in detecting the type of failure, that is, whether it is accidental or deliberate, and what should be done in response. Extensive recent research in securing routing in mobile ad hoc networks has considered enforcing cooperation and authentication. Another key area for research, fundamental to the security of mobile ad hoc networks, is key management. Many research problems relating to the security of ad hoc networks remain, and no single solution is ever likely to address all problems.

Notes

1 The work reported in this chapter has formed part of the Networks & Services area of the Core 2 Research Programme of the Virtual Centre of Excellence in Mobile & Personal Communications, Mobile VCE, www.mobilevce.com, whose funding support, including that of EPSRC, is gratefully acknowledged. Fully detailed technical reports on this research are available to Industrial Members of Mobile VCE.

2 www.minorityreport.com

3 The current status of MANET can be found at www.ietf.org/html.charters/manet-charter.html

4 For details see www.etsi.org/frameset/home.htm?/technicalactiv/Hiperlan/hiperlan2.htm

5 In the AODV literature route requests are known as RREQs, and route replies are known as RREPs.

6 www.terminodes.org

References

1 JUBIN, J. and TORNOW, J.: 'The DARPA packet radio network protocols', *Proceedings of the IEEE*, 1987, **75**, pp. 21–32

2 STAJANO, F. and ANDERSON, R.: 'The resurrecting duckling: security issues for ad-hoc wireless networks', *in* CHRISTIANSON, B., CRISPO, B. and ROE, M. (Eds): 'Security protocols', 7th international workshop, 19–21 April, 1999, Cambridge, UK, *Lecture notes in computer science*, vol. 1796 (Springer, Berlin, 2000) pp. 172–94

3 KENT, S. and ATKINSON, R.: 'Security architecture for the internet protocol'. RFC 2401, Internet Engineering Task Force, November 1998

4 NOBLE, I.: 'Wireless networks wide open', on BBC Interactive web site, November 2001

5 Institute of Electrical and Electronics Engineers.: 'Wireless LAN medium access control (MAC) and physical layer (PHY) specifications IEEE 802.11-1999', March 1999

6 MAXIM, M. and POLLINO, D.: 'Wireless security' (RSA Press, McGraw-Hill/Osborne, 2002)

7 PERKINS, C. and ROYER, E.: 'The ad hoc on-demand distance-vector protocol', *in* PERKINS, C. (Ed.): 'Ad hoc networking', chapter 6 (Addison-Wesley, Reading, MA, 2001) pp. 173–219

8 JOHNSON, D., MALTZ, D. and BROCH, J.: 'DSR – the dynamic source routing protocol for multihop wireless ad hoc networks', *in* PERKINS, C. (Ed.): 'Ad hoc networking', chapter 5 (Addison-Wesley, Reading, MA, 2001) pp. 139–72

9 TSENG, Y., WU, S., LAIO, W. and CHAO, C.: 'Location awareness in ad hoc wireless mobile networks', *IEEE Computer*, 2001, **34**, (6), pp. 46–52

10 CLAUSEN, T., HANSEN, G., CHRISTENSEN, L. and BEHRMANN, G.: 'The optimized link state routing protocol, evaluation through experiments and simulation'. Proceedings of the 4th international symposium on *Wireless personal multimedia communications*, 9–12 September, 2001, Aalborg, Denmark (IEEE Press, 2001) pp. 841–6

11 BELLUR, B. and OGIER, R.: 'A reliable, efficient topology broadcast protocol for dynamic networks'. Proceedings IEEE INFOCOM '99, The conference on computer communications, eighteenth annual joint conference of the IEEE computer and communications Societies, *The Future Is Now*, vol. 1, 21–25 March, New York, NY, USA (IEEE Press, 1999) pp. 178–186

12 PEI, G., GERLA, M. and CHEN, T.-W.: 'Fisheye state routing: a routing scheme for ad hoc wireless networks'. IEEE international conference on *Communications, ICC 2000, Global convergence through communications*, vol. 1, 18–22 June, 2000, New Orleans, USA (IEEE Press, 2000) pp. 70–4

13 GERLA, M., HONG, X. and PEI, G.: 'LANMAR: landmark routing for large scale wireless ad hoc networks with group mobility', *in* VAIDYA, N., CORSON, M. and DAS, S. (Eds): Proceedings of the first ACM international symposium on *Mobile ad hoc networking and computing*, 11 August, 2000, Boston, MA, USA, 150 (ACM Press, 2000) pp. 11–18

14 HAAS Z. and PEARLMAN, M.: 'The performance of query control schemes for the zone routing protocol', *in* SEN, A. and VAIDYA, N. (Eds): Proceedings of the third international workshop on *Discrete algorithms and methods for mobile computing and communications*, 2 September, 1999, Seattle, WA, USA, 92 (ACM Press, 1999) pp. 23–9

15 HAAS, Z. and PEARLMAN, M.: 'ZRP – a hybrid framework for routing in ad hoc networks', *in* PERKINS, C. (Ed.): 'Ad hoc networking', chapter 7 (Addison-Wesley, Reading, MA, 2001) pp. 221–53

16 ZHOU, L. and HAAS, Z.: 'Securing ad hoc networks', *IEEE Network*, 1999, **13**, (6), pp. 24–30

17 HUBAUX, J., BUTTYAN, L. and CAPKUN, S.: 'The quest for security in mobile ad hoc networks', *in* VAIDYA, N., CORSON, M. and DAS, S. (Eds): Proceedings of the ACM symposium on *Mobile ad hoc networking and computing*, 4–5 October, 2001, Long Beach, CA, USA, 356 (ACM Press, 2001) pp. 146–55

18 Bluetooth SIG.: 'Specification of the Bluetooth system', v1.1 edn., February 2001

19 HASSAN, A., STARK, W., and HERSHEY, J.: 'Frequency-hopped spread spectrum in the presence of a follower partial-band jammer', *IEEE Transactions on Communications*, 1993, **41**, (7), pp. 1125–31

20 PICKHOLTZ, R., SCHILLING, D. and MILSTEIN, B.: 'Theory of spread-spectrum communications – a tutorial', *IEEE Transactions on Communications*, 1982, **30**, (5), pp. 855–84

21 LUNDBERG, J.: 'Routing security in ad hoc networks', *in* PEHU-LEHTONEN, H. and LIPMAA, H. (Eds): Proceedings of the Helsinki University of Technology seminar on *Network Security*, Fall 2000, Helsinki, Finland. Helsinki

University of Technology, 2000. Proceedings are only available online at http://www.tcm.hut.fi/Opinnot/Tik-110.501/2000/papers/

22 HU, Y., PERRIG, A. and JOHNSON, D.: 'Packet leashes: a defense against wormhole attacks in wireless ad hoc networks'. Proceedings of the twenty-second annual joint conference of the *IEEE computer and communications societies*, 1–3 April, 2003, San Fransisco, CA, USA (IEEE Press, 2003) (in press)

23 International Organisation for Standardization: 'ISO 7498-2: information processing systems – open systems interconnection – basic reference model – part 2': Security architecture, 1st edn, 1989

24 BUCHEGGER, S. and LE BOUDEC, J.-Y.: 'Performance analysis of the CON-FIDANT protocol (cooperation of nodes: fairness in dynamic ad-hoc networks)', *in* HUBAUX, J., GARCIA-LUNA-ACEVES, J. J. and JOHNSON, D. (Eds): Proceedings of the third ACM international symposium on *Mobile ad hoc networking and computing*, 9–11 June, 2002, Lausanne, Switzerland, 238 (ACM Press, 2002) pp. 226–36

25 MICHIARDI, P. and MOLVA, R.: 'CORE: a collaborative reputation mechanism to enforce node cooperation in mobile ad hoc networks', *in* JERMAN-BLAZIC, B. and KLOBUCAR, T. (Eds): *Communications and multimedia security, IFIP TC6/TC11* Sixth joint working conference on *Communications and multimedia security*, 26–27 September, 2002, Portoroz, Slovenia, vol. 228 of IFIP Conference Proceedings (Kluwer Academic, Dordrecht, 2002) pp. 107–21

26 YANG, H., MENG, X. and LU, S.: 'Self-organized network-layer security in mobile ad hoc networks', *in* MAUGHAN, D. and VAIDYA, N. (Eds): Proceedings of the ACM workshop on *Wireless security*, 28 September, 2002, Atlanta, GA, USA, 94 (ACM Press, 2002) pp. 11–20

27 BUTTYAN, L. and HUBAUX, J.: 'Report on a working session on security in wireless ad hoc networks', *ACM Mobile Computing and Review*, 2002, **7**, (1) pp. 74–91

28 BUTTYAN, L. and HUBAUX, J.: 'Stimulating cooperation in self-organising mobile ad hoc networks' *ACM/Kluwer Mobile Networks and Applications (MONET)*, 2003, **8**, (5) (in press)

29 YOON, J., LIU, M. and NOBLE, B.: 'Random waypoint considered harmful'. Proceedings of the twenty-second annual joint conference of the *IEEE computer and communications societies*, **2**, 1–3 April, 2003, San Fransisco, CA, USA (IEEE Press) pp. 1312–1321

30 YI, S., NALDURG, P. and KRAVETS, R.: 'Security-aware ad hoc routing for wireless networks', *in* VAIDYA, N., CORSON, M. and DAS, S. (Eds): Proceedings of the 2001 ACM international symposium on *Mobile ad hoc networking and computing*, 4–5 October, 2001, Long Beach, CA, USA, 302 (ACM Press, 2001) pp. 299–302

31 HU, Y., JOHNSON, D. and PERRIG, A.: 'SEAD: secure efficient distance vector routing for mobile wireless ad hoc networks'. Proceedings of the fourth IEEE workshop on *Mobile computing systems and applications*, 20–21 June, 2002, Calicoon, New York, USA (IEEE Press, 2002) pp. 3–13

32 HU, Y., PERRIG, A. and JOHNSON, D.: 'Ariadne: a secure on-demand routing protocol for ad hoc networks', *in* AKYILDIZ, I., LIN, J., JAIN, R., BHARGHAVAN, V., and CAMPBELL, A. (Eds): Proceedings of the eighth annual international conference on *Mobile computing and networking*, 28 September, 2002, Atlanta, GA, USA, 286 (ACM Press, 2002) pp. 12–23

33 ZAPATA, M. and ASOKAN, N.: 'Securing ad hoc routing protocols', *in* MAUGHAN, D. and VAIDYA, N. (Eds): Proceedings of the ACM workshop on *Wireless security*, 28 September, 2002, Atlanta, GA, USA, 94 (ACM Press, 2002) pp. 1–10

34 PAPADIMITRATOS, P. and HAAS, Z.: 'Secure routing for mobile ad hoc networks'. *SCS communication networks and distributed systems modeling and simulation conference*, 27–31 January, 2002, San Antonio, TX, USA, The Society of Computer Simulation International, 2002, pp. 27–31

35 SHAMIR, A.: 'How to share a secret', *Communications of the ACM*, 1979, **22**, (11), pp. 612–13

36 LAMPORT, L., SHOSTAK, R. and PEASE, M.: 'The byzantine generals problem', *ACM Transactions on Programming Languages and Systems (TOPLAS)*, 1982, **4**, (3), pp. 382–401

37 SHAMIR, A.: 'Identity-based cryptosystems and signature schemes', *in* BLAKLEY, G. and CHAUM, D. (Eds): 'Advances in cryptology', Proceedings of *CRYPTO 84*, 19–22 August, 1984, Santa Barbara, CA, USA, *Lecture notes in computer science*, vol. 196 (Springer, Berlin, 1984) pp. 47–53

38 BONEH, D. and FRANKLIN, M.: 'Identity-based encryption from the weil pairing', *in* KILIAN, J. (Ed.): 'Advances in Cryptology', Proceedings of *CRYPTO 2001*, 21st annual international cryptology conference, 19–23 August, 2001, Santa Barbara, CA, USA, *Lecture notes in computer science*, vol. 2139 (Springer, Berlin, 2001) pp. 213–29

39 COCKS, C.: 'An identity based encryption scheme based on quadratic residues', in HONARY, B. (Ed.): 'Cryptography and coding', Proceedings of the 8th IMA international conference, 17–19 December, 2001, Cirencester, UK, *Lecture notes in computer science*, vol. 2260 (Springer, Berlin, 2001) pp. 360–3. See also www.cesg.gov.uk/technology/idpkc/index.htm

40 DIERKS, T. and ALLEN, C.: 'The TLS protocol', RFC 2246, Internet and Engineering Task Force, January 1999

41 PERKINS, C.: 'IP mobility support', RFC 2002, Internet and Engineering Task Force, October 1996

Chapter 11

Security issues in a MobileIPv6 network

Robert Maier, Vaia Sdralia, Joris Claessens and
Bart Preneel

Introducing mobility into a communications environment that was initially designed for fixed nodes brings up many challenges that come from a technical as well as from a security point of view. The challenges posed by introducing mobility for seamlessly roaming between IP networks are addressed by the MobileIP standard. This chapter describes the solution proposed by MobileIP with a focus on the security mechanisms used in version 6 of this protocol.

11.1 Introduction

Mobility is an increasing trend in today's computing environments. Environments that were not designed to be mobility-aware are now faced with the challenge of adapting to a more mobile world. The Internet is a key example of such a global communications environment where mobile devices like laptops and PDAs can already operate as nodes while devices such as watches, cars and home appliances are likely to connect in the near future. However, the Internet was originally designed for fixed nodes. Connecting mobile devices to it was not an envisioned development direction at the moment of the design and its initial deployment. Therefore, using the standard Internet stack of protocols for connecting mobile devices to the Internet would limit their mobility.

The IP address assigned to each node is used for identifying the peers of a connection. The routing mechanisms on the Internet use the IP address in order to deliver packets to the right destination. For more information on the inner workings of the Internet, like the IP address and the routing process, the reader is referred to Reference 1. When the IP address of a node is changed, all the connections that were already established at that time will be dropped. Moreover, the IP address contains information on the location of the node.[1] This means that if a node changes its location without changing its IP address, it will lose its connection to the Internet. Without updating the affected routers with the path to the node's new location, every packet destined to it will end up in the previous location of the node, hence it will be lost. Thus, in order to provide mobility in the Internet it seems that two contradictory conditions need to be fulfilled as a location change would require an IP change and the IP change would result in disrupting at least the immediate connectivity of the node.

This chapter focuses on the MobileIPv6 protocol. In addition to the technical details relevant for understanding the protocol, we particularly address the security mechanisms that are used to counter different types of threats. These security mechanisms are the key elements of MobileIPv6 security. To put the subject in a broader perspective, we also describe known open problems and related security issues. An introduction to the MobileIPv6 protocol itself is given in the next section. Section 11.3 deals with the security mechanisms of MobileIPv6. These mechanisms protect against generic threats (threats that do not arise because of the mobility factor) and against specific mobility related threats. The same section also includes a comparison of the security offered by MobileIPv4 and MobileIPv6. Open problems, such as location privacy and related security issues (AAA), are described in Sections 11.4 and 11.5.

11.2   An introduction to MobileIP

In order to address the challenges raised by introducing mobility into the Internet, a working group has been established by the Internet Engineering Task Force (IETF) in 1991. This group develops the MobileIP protocol and also works on several extensions to it. In this chapter, we address the extensions that are relevant from a security point of view. The MobileIP solution constitutes a protocol situated at the network layer in the Internet protocol stack. It is necessary to address the problem at the network layer in order to provide transparency for the transport layer. Note that if this transparency is not needed, there are other usable solutions, like DHCP [2] and dynamic DNS updates [3]. However, these solutions disrupt the currently established transport layer connections, that is, they do not provide transparency for the transport layer.

This section presents the solution adopted by MobileIP. Conceptually, there is only one MobileIP protocol. However, due to the fact that there are two versions of the IP protocol (IPv4 [4] and IPv6 [5]), two different versions of MobileIP evolved from the common idea. Thus, MobileIPv4 and MobileIPv6 have many common features, but MobileIPv6 offers many improvements compared to MobileIPv4. In contrast to MobileIPv4, which is, in fact, a 'patch' for the IP protocol, MobileIPv6 is completely integrated into IPv6. Herein we focus on MobileIPv6. Complete descriptions of the protocol are given in References 6 (for MobileIPv4) and 7 (for MobileIPv6).

The main challenge addressed by the MobileIP protocol is delivering a packet to its destination, regardless of the end node's point of connection. This needs to be done in a transparent way, so that neither the routing mechanism nor the source/destination

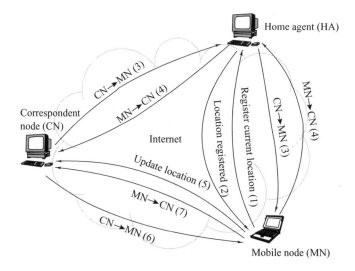

Figure 11.1 Traffic flow in a MobileIP enabled environment

node of the packet are affected. This means that the traffic targeted to a mobile node arrives at the same IP address, regardless of the actual location of the node. In order to accomplish this, MobileIP introduces a number of additional participants to the protocol (see Figure 11.1). A node that has the capability of moving from one network or subnetwork to another is called a Mobile node and is denoted by MN. The MN is initially connected to its home network, which is its publicly known location. When the MN leaves its home network and attaches to another network (visited network), the traffic destined for the MN is handled by the MN's Home agent (denoted by HA). In particular, the HA intercepts the traffic and forwards it via tunnelling to the MN's actual location. A peer node with which an MN communicates is called a Correspondent node (denoted by CN).

In MobileIP, an MN has two IP addresses.

- A *home IP address*, which is the IP address that the MN is using when it is connected to the home network. This IP address remains unchanged when the MN is visiting other networks.
- A *care-of IP address*, which is a temporary IP address that the MN is using while it is attached to a visited network. In particular, every time the MN connects to a different visited network it gets a different care-of address, specific to that network.

A CN can use both IP addresses to reach the MN and does not need any additional information to communicate with the MN. Even if the CN does not know the MN's location or the MN's care-of address, it can still reach the MN by using its home address.

A more detailed description of the MobileIP protocol, containing the different steps and messages exchanged by the participants in the protocol (as shown in Figure 11.1), is given below.

1. When an MN leaves its home network, it acquires an IP address from the visited network (using, e.g. DHCPv6). This is the care-of IP address, which identifies its current point of attachment. It then registers this IP address with its HA. The association between its home IP address and the care-of IP address is called binding and the registration request is performed using a *Binding Update* message. The registration process permits the HA to create a binding and thus to become aware of the current location of the MN.
2. Optionally, the HA may acknowledge the registration process, using a *Binding Acknowledge* message.
3. After the registration process, the HA has a binding for the MN, so it can start forwarding traffic to and from it. When a CN sends a packet for the first time to the MN, it will use the MN's home address and the packet will end up on the MN's home network. If the MN is not 'at home' the packet gets intercepted by the MN's HA and it is tunnelled to the actual location of the MN.
4. The reply packet can be reverse-tunnelled, through the HA, to the CN.
5. If the CN supports MobileIP, it can be informed by the MN of its actual location. However, this will not affect the transparency that MobileIP provides for the upper layers (transport and application) in the TCP/IP stack. Only the network layer of the CN will be informed of the MN's actual address and it will transform the packets coming from the MN in such away that they appear to originate from the MN's home address.
6. If the CN supports MobileIP, subsequent packets sent by the CN to the MN can be targeted directly to the care-of address of the MN, thus bypassing the HA.
7. Finally, packets from the MN to the CN can either be sent through the reverse tunnel, like in step 4, or directly to the CN.

11.3   MobileIPv6 security mechanisms

This section deals with the specific security aspects of MobileIPv6. It describes the threats to MobileIPv6 and the security mechanisms adopted by the MobileIPv6 protocol, as discussed in Reference 7, in order to address these threats. A comparison between the security provided by MobileIPv6 and MobileIPv4 is then presented.

11.3.1   MobileIPv6 threats and solutions

Even without support for mobility, IPv6 packets face several generic threats. In particular, there are threats to the confidentiality, integrity and availability of the IPv6 traffic. These extend to MobileIPv6 packets and affect both payload and signalling. As most of these threats are not specific to MobileIP, the protocol does not offer new security mechanisms to address them. They must be addressed using existing generic security mechanisms at the network layer, that is, IPSEC [8–10] or at other layers in the Internet protocol stack. However, some of them (especially the ones affecting MobileIPv6 signalling packets) introduce new threats that are specific to MobileIPv6. These together with the solutions proposed by MobileIPv6 are listed and analysed in the following subsections.

11.3.1.1 Threats against *binding updates*

A binding procedure is the process of informing the HA or the CN of the MN's current binding. It is initiated by the MN and is performed using a *binding update* (BU) message. This message has to be authentic so that an attacker will not be able to impersonate the MN and falsely update the MN's location. Depending on the target, there are two possible types of binding procedures:

- the *home binding procedure*, when the MN informs the HA of its current binding (cf. step 1 of the protocol in Figure 11.1);
- the *correspondent binding procedure*, when the MN informs a MobileIPv6-aware CN of its current binding (cf. step 5 of the protocol in Figure 11.1).

These procedures use different security mechanisms to protect against the threats that they face.

11.3.1.1.1 Threats against home binding procedure

If there is no mechanism to check the authenticity of the BU message sent by the MN to the HA, an attacker could impersonate the MN and change its binding with the HA. This would lead to the HA assuming that the MN has moved and would forward the MN's traffic to a false location. This event could result in a Denial of Service (DoS) attack against the MN. It could also be used by the attacker to mount a man-in-the-middle attack. The attacker could send a spoofed BU message to the HA specifying that the binding needs to be updated without even having access to the link between the MN and the HA.

In order to address this threat, MobileIPv6 makes use of IPSEC [8–10]. Since the MN and the HA are generally administered by the same organisation, sharing a Security Association (SA) should not be difficult, thus enabling the use of IPSEC for the provision of message authentication (origin authentication and message integrity). For more information on how IPSEC is used to protect the BU messages between the MN and HA, the reader is referred to Reference 11.

Another threat against the home binding procedure is introduced by the fact that the attacker could replay authentic and valid BU messages. This would result in the attacker gaining the same advantages as before. For mounting a man-in-the-middle attack using this method, the attacker would need access to one of the previous locations (care-of addresses) of the MN (namely, the one for which he can replay a BU message). IPSEC protects against replay attacks, but only if a dynamic SA establishment is used rather than a static one. However, there are cases where manual establishment of the SA is desired. In order not to enforce this requirement, IPSEC is not chosen as the protection mechanism against this threat. Sequence numbers are used to prevent against replay attacks in the home binding procedure. The sequence numbers also provide for a correct ordering of the messages that IPSEC anyway does not provide. The integrity of the sequence numbers, is then protected using IPSEC. This mechanism does not protect against replay attacks targeted to a HA that has just rebooted and lost its state with respect to sequence numbers.

Third, a malicious MN could lie about its care-of address and in this way mount a DoS attack against another node in the Internet, by redirecting its traffic to the victim node. For example, an MN could request several streams of data and then pretend that it has moved and during the Registration Procedure lie about its care-of address. The result would be that the traffic requested by the MN will be redirected to the victim node. This can be used to mount a DoS attack against another node in the Internet. If the targeted node cannot handle large bandwidth traffic a malicious MN would be enough to mount such an attack. Otherwise, more MNs would be needed in order to mount a Distributed DoS attack against the target node. However, such an attack is very easily traced since the MN needs to use its SA while registering its new location with the HA. The HA can then stop providing services to the misbehaving node. Thus, no additional security mechanism is put in place to address this threat.

11.3.1.1.2   Threats against correspondent binding procedure

The correspondent binding procedure is similar to the home binding procedure, but it is performed between the MN and a CN. Hence, most of the threats presented for the home binding procedure also apply in this case.

In particular, an attacker could send fake BU messages to a CN, thus hijacking the connection with the MN. This could again result in mounting a man-in-the-middle attack or a DoS attack against the MN or the CN. Thus, as in the previous case, message authentication is needed for the BU message that is sent to the CN. However, in contrast to the previous case, the MN and the CN would not normally share a SA. If such a SA between CN and MN exists (established either statically or dynamically[2]), IPSEC can be used to protect the BU messages. If a SA between the CN and the MN does not exist, IPSEC cannot be used to address these threats. The security mechanism used in this case is the *Return Routability* (RR) procedure.

This procedure was developed to provide adequate authentication between the MN and the CN. Specifically, it ensures that the MN is able to receive messages at its home and care-of addresses and it protects the binding messages between the MN and the CN. It consists of two pairs of messages, each pair travelling on a different path between the MN and CN (see Figure 11.2). The reason for the two-path communication between the MN and the CN in the RR procedure is that in this way both the home address and the current care-of address claimed by the MN are verified to be attainable. Within the RR process, the CN keeps a secret key (K_{CN}) and a random number (N_i), which it renews at regular intervals. During these time intervals the CN uses the same K_{CN} and N_i with all the MNs so as not to have to generate and store a new random number every time a new MN contacts it. Each value of N_i is identified by i, which is communicated in the protocol, so when N_i is replaced by N_{i+1}, the CN can distinguish messages that should be checked against

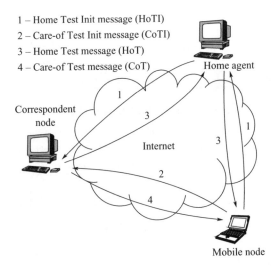

1 – Home Test Init message (HoTI)
2 – Care-of Test Init message (CoTI)
3 – Home Test message (HoT)
4 – Care-of Test message (CoT)

Figure 11.2 Return Routability procedure

the old random number rather than the new one. Therefore, the CN has to keep the current N_i as well as a small set of its previous values in memory. Older values are discarded and messages using them can be rejected as replays. Regarding the secret key K_{CN}, it can be a fixed value or regularly updated. In the case of an update of K_{CN}, it can be done at the same time as the N_i update, so that i identifies both the random number and the key. Finally, a new K_{CN} is generated each time the CN boots so as to avoid the need for secure persistent storage of the key. The K_{CN} is known only to the CN and is used to generate the two parts of the binding management shared key. The mechanism uses a HMAC_SHA1 [12] algorithm to generate the two parts of the shared key. The outputs of this algorithm are never verified by another participant in the protocol, this algorithm is only used to avoid state management by the CN.

The outcome of this procedure is the establishment of a shared cryptographic key between the two parties. This cryptographic key is then used to protect the BU message that follows the RR procedure. The integrity of the BU message is protected by an 'authenticator', that is, a Message Authentication Code, more precisely the first 96 bits of a HMAC [12, 13] computed over the BU message and keyed with the previously established key. Schematically, this procedure can be represented as in Figure 11.3.

Another possible threat arises from the case of a malicious MN that sends a large number of invalid RR initiation messages in order to mount a DoS attack against the CN. If no measure is taken against this attack, the CN could be deceived into generating and storing a large number of tokens for the falsely initiated RRs. In order to address this threat, the CN does not keep state over the initiated RRs, nor it stores the generated tokens, but it recomputes them dynamically as soon as it needs to verify the authenticity of an incoming BU message.

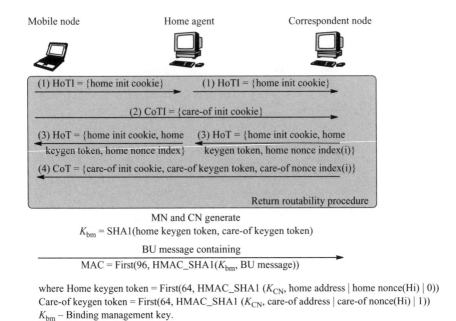

Mobile node   Home agent   Correspondent node

(1) HoTI = {home init cookie}   (1) HoTI = {home init cookie}

(2) CoTI = {care-of init cookie}

(3) HoT = {home init cookie, home   (3) HoT = {home init cookie, home
keygen token, home nonce index}   keygen token, home nonce index(i)}

(4) CoT = {care-of init cookie, care-of keygen token, care-of nonce index(i)}

Return routability procedure

MN and CN generate
K_{bm} = SHA1(home keygen token, care-of keygen token)

BU message containing
MAC = First(96, HMAC_SHA1(K_{bm}, BU message))

where Home keygen token = First(64, HMAC_SHA1 (K_{CN}, home address | home nonce(Hi) | 0))
Care-of keygen token = First(64, HMAC_SHA1 (K_{CN}, care-of address | care-of nonce(Hi) | 1))
K_{bm} – Binding management key.

Figure 11.3   Return Routability procedure, followed by the binding procedure (schematic view)

However, the reader should note that the RR procedure does not protect against an attacker who has access to the two links, the one between the CN and the HA and the one between the CN and the MN (at its current location). Such an attacker would have access to the information used to generate the shared key and could use it to send fake BU messages pretending to be the MN. There are proposed alternative mechanisms that address this problem; for example, by using public key signatures for binding message authentication.

These mechanisms are detailed in Section 11.4. Also, the reasons that are at the base of this method's choice as the default mechanism by the MobileIPv6 specification are given in Reference 14.

11.3.1.2   Protecting the location of the MN – location privacy

The location of the nodes becomes an important characteristic in a mobile Internet environment as certain information can be deduced based on their IP address.

One of the mechanisms used by MobileIPv6 to provide transparency for the upper layers in the Internet protocol stack uses the Home Address Destination option in the headers of the payload packets that are sent from the MN to the CN. The payload packets sent by the MN have as source address the care-of address of the MN, while its home IP address is placed in a field inside the IPv6 header. Before processing the incoming packet, the IPv6 protocol from the network layer of the CN replaces the source address of the packet with the actual home address of the MN, found in the Home Address Destination field of the IPv6 header. After this processing, the packet seems to come from the home IP address of the MN. However, this mechanism does not protect the care-of address (i.e. the reference to the actual location of the MN) from a malicious CN. The MN's home IP address is sent in clear, so any eavesdropper on the route between the MN and CN (possibly via the HA) can easily track the movements of the MN by capturing and analysing the traffic.

When using MobileIP, the actual location of the MN can be protected to a certain level, by using the HA as a 'proxy' for MN's traffic. If the attacker (the CN or any other node on the Internet) does not have access to the MN–HA link, it will not notice when the MN moves 'away from home'. However, MobileIP permits a node to initiate a correspondent binding procedure, thus revealing its actual location to the CN. This improves efficiency of the communication between the MN and CN since the packets do not have to pass through the HA. Nevertheless, if the MN needs to keep its actual location private from a CN, it will not initiate a correspondent binding procedure and will use the bi-directional tunnelling route (via the HA). In this case, the CN will not be aware that the MN moved away from its home network. The CN will keep sending packets to the MN's home location, which will eventually arrive to the MN tunnelled by the HA. In the reverse direction, the communication from the MN goes through the same tunnel, hence, the CN receives packets from the MN's home location that have as source address the home IP address of the MN. However, this method does not protect against attackers that can do traffic analysis on the MN–HA link.

Thus, information on the location of an MN in MobileIPv6 should be protected if the user desires to do so for privacy reasons. However, no mechanism exists in the MobileIPv6 specification to address the threats faced by the location information. The generic solution is to use bi-directional tunnelling (traffic between the CN and the MN is tunnelled via the HA). For this solution to be effective against an attacker with traffic analysis capabilities, the traffic between the MN and HA should be encrypted. Alternative mechanisms to deal with the location privacy problem are presented in Section 11.4.

11.3.2   Security of MobileIPv6 versus MobileIPv4

MobileIPv6 is based on the main ideas used in MobileIPv4. However, it is better integrated with IPv6 and this leads to a more robust protocol. There are several differences between MobileIPv6 and MobileIPv4, most of them affecting the protocol's robustness, efficiency and security. The most notable differences are listed in this section and the ones affecting the protocol's security are detailed.

In MobileIP there are several paths for a packet to be routed between the CN and the MN. The packet can travel both ways via the HA, this being the most inefficient route. Alternatively, it can be routed via the HA when travelling from the CN to the MN and directly when travelling in the reverse direction. This is called *triangular routing* and is the default routing method for MobileIPv4. The most efficient solution is to have the packet routed directly between the CN and the MN. The latter route optimisation support is included by default in the MobileIPv6 protocol specification and is only an extension to the MobileIPv4 protocol. We will refer to this latter technique as *Route Optimisation* (RO). In order to use RO in MobileIPv6, the MN signals to the CN that its location has changed and keeps the CN informed of its actual location. This is implemented by sending BU messages also to the CN. Since, generally, the MN and CN do not share a SA, MobileIPv6 proposes a security mechanism (the RR procedure) for protecting these BU messages. This mechanism has been chosen for the basic specification of the MobileIPv6 protocol, but alternative mechanisms can also be used. Having such a mechanism that will provide more security than the current one while it does not bring any new disadvantages (from a security or performance point of view) is an open problem and is detailed in Section 11.4. This security mechanism, together with the CN's capability to understand BU messages is lacking in the base specification of MobileIPv4. There, by default, the MN can choose between triangular routing and no routing optimisations.

In MobileIPv4, when using triangular routing, the MN places its home IP address as the source address for the packet. This can lead to the packet being filtered by firewalls that guard the visited network. Such an incident is called *ingress filtering* and it occurs when firewalls are configured to filter packets originating from inside the guarded network that have a source IP address not belonging to that network. In this case, MobileIPv4 proposes reverse tunnelling as the solution, while in MobileIPv6 the problem does not occur at all, since the source IP address of the packet will always belong to the visited network. Whenever an MN in MobileIPv6 sends a packet directly to the CN, the packet will have as source IP address the care-of address of the MN, an IP address belonging to the visited network and hence it will pass the packet filters enforcing the ingress filtering policy. The actual home address of the MN is carried in a field of the IPv6 header and interchanged with the actual source IP address only at the CN's network layer. Hence, route optimisation techniques (triangular routing and RO) in MobileIPv6 are still possible even if ingress filtering is enforced and, thus, the integration of the MobileIPv6 protocol with IPv6 results in a better coexistence with the ingress filtering policy.

Another difference that is reflected in the robustness and security of MobileIPv6 is the fact that MobileIPv6 does not depend on the Address Resolution Protocol (ARP) [15]. In MobileIPv4, the HA made use of ARP to capture the traffic of the MN while the MN was 'away from home'. ARP is inherently an insecure protocol, not providing any mechanism to enforce security. MobileIPv6 uses the IPv6 replacement of ARP, the Neighbor Discovery Protocol [16]. Additionally, IPv6 provides for a better integration with the MobileIPv6 protocol and thus reduces the overhead of the encapsulation mechanism used in MobileIPv4. This is achieved via the Mobility Header, which offers better support to mobility than IPv4 and thus does not require the use of the encapsulation mechanisms for redirecting the traffic to the new location of the MN.

In MobileIPv4, Foreign Agents (FA) are protocol participants used to provide a care-of address to the MN and also to encapsulate/decapsulate the traffic to/from the MN. Due to new IPv6 mechanisms (Neighbor Discovery [16], Stateless Address Autoconfiguration [17]) the MN is able to obtain its care-of address without the help of the foreign agents and hence they are no longer needed in MobileIPv6. This protocol modification solves several security issues, since the MN does not need to depend on an unknown (and possibly malicious) Foreign Agent while being in the visited network.

11.4 Open issues

Herein we detail some of the open problems of MobileIPv6 with respect to location privacy and the route optimization issue. Solutions for these problems are being actively pursued by the IETF and the research community. Some generic solutions have already been proposed, but they do not always offer the best protection while they can prompt other disadvantages (e.g. in terms of performance).

One such problem is the MN's location privacy. The MobileIPv6 protocol does not address this issue in a specific manner. Although it is acknowledged and a generic solution to it is provided, if used, it would result in a possible degradation of the protocol's efficiency. In particular, the MN can choose not to use route optimization mechanisms, but send and receive traffic through a bidirectional tunnel, via its HA, thus obtaining a certain level of protection for its location. Nevertheless, it loses the advantages brought by a more efficient routing of traffic. In order to protect against an attacker that has access to the MN–HA link and is capable of performing traffic analysis on this link, the traffic between the MN and the HA needs to be (as a first line of defense) at least encrypted. Location privacy in MobileIPv6 has been the subject of research and mechanisms that address it have been proposed and they are mostly based on hiding the location information. This is attained either by dynamically replacing the MN's home IP address by a Temporal Mobile Identifier which would identify the MN [18] or by generating random interface identifiers periodically for the MN [19]. These methods are described briefly and an enhancement to the latter method is proposed in Reference 20.

Another problem comes from the fact that IPv6 security mechanisms (more specifically, IPSEC) are not adequate against every possible attack. There are two main reasons why IPSEC cannot be used in some cases and thus alternative security mechanisms are required.

- IPSEC relies on a public key infrastructure that has not yet been deployed and will probably not be deployed in the near future.
- The key management mechanism that IPSEC provides, IKEv2 [21], is too costly for many nodes with respect to performance. This mechanism might not be easily applied to nodes that have a very limited computing power or memory.

One point where this problem materialises is the protection of the signalling between the MN and the CN, when route optimisation is desired. The first version of the MobileIPv6 protocol was using IPSEC to protect these messages. The Internet Engineering Steering Group (IESG) decided that IPSEC is not really suited to protect MN–CN signalling with respect to route optimisation, because of the reasons previously listed and hence rejected that version of the MobileIPv6 protocol. The IESG then proposed that an alternative protection scheme that would not be based on IPSEC should be used. The current mechanism, the RR procedure, does not provide a high level of security, but it tries to comply with the design requirement stating that it should provide at least the level of security that IPv4 provides without mobility support. However, there are certain attacks that are easier to mount in the MobileIP environment by using the RR mechanism [22]. Finding a suitable alternative mechanism is still an open issue. Mechanisms that try to address these issues have been proposed, both before and after the adoption of the RR procedure as the default mechanism. We briefly describe and reference some of them.

A generic mechanism was proposed by the IESG for consideration when designing the non-IPSEC solution for protection of the MN–CN signalling. This generic mechanism, namely the Purpose-Built Keys (PBK) [23], is still actively developed by the Networking WG of the IETF. PBK provides a framework to obtain data (and not entity) authentication for a node in the Internet. It was designed for the Internet in general, to offer a 'good enough' security and it can be easily applied to the MobileIP environment. The overall security of the scheme depends on the security of the first step. Basically, the scheme provides a smaller vulnerability window than if no security scheme is used at all. The steps describing the PBK protocol are as follows.

1. The MN generates a temporary public/private key pair and sends to the CN the hash of the public key (PBID) at the beginning of the communication session. The public key is also sent to the CN later during the established communication session by attaching the information to the normal IP packets, either using an extra header or existing header fields.
2. When the MN wants to authenticate a BU message (data authentication) to the CN, it appends the PBID to the message and signs the BU message using its private key. Replay protection can be addressed by using nonces or timestamps.
3. The CN verifies the signature and sends a challenge to the source IP address of the BU message.
4. The MN receives the challenge at its new address and sends back a response consisting of the encryption of the challenge using its private key.

The obvious weak point of this protocol is in its first step. A man-in-the-middle attack can be mounted and if the first step of the protocol is attacked, the security of the entire protocol can be compromised. The RR procedure (the current mechanism used to provide authentication of the BU messages) suffers from similar shortcomings. If the attacker has access to the CN–MN and CN–HA links, it will be able to recreate the key and falsify BU messages. However, one advantage over the PBK is that it uses symmetric cryptography and was probably chosen because of its performance as opposed to the performance of the public key cryptography used in the PBK. The RR is the current mechanism used by the MobileIPv6 protocol and was described in Section 11.3.

Another mechanism that has been proposed for the authentication between MN and CN is the Cryptographically Generated Addresses (CGA). This mechanism was first proposed in Reference 24 and it has been taken into consideration by the MobileIP WG. This mechanism is complementary to the RR procedure and it addresses the existing security issue in the RR. According to the CGA, a part of the MN's home

IPv6 address is derived from its public key. An IPv6 address (128 bits) consists of a 64-bit network prefix and a 64-bit interface identifier. The network prefix is used for routing while the interface identifier specifies the unique node. Although the interface identifier is normally derived from the link layer address of the node, almost any value is valid as only two bits must have fixed values. Hence, the remaining 62 bits of the interface identifier can be generated by the CGA method using the node's public key. The uniqueness of the address is then ensured by the Duplicate Address Detection algorithm of the IPv6 using a hash function. According to CGA, the CN can be certain that a message is originating from the MN that owns the specific public key by verifying that the home IP address was really derived from that key. In particular, the CN recreates the CGA address from the public key and then it compares the received home IP address with the recreated address. Afterwards, the CN verifies the validity of the BU message by calculating the hashed value of the signature and comparing it to the one that is recovered from the message by using the public key.

The presented mechanisms suffer from several problems, from a security as well as from a practical implementation point of view (with respect to performance). A solution that addresses these shortcomings is presented in Reference 22. The proposed mechanism uses public key cryptography, as the CGA does, but it does not request the MN to perform any computationally intensive operations, these being delegated to the HA.

A more complete analysis on the authentication protocols proposed for the BU message can be found in Reference 25. Finding better solutions for the open problems identified in this section is the subject of present research. However, since current solutions adhere to the design requirements they are already taken into account for standardisation. As soon as better solutions emerge, they may be used to extend the MobileIPv6 protocol.

11.5   AAA requirements for MobileIP

The ability of movement between networks that is provided by MobileIP requires authorisation, and therefore authentication and accounting (AAA) since these three functions are closely co-dependent. This fact creates the need for the design and specification of AAA protocols. In particular, when an MN wants to access a resource within a visited network, the FA or in the case of MobileIPv6 the local router or the address allocation entity will probably request or require the provision of credentials that can be authenticated before access to resources is permitted. Once authenticated, the MN may be authorised to access services within the visited network while an accounting of the actual resources used may then be assembled. Herein we briefly describe the functional and performance requirements of AAA protocols within MobileIP as presented in Reference 26.

The basic model for the operation of AAA protocols within MobileIPv6 includes an AAA server at the home network, which can check the credentials of all MNs that are administered by it, and a local AAA server at the visited network, which has to dynamically establish security relationships with the home AAA in order to be able to authenticate each visiting MN. Furthermore, the functionality of the AAA servers should include the authentication or reauthentication of the MN for MobileIP registration, the authorisation of the MN to use at least the set of resources for minimal MobileIP functionality as well as other requested services and finally the initialisation of the accounting for service utilisation.

In order to make AAA protocols flexible within MobileIP, the MN can use a resource at the visiting network without requiring further involvement by the AAA servers after the initial registration has been completed. This means that the initial registration process will take longer than any subsequent registrations. In order to reduce that time an integration between the AAA functions and the MobileIP entities is needed. Furthermore, not only the MN must be able to provide complete credentials without ever having been in touch with its home network but any nodes in the path between the home and the visited network must not be able to acquire and reuse the MN's credentials.

Finally, as there can be a need for subsequent registrations, the AAA servers must also be able to perform a key distribution functionality during the initial registration process and validate the MN's certificates. This key distribution must be able to identify or create a security association between the MobileIP entities (MN, HA) and participate in the distribution of these associations. However, the created security associations should be long enough in order to avoid frequent initiation of the AAA key distribution, as this will lead to lengthy delays between [re]registrations.

11.6   Conclusions

MobileIP introduces mobility into the Internet, at the network layer and provides transparency to the upper layers of the TCP/IP stack. Using MobileIP, a node can maintain its currently established connections even if it changes its point of attachment to the Internet. Having mobility as a new characteristic of the Internet generates new security threats and influences the existing ones. There are several threats that MobileIPv6 faces, both generic (originating from the Internet environment) and specific, introduced by the protocol. The current specification of the MobileIPv6 protocol employs mechanisms to address these threats. However, these solutions do not always offer the best possible solution, they sometimes only satisfy a design requirement. Alternative mechanisms have been proposed by the research community and are actively being pursued by the IETF. There are certain issues, like location privacy of the MN for instance, that need further research, even if for some of them the specification already provides a solution. The proposed solution is not always efficient either from a security or from a performance point of view. We described threats related to the binding update procedure, both with respect to the HA and for route optimisation. In an envisioned Mobile Internet, location privacy of the MNs becomes an important issue that needs addressing. In MobileIPv6, location privacy is still an open issue and needs further research.

Notes

1 By location of the node we understand the point of connection to the Internet.
2 Mechanisms for this exist, but they need a Public Key Infrastructure put in place.

References

1 TANENBAUM, A. S.: 'Computer networks' (Prentice Hall, 2002, 4th edn.)
2 DROMS, R.: 'Dynamic Host Configuration Protocol', RFC 2131, March 1997, available at: http://www.ietf.org/rfc/rfc2131.txt
3 VIXIE, P., THOMSON, S., REKHTER, Y. and BOUND, J.: 'Dynamic updates in the domain name system (DNS UPDATE)', RFC 2136, April 1997, available at: http://www.ietf.org/rfc/rfc2136.txt
4 Internet Engineering Task Force: 'Internet Protocol', RFC 791, September 1981, available at: http://www.ietf.org/rfc/rfc0791.txt
5 DEERING, S. and HINDEN, R.: 'Internet Protocol, Version 6 (IPv6)', RFC 2460, December 1998, available at: http://www.ietf.org/rfc/rfc2460.txt
6 PERKINS, C.: 'IP mobility support for IPv4', RFC 3344, August 2002, available at: http://www.ietf.org/rfc/rfc3344.txt
7 JOHNSON, D., PERKINS, C. and ARKKO, J.: 'Mobility support in IPv6', February 2003, available at: http://www.ietf.org/internet-drafts/draft-ietf-mobileip-ipv6-21.txt
8 KENT, S.: 'IP Authentication Header', RFC 2402, November 1998, available at: http://www.ietf.org/rfc/rfc2402.txt
9 KENT, S.: 'IP Encapsulating Security Payload (ESP)', RFC 2406, November 1998, available at: http://www.ietf.org/rfc/rfc2406.txt
10 HARKINS, D. and CARREL, D.: 'The Internet Key Exchange (IKE)', RFC 2409, November 1998, available at: http://www.ietf.org/rfc/rfc2409.txt
11 ARKKO, J., DEVARAPALLI, V. and DUPONT, F.: 'Using IPSEC to protect Mobile IPv6 signaling between mobile nodes and home agents', March 2003, available at: http://www.ietf.org/internet-drafts/draft-ietf-mobileip-mipv6-ha-ipsec-04.txt
12 KRAWCZYK, H., BELLARE, M. and CANETTI, R.: 'HMAC: Keyed-Hashing for Message Authentication', RFC 2104, February 1997, available at: http://www.ietf.org/rfc/rfc2104.txt
13 National Institute of Standards and Technology: 'Secure Hash Standard', FIPS PUB 180-1, April 1995, available at: http://www.itl.nist.gov/fipspubs/fip180-1.htm
14 NIKANDER, P., AURA, T., ARKKO, J. and MONTENEGRO, G.: 'MobileIP version 6 route optimization security design background', April 2003, available at: http://www.ietf.org/internet-drafts/draft-nikander-mobileip-v6-ro-sec-00.txt
15 PLUMMER, D.: 'Ethernet Address Resolution Protocol: or converting network protocol addresses to 48.bit Ethernet address for transmission on

Ethernet hardware', STD 37, RFC 826, November 1982, available at: http://www.ietf.org/rfc/rfc826.txt

16 NARTEN, T., NORDMARK, E. and SIMPSON, W.: 'Neighbor discovery for IP Version 6 (IPv6)', RFC 2461, December 1998, available at: http://www.ietf.org/rfc/rfc2461.txt

17 THOMSON, S. and NARTEN, T.: 'IPv6 Stateless Address Autoconfiguration', RFC 2462, December 1998, available at: http://www.ietf.org/rfc/rfc2462.txt

18 CASTELLUCCIA, C. and DUPONT, F.: 'A simple privacy extension for Mobile IPv6', February 2001, available at: http://www.inrialpes.fr/planete/people/ccastel/draft-castelluccia-mobileip-privacy-00.txt

19 NARTEN, T. and DRAVES, R.: 'Privacy extensions for stateless address autoconfiguration in IPv6', RFC 3041, January 2001, available at: http://www.ietf.org/rfc/rfc3041.txt

20 ESCUDERO PASCUAL, A.: 'Location privacy in IPv6 – tracking the binding updates', 8th International Workshop on *Interactive Distributed Multimedia Systems*, September 2001.

21 KAUFMAN, C.: 'Internet Key Exchange (IKEv2) Protocol', Internet Draft, February 2003, available at: http://www.ietf.org/internet-drafts/draft-ietf-ipsec-ikev2-05.txt

22 DENG, R. H., ZHOU, J. and BAO, F.: 'Defending against redirect attacks in MobileIP'. Proceedings of the 9th ACM Conference on *Computer and Communications security* (ACM Press, 2002), pp. 59–67

23 BRADNER, S., MANKIN, A. and SCHILLER, J. I.: 'A framework for purpose built keys (PBK)', Internet Draft, January 2003, available at: http://www.ietf.org/internet-drafts/draft-bradner-pbk-frame-04.txt

24 O'SHEA, G. and ROE, M.: 'Child-Proof Authentication for MIPv6 (CAM)', *Computer Communications Review*, April, 2001

25 AURA, T., ROE, M. and ARKKO, J.: 'Security of Internet location management'. Proceedings of the 18th *Annual Computer Security Applications* Conference, IEEE Press, December 2002, pp. 78–87

26 GLASS, S., HILLER, T., JACOBS, S. and PERKINS, C.: 'Mobile IP Authentication, Authorization and Accounting Requirements', RFC 2977, October 2000, available at: http://www.ietf.org/rfc/rfc2977.txt

Part III

Mobile code issues

The two chapters in Part III consider software security issues arising in future mobile systems. Multi-agent systems, based on *agents*, that is, pieces of software that can act autonomously, appear to be a promising technology in a variety of application domains, including middleware for mobile systems. When agents themselves are mobile, a variety of significant security issues arise, which are the focus of Chapter 12.

Chapter 13 is concerned with the major security issues associated with the use of mobile code within the context of the mobile phone, including Software Defined Radio. Whilst some of the security issues are the same as for the existing PC environment, new issues arose because of the need for mobile devices to have very simple user interfaces and yet retain the reliability which their users need and expect of them.

Chapter 12

Security for agent systems and mobile agents

Niklas Borselius

The agent paradigm is currently attracting much research. A mobile agent is a particular type of agent with the ability to migrate from one host to another where it can resume its execution. In this chapter, we consider security issues that need to be addressed before multi-agent systems in general, and mobile agents in particular, can be a viable solution for a broad range of commercial applications. We do this through considering the implications of the characteristics given to agents and general properties of open multi-agent systems. We then look in more detail at technology and methods applicable to mobile agent systems.

12.1 Introduction

Agents are independent pieces of software capable of acting autonomously in response to input from their environment. Agents can be of differing capabilities, but typically possess the required functionality to fulfil their design objectives. To be described as 'intelligent', software agents should also have the ability to act autonomously, that is without direct human interaction, be flexible, and in a multi-agent system, be able to communicate with other agents, that is, to be social. Agents are, to various degrees, aware of their environment, which often also can be affected by the agents' actions.

A mobile agent is a particular class of agent with the ability during execution to migrate from one host to another where it can resume its execution. It has been suggested that mobile agent technology, amongst other things, can help to reduce network traffic and to overcome network latencies [1]. An agent's ability to move does, however, introduce significant security concerns.

The concept of an agent originates from the area of Artificial Intelligence (AI) but has now gained more widespread acceptance in mainstream computer science [2]. The term 'agent' has become rather fashionable, and a more mature technology than currently available is often implied. This is, in particular, true for security in multi-agent systems. Over-simplified assumptions and non-applicable references to security solutions are not uncommon in the literature. Naturally, security is not a driving force for research and development of multi-agent systems, and therefore has not received much attention from the agent community. Nevertheless, in order for agent technology to gain widespread use and provide viable solutions on a wider scale for commercial applications, security issues need to be properly addressed.

Autonomous agents and multi-agent systems represent a relatively new way of analysing, designing and implementing complex software systems. In this chapter, we are only concerned with the security of the system and its components (leaving design methodologies to others). Several multi-agent systems are available as commercial products and many more have been implemented in various research projects, with varying success. Recent standardisation efforts [3, 4] have proven rather successful and are still evolving. Today there is growing interest and research in implementing and rolling out (open) multi-agent systems on a wider scale.[1] Mobile VCE (www.mobilevce.com) is undertaking one such project where the agent paradigm is researched in a mobile telecommunications setting.

The remainder of this chapter is organised as follows. Section 12.2 briefly describes those characteristics of agents and multi-agent systems of relevance to security. In Section 12.3, we will discuss some security implications of these characteristics. Section 12.4 considers security technology and some recent research aimed at enhancing the security of mobile agent systems.

12.2 Agents and multi-agent systems

In this section, we will briefly describe some properties of agents and multi-agent systems (MAS). This is not intended to be a complete description of agents or MAS (we are, e.g. not concerned with AI properties for agents here). We try to focus on issues with possible security implications. For broader and more general introductory material to agents and MAS see, for example, References 5 and 6.

Agents are software entities that have some kind of autonomy and certain 'intelligence'. An agent is often assumed to represent another entity, such as a human or an organisation on whose behalf it is acting. No single universal definition of agents exists, but there are certain widely agreed universal characteristics of agents; these include situatedness, autonomy and flexibility [7].

- *Situatedness* means that the agent receives sensory input from its environment and that it can perform actions which change the environment in some way.
- *Autonomy* means that an agent is able to act without the direct intervention of humans (or other agents), and that it has control over its own actions and internal state.
- *Flexibility* can be defined to include the following properties:
  - *responsive*: refers to an agent's ability to perceive its environment and respond in a timely fashion to changes that occur in it;
  - *proactive*: agents are able to exhibit opportunistic, goal-driven behaviour and take the initiative where appropriate;

o *social*: agents should be able to interact, when appropriate, with other agents and humans in order to solve their own problems and to help others with their activities.

A number of other attributes are sometimes discussed in the context of agency. These include but are not limited to [8] the following.

- *Rationality*: the assumption that an agent will not act in a manner that prevents it from achieving its goals and will always attempt to fulfil those goals.
- *Veracity*: the concept that an agent will not knowingly communicate false information.
- *Benevolence*: an agent cannot have conflicting goals that either force it to transmit false information or to effect actions that cause its goals to be unfulfilled or impeded.
- *Mobility*: the ability for an agent to move across networks and between different hosts to fulfil its goals.

An MAS is a system composed of multiple autonomous agents with the following characteristics [7]:

- each agent cannot solve a problem unaided,
- there is no global system control,
- data is decentralised, and
- computation is asynchronous.

Computer hosts, or platforms, provide agents with environments in which they can execute. A platform typically also provides additional services, such as communication facilities, to the agents it is hosting. In order for agents to be able to form a useful open MAS where they can communicate and cooperate, certain functionality needs to be provided to the agents. This includes the functionality to find other agents or find particular services. This additional functionality can either be implemented as services offered by other agents or as services more integrated with the MAS infrastructure itself. Examples of such services include facilitators, matchmakers, mediators and blackboards.

Open MAS are usually envisioned as systems, communicating over the Internet, allowing anybody to connect to a platform on which agents are running. This means that the MAS lacks a global system control and that information in general is highly decentralised.

12.3 Security implications

In this section we will discuss agent security issues based on the characteristics described in the previous section.

12.3.1 Agent execution

Naturally, agents need to execute somewhere. A computer host, the immediate environment of an agent, is ultimately responsible for the correct execution and protection of the agent. This leads us to the question of where access control decisions should be performed and enforced. Does the agent contain all necessary logic and information required to decide if an incoming request is authentic (originating from its claimant) and if so, is it authorised (has the right to access the requested information or service)? Or, can the agent rely on the platform for access control services?

The environment might also need certain protection from the agents that it hosts. An agent should, for example, be prevented from launching a denial of service attack through consuming all resources on a host, thus preventing the host from carrying out other things (such as executing other agents). Security issues related to the executing host become even more apparent for agents that are mobile, further described in Section 12.3.5.

12.3.2   Situatedness

The meaning of the term 'environment' depends on the application and appears to vary somewhat arbitrarily in the agent literature; it can, for example, be the Internet or the host on which the agent is executing. An agent is assumed to be 'aware' of certain states or events in its environment. Depending on the nature and origin of this information, its authenticity and availability need to be considered (confidentiality of such information might also be relevant). If an agent's 'environment' is limited to the host on which it is executing, no specific security measures might be necessary (assuming the host environment cannot be spoofed). The situation is, however, likely to be different if the agent receives environment information from, or via, the Internet. (Security of communication is further explored in the following text.)

12.3.3   Autonomy

Autonomy, when combined with other features given to agents, can introduce serious security concerns. If an agent, for example, is given authority to buy or sell things, it should not be possible for another party to force the agent into committing to something it would not normally commit to. Neither should an agent be able to make commitments it cannot fulfil. Hence, issues related to delegation needs to be considered for agents.

The autonomy property does not necessarily introduce any 'new' security concerns; this property is held by many existing systems. It is worth mentioning that Internet worms (often referred to as viruses) also hold this property, which enables them to spread efficiently without requiring any (intentional or unintentional) human interaction. The lesson to learn from this is that powerful features can also be used for malicious purposes if not properly controlled.

12.3.4   Communication

Of the flexibility properties, social behaviour is certainly interesting from a security point of view. This means that agents can communicate with other agents and humans. Just as an agent's communication with its environment needs to be protected, so does its communication with other agents and humans. The following security properties should be provided:

- *confidentiality*: assurance that communicated information is not accessible to unauthorised parties;
- *data integrity*: assurance that communicated information cannot be manipulated by unauthorised parties without being detected;
- *authentication of origin*: assurance that communication originates from its claimant;
- *availability*: assurance that communication reaches its intended recipient in a timely fashion;
- *non-repudiation*: assurance that the originating entity can be held responsible for its communications.

Fundamental to the above-mentioned communication security properties are issues relating to the identification and authentication of the sending and receiving parties. These issues are further discussed in Section 12.3.7.

It should be noted that security usually comes at a cost. Additional computing resources as well as communication resources are required by most solutions to the above-mentioned security requirements. Therefore, security needs to be dynamic. Sometimes it makes sense to protect all communication within a system to the same degree, as the actual negotiation of security mechanisms can then be avoided. However, in a large-scale open MAS, security services and mechanisms need to be able to fit the purpose and nature of the communications of various applications with differing security requirements.

Some implementations of MAS assume that security is provided transparently by a lower layer. This approach might be sufficient in a closed system where the agents can trust each other and the only concern is external malicious parties. However, we believe that agents in an open system may often need to be 'security aware', that is, they need to be able to make decisions based on where information is originating from and how well protected it is. As suggested elsewhere [14], public key cryptography and a supporting public key infrastructure can be used as important parts of inter-agent communication.

With a public key infrastructure in place, security protocols and mechanisms already developed for other applications can be made to fit the requirements of MAS to provide authentication, confidentiality and data integrity.

12.3.5 Mobility

The use of mobile agents raises a number of security concerns. Agents need protection from other agents and from the hosts on which they execute. Similarly, hosts need to be protected from agents and from other parties that can communicate with the platform. The problems associated with the protection of hosts from malicious code are quite well understood.

The problem posed by malicious hosts to agents seems more complex to solve. Since an agent is under the control of the executing host, the host can, in principle, do anything to the agent and its code. The particular attacks that a malicious host can make can be summarised as follows:

- observation of code, data and flow control,
- manipulation of code, data and flow control – including manipulating the route of an agent,
- incorrect execution of code – including re-execution,
- denial of execution – either in part or whole,
- masquerading as a different host,
- eavesdropping on agent communications,
- manipulation of agent communications,
- false system call return values.

In Section 12.4, we describe countermeasures addressing security issues arising from the mobility property.

12.3.6   Rationality, veracity and benevolence

These properties could at a first glance appear to be very security relevant. However, on closer consideration they seem to be too abstract for us to consider as practical security concerns. The meaning (from a security point of view) of these properties seems to be: 'Agents are well behaved and will never act in a malicious manner'. If we make this a genuine requirement, then the required redundancy for such a system is likely to make the system useless. It would, of course, be valuable to have a system where agents can be assumed to behave truthfully and honestly in every situation. This does not seem a likely scenario for an MAS that is not under very strict control and under a single authority, and would not correspond to the assumed open system scenario. However, measures can be taken to limit maliciously behaving agents. Assurance that only information from trusted sources is acted upon and that agents (or their owner) can be held responsible for their actions, as well as monitoring and logging of agent behaviour, are mechanisms that can help in creating a system where the actions of malicious agents can be minimised.

12.3.7   Identification and authentication

Identification is not primarily a security issue in itself; however, the means by which an agent is identified are likely to affect the way an agent can be authenticated. For example, an agent could simply be identified by something like a serial number, or its identity could be associated with its origin, owner, capabilities or privileges. As mentioned in Section 12.3.4, authentication is often fundamental to secure communication. If identities are not permanent, security-related decisions cannot be made on the basis of an agent's identity.

Connected with identification and authentication is anonymity. While an entity's identity is of major importance to certain applications and services, it is not needed in others. An open MAS would probably require some sort of anonymity service to acquire great acceptance today. In fact, agents are likely to be ideal for providing anonymity to their owners as they are independent pieces of code, possess some degree of autonomy, and do not require direct user interaction.

12.3.8   Trust

Agents need to be able to make decisions based on information received and collected from other entities. In order to make these decisions they need to be able to evaluate the trustworthiness of the information or the information source. For mobile agents, decisions need to be made whether to transfer and execute on a particular host or not.

The issues surrounding trust within agent systems are currently attracting much research within the agent community. Various mechanisms for agents to reason about trust have been proposed, see, for example, Reference 10. Trust mechanisms based on reputation is one approach suggested by many [11, 30].

Creating trust between entities without any, or very limited, common history or knowledge of each other, which would be the case in an open MAS, is a non-trivial task. Even though public key infrastructure (PKI) technology still has to prove itself viable for an open system on a global scale, a PKI may well be the best available solution for distribution of trust in an open MAS [13].

12.3.9   Authorisation and delegation

Authorisation and delegation are important issues in MAS. Not only do agents need to be granted rights to access information and other resources in order to carry out their tasks, but they will also be acting on behalf of a person, organisation or other agents, requiring transfer of access rights between different entities. With a PKI in place, delegation can be done through various types of certificate, including attribute certificates for delegation of rights and issuing of 'traditional' public key certificates for delegation of signing rights.

12.4   Security measures for mobile agents

Many commercial and research MAS architectures have been implemented and many are still under development.[2] Several of these recognise security as an issue to be taken care of in the future, whilst others imply that security is provided for. It is common for MAS implementations to assume a VPN-like (Virtual Private Network) underlying network to provide security services. This approach usually does not provide for much flexibility, since secure communication between parties without pre-established relationships becomes cumbersome. Nevertheless, this solution can use well-established security protocols and be adequate for applications where all communication is protected to the same degree. Such an approach usually leaves the agents completely unaware of security services as this is handled between agent platforms (or perhaps even at the link level). The agents themselves are also unprotected from malicious hosts if no other security measures are applied.

In this section, we will describe available technologies and research efforts addressing the security issues arising from the mobility property of mobile agents.

First, we consider mechanisms addressing various security aspects of the mobile agent, and second we examine technologies protecting the executing hosts from agents.

12.4.1 Protecting agents

The security issues that arise with mobile agents are relatively well understood by the security community and hence much research is being devoted to the area. There have been many attempts to address the threats posed to mobile agents, most addressing a particular part of the problem.

As stated before, once an agent has arrived at a host, little can be done to stop the host from treating the agent as it likes. The problem is usually referred to as the *malicious host problem*. A simple example often used to illustrate how a malicious host can benefit from attacking a mobile agent is the shopping agent. An agent is sent out to find the best airfare for a flight with a particular route. The agent is given various requirements, such as departure and destination, time restrictions, etc., and sent out to find the cheapest ticket before committing to a particular purchase. The agent will visit every airline and query their databases before committing to a purchase and reporting back to the agent owner (see Figure 12.1). A malicious host can interfere with the agent execution in several ways in order to make its offer appear most attractive. For example, a malicious host could try to: (i) erase all information previously collected by the agent – in this way the host is guaranteed at least to have the best current offer; (ii) change the agent's route so that airlines with more favourable offers are not visited; (iii) simply terminate the agent to ensure that no competitor gets the business either; (iv) make the agent execute its commitment function, ensuring that the agent is committing to the offer given by the malicious host (if the agent is carrying electronic money it could instead take it from the agent directly). In addition to this, the agent might be carrying information that needs to be kept secret from the airlines (e.g. maximum price).

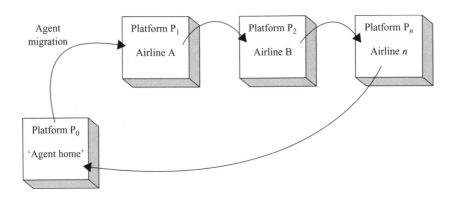

Figure 12.1   Shopping agent, sent out to find best airfare

There is no universal solution to the malicious host problem, but some partial solutions have been proposed. Many of the security mechanisms are aimed at detecting, rather than preventing, misbehaving hosts. In the following sections, we will describe some of the mechanisms proposed to address the malicious host problem.

12.4.1.1   Contractual agreements

The simplest solution (at least from a technical perspective) is to use contractual means to tackle the malicious host problem. Operators of agent platforms guarantee, via contractual agreements, to operate their environments securely and not to violate the privacy or the integrity of the agent, its data, and its computation. However, to prove that such an agreement has been broken might be a non-trivial task.

12.4.1.2   Trusted hardware

If the operators of the available execution environments cannot be trusted, one obvious solution is to let a trusted third party supply trusted hardware, in the form of tamper-resistant devices, that are placed at the site of the host and interact with the agent platform [14]. A tamper-resistant device can, for example, come in the form of a smartcard. Such trusted hardware can then either protect the complete execution environment of the agent or perform certain security sensitive tasks. However, such trusted hardware must be used carefully and might appear to offer more security than it really does. The agent must still be able to communicate with resources at the local platform (the part under control of an untrusted party), for example, to interact with a local database. All such interactions can still be intercepted by the untrusted party.

If the trusted hardware is only used to protect security sensitive actions this might be even more vulnerable. It might, for example, be tempting to let the agent's private signature key be protected such that it only will be available when decrypted inside the trusted device. A signature algorithm can then be executed within the device using the agent's private key. In this way, the private signature key is never exposed to the host. However, the host might be able to interfere with the communication taking place between the agent residing on the host and the trusted device in such a way that a correct signature is produced on information falsely manufactured by the host.

Above all else, the major drawback of trusted hardware is the cost of such a solution.

12.4.1.3   Trusted nodes

By introducing trusted nodes into the infrastructure to which mobile agents can migrate when required (illustrated in Figure 12.2), sensitive information can be prevented from being sent to untrusted hosts, and certain misbehaviours of malicious hosts can be traced. The owner's host, that is, the platform from where the mobile agent first is launched, is usually assumed to be a trusted node. In addition to this, service providers can operate trusted nodes in the infrastructure.

In our example with the shopping agent, the mobile agent can be constructed so that the commitment function (e.g. the agent's signature key) is encrypted such that it can only be decrypted at a trusted host. Once the agent arrives at the trusted host,

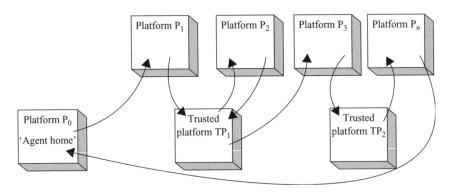

Figure 12.2   Trusted nodes can be used to protect mobile agents

it can compare the collected offers and commit to the best offer. Alternatively [15], one agent containing the ability to commit to a purchase can be sent to a trusted node. From this node, one or several sub-agents are sent to the airline hosts to collect offers. Depending on the threat scenario, single-hop agents can be used, that is, agents only visiting one host before returning back, or one or several multi-hop agents can be used. Once the sub-agent or agents have returned to the trusted node, the best offer is selected and the agent commits to a purchase. This last alternative does limit the agent's mobility, but may be beneficial in certain scenarios.

12.4.1.4   Cooperating agents

By using cooperating agents, a similar result to that of trusted nodes can be achieved [16]. Information and functionality can be split between two or more agents in such a way that it is not enough to compromise only one (or even several) agents in order to compromise the task. An identical scenario to that described using trusted nodes can, for example, be achieved by letting the agent residing on the trusted host be executed on any host that is assumed not to be conspiring with any of the airlines.

By applying fault-tolerant techniques, the malicious behaviour of a few hosts can be countered. One such scheme for ensuring that a mobile agent arrives safely at its destination has been proposed in Reference 12. Although a malicious platform may cause an agent to operate incorrectly, the existence of enough replicates ensures the correct end result.

Again, referring to the shopping agent, several mobile agents can be used, taking different routes, and before deciding on the best offer the agents communicate their votes amongst each other.

12.4.1.5   Execution tracing

Execution tracing [17] has been proposed for detecting unauthorised modifications of an agent through the faithful recording of the agent's execution on each agent platform. Each platform is required to create and retain a non-repudiable log of the operations performed by the agent while executing on the platform. The major drawbacks of this approach are not only the size of the logs created, but also the necessary management of created logs.

Partial result authentication codes (PRACs) were introduced by Yee [18]. The idea is to protect the authenticity of an intermediate agent state or partial result that results from running on a server. PRACs can be generated using symmetric cryptographic algorithms. The agent is then equipped with a number of encryption keys. Every time the agent migrates from a host, the agent's state or some other result is processed using one of the keys, producing a message authentication code (MAC) on the message. The key that has been used is then disposed of before the agent migrates. The PRAC can be verified at a later point to identify certain types of tampering. A similar functionality can be achieved using asymmetric cryptography by letting the host produce a signature on the information instead.

12.4.1.6 Encrypted payload

Asymmetric cryptography (also known as public key cryptography) is well suited for a mobile agent that needs to send back results to its owner or which collects information along its route before returning with its encrypted payload to its owner. This is due to the fact that the encryption key does not need to be kept secret. However, to encrypt very small messages is either very insecure or results in a large overhead compared with the original message. A solution called sliding encryption [19] has been proposed that allows small amounts of data to be encrypted, and consequently added to the cryptogram, such that the length of the resulting ciphertext is minimised. Due to the nature of asymmetric cryptography the agent is not able to access its own encrypted payload until arriving at a trusted host where the corresponding decryption key is available.

12.4.1.7 Environmental key generation

Environmental key generation [20] allows an agent to carry encrypted code or information. The encrypted data can be decrypted when some predefined environmental condition is true. Using this method, an agent's private information can be encrypted and only revealed to the environment once the predefined condition is met. This requires that the agent has access to some predictable information source. Once the private information has been revealed, it would, of course, be available also to the executing host. However, if the condition is not met on a particular host, the private information is not revealed to the platform.

12.4.1.8 Computing with encrypted functions

Sander and Tschudin [21] have proposed a scheme whereby an agent platform can execute a program embodying an enciphered function without being able to discern the original function. For example, instead of equipping an agent with function f, the agent owner can give the agent a program $P(E(f))$ which implements $E(f)$, an encrypted version of f. The agent can then execute $P(E(f))$ on x, yielding an encrypted version of $f(x)$. With this approach, an agent's execution would be kept secret from the executing host as would any information carried by the agent.

For example, the means to produce a digital signature could thereby be given to an agent without revealing the private key. However, a malicious platform could still use the agent to produce a signature on arbitrary data. Sander and Tschudin therefore suggest combining the method with undetachable signatures (see following text).

Although the idea is straightforward, the trick is to find appropriate encryption schemes that can transform functions as intended; this remains a research topic. Recently Barak *et al.* [22] have shown that it is unlikely that obtaining theoretical justification for a program's ability to completely hide its information appears infeasible.

12.4.1.9 Obfuscated code

Hohl [23] proposes what he refers to as Blackbox security to scramble an agent's code in such a way that no one is able to gain a complete understanding of its function. However, no general algorithm or approach exists for providing Blackbox security. A time-limited variant of Blackbox protection is proposed as a reasonable alternative. This could be applicable where an agent only needs to be protected for a short period. One serious drawback of this scheme is the difficulty of quantifying the protection time provided by the obfuscation algorithm.

12.4.1.10 Undetachable signatures

By binding usage restrictions to a signature key given to the agent, we can potentially limit the damage a malicious host can do. Sander and Tschudin [21] proposed one such scheme, which they refer to as undetachable signatures. Their original scheme has since been improved [24]. The idea is to encode constraints into the signature key. If the constraints are not met, a valid signature is not produced, thus preventing arbitrary messages from being signed.

An alternative to undetachable signatures is to use digital certificates to regulate the validity of digital signatures [25]. Digital certificates are used to let a verifier check the validity of a digital signature. Certificates usually include a validity period under which valid signatures can be produced. By extending the constraints included in the certificate to context-related values such as executing host, maximum value of a purchase and so on, certificates can be used to further restrict the usage of signature keys and thereby decrease the involved risks regarding improper use of the signature key. One advantage with this scheme over undetachable signatures is that it relies on already well-established cryptographic techniques.

12.4.2 Protecting the agent platform

More mature technology is available to address the problem of protecting the agent platform from malicious agents. Techniques similar to those used to address security issues associated with downloading software from the Internet can be applied to the mobile agent scenario.

12.4.2.1 Sandboxing and safe code interpretation

Sandboxing isolates applications (or in our case agents) into distinct domains enforced by software. The technique allows untrusted programs to be executed within their virtual address space thereby preventing them from interfering with other applications. Access to system resources can also be controlled through a unique identifier associated with each domain.

Agents are usually developed using an interpreted script or programming language. The main motivation for this is to support agent platforms on heterogeneous computer systems. The idea behind safe code interpretation is that commands considered insecure can either be made safe or denied to the agent. Java is probably the most widely used interpretative language used today. Java also utilises sandboxing and signed code (described below); this makes Java well suited for development of mobile agents.

12.4.2.2 Proof carrying code

Proof carrying code [26] requires the author of an agent to formally prove that the agent conforms to a certain security policy. The execution platform can then check the agent and the proof before executing the agent. The agent can then be run without any further restrictions. The major drawback of this approach is the difficulty in generating such formal proofs in an automated and efficient way.

12.4.2.3 Signed code

By digitally signing an agent its authenticity, origin and integrity can be ensured. Typically, the code signer is either the creator of the agent, the agent owner (on whose behalf the agent is acting) or some party that has reviewed the agent. The security policy at the platform, perhaps in conjunction with attribute certificates supplied with the signed code, would then decide if a particular signature means that the code should be executed.

12.4.2.4 Path histories

The idea behind path histories [27] is to let a host know where a mobile agent has been executed previously. If the agent has been running on a host that is not trusted, the newly visited host can decide not to let the agent execute or restrict the execution privileges. Path histories require each host to add a signed entry to the path, indicating its identity and the identity of the next platform to be visited, and to supply the complete path history to the next host.

12.4.2.5 State appraisal

State appraisal [28] attempts to ensure that an agent's state has not been tampered with and that the agent will not carry out any illegal actions through a state appraisal function which becomes part of the agent code. The agent author produces the appraisal function, which is signed by the author, together with the rest of the agent. An agent platform uses the function to verify that an incoming agent is in a correct state and to determine what privileges an agent can be granted during execution. The theory, which is still to be proven in practice, requires that the legal states can be captured and described in an efficient and secure way.

12.5   Conclusions

The security issues for non-mobile agents can, at least in theory, to a great extent be tackled through existing security technology and protocols. However, issues related to trust and delegation in a large scale multi agent system are non-trivial to solve. Although a PKI is likely to be an important part of the solution, agents need to be able to reason and make decisions based on various security parameters. Execution of agents (mobile as well as non-mobile) on untrusted platforms is another factor introducing non-trivial security concerns, in particular related to correct agent execution and confidentiality of agent data. The required security level and security measures must, as always, depend on the application. Current standardisation efforts and large-scale agent projects are likely to facilitate greater use of agent technology in the future. One such research effort is ongoing within the Mobile VCE Core II Research Programme, which has recently proposed a security architecture for agent-based mobile middleware [29].

There does not seem to be a single solution to the security problems introduced by mobile agents unless trusted hardware is introduced, which is likely to prove too expensive for most applications. The way forward appears to lie in a range of mechanisms aimed at solving particular (smaller) problems. This could, for example, include mechanisms that depend on agents executing on several hosts rather than on only one host, mechanisms and protocols binding agent actions to hosts, generation of various types of audit information that can be used in case of disputes, and so on. Solutions to certain problems do exist, but for mobile agents to be more widely adopted this is an area that requires further research.

Acknowledgements

The work reported in this paper has formed part of the Software Based Systems area of the Core II Research Programme of the Virtual Centre of Excellence in Mobile & Personal Communications, Mobile VCE, www.mobilevce.com, whose funding support, including that of the EPSRC, is gratefully acknowledged. Fully detailed technical reports on this research are available to Industrial Members of Mobile VCE.

Notes

1   See http://www.agentcities.org for an example of an ongoing effort to implement large-scale multi-agent systems.
2   See http://www.agentbuilder.com/AgentTools for a list of available systems.

References

1 HARRISON, C. G., CHESS, D. M., and KERSHENBAUM, A.: 'Mobile agents: are they a good idea?', Technical Report, IBM Research Division, 1995

2 LUCK, M. and D'INVERNO, M.: 'A conceptual framework for agent definition and development', *The Computer Journal*, 2001, **44**, (1), pp. 1–20

3 GENESERETH, M. and FIKES, R.: 'Knowledge interchange format, version 3.0 reference manual'. Technical Report Logic-92-1, Computer Science Department, Stanford University, USA, 1992

4 BURG, B.: 'Towards the deployment of an open agent world', *in* HERMES (Ed.): 'Journées Francophones d'Intelligence Artificielle Distribuée et de Systèmes Multi-Agents' (JFIADSMA 2000), October 2001

5 HAYZELDEN, A. L. G. and BOURNE, R. A. (Eds): 'Agent technology for communication infrastructures' (John Wiley & Sons, Chichester, UK, 2000)

6 WOOLDRIDGE, M.: 'An introduction to multiagent systems' (John Wiley & Sons Chichester, UK, 2002)

7 JENNINGS, N. R., SYCARA, K., and WOOLDRIDGE, M.: 'A roadmap of agent research and development', *Autonomous Agents and Multi-Agent Systems*, 1998, **1**, (1), pp. 275–306

8 JENNINGS, N. R. and WOOLDRIDGE, M.: 'Intelligent agents: theory and practice', *The Knowledge Engineering Review*, 1995, **10**, (2), pp. 115–152

9 HE, Q., SYCARA, K. P. and FININ, T. W.: 'Personal security agent: KQML-Based PKI', *in* SYCARA, K. P. and WOOLDRIDGE, M. (Ed.): Proceedings of the 2nd international conference on *Autonomous Agents* (ACM Press, New York, USA, 1998) pp. 377–84

10 CASTELFRANCHI, C. and TAN, Y.-H. (Eds): 'Trust and deception in virtual societies' (Kluwer Academic Publishers, Dordrecht, The Netherlands, 2001)

11 BIRK, A.: 'Learning to trust', *in* FALCONE, R., SINGH, M. and TAN, Y. H. (Eds): 'Trust in cyber-societies', LNAI, vol. 2246 (Springer-Verlag, Berlin, 2001) pp. 27–54

12 SCHNEIDER, F. B.: 'Towards fault-tolerant and secure agentry', *in* MAVRONICOLAS, M. and TSIGAS, P. (Eds): 'Eleventh international workshop on *Distributed Algorithms*', LNCS, vol. 1320 (Springer-Verlag, Berlin, 1997) pp. 1–14

13 BOSWORTH, K. P. and TEDESCHI, N.: 'Public key infrastructures – the next generation', *in* TEMPLE, R. and REGNAULT, J. (Eds): 'Internet and wireless security', BTexact Communications Technology Series 4 (IEE, London, 2002) pp. 95–120

14 WILHELM, U. G., STAAMANN, S. and BUTTYÁN, L.: 'Introducing trusted third parties to the mobile agent paradigm', *in* VITEK, J. and JENSEN, C. (Eds): 'Secure Internet programming', LNCS, vol. 1603 (Springer-Verlag, New York, 1999) pp. 471–91

15 BORSELIUS, N., MITCHELL, C. J. and WILSON, A. T.: 'On mobile agent based transactions in moderately hostile environments', *in* DE DECKER, B., PIESSENS, F., SMITS, J. and VAN HERREWEGHEN, E. (Eds):

'Advances in network and distributed systems security'. Proceedings of IFIP TC11 WG11.4 first annual working conference on *Network Security*, KU Leuven, Belgium, November 2001 (Kluwer Academic Publishers, Boston, MA, 2001) pp. 173–86

16 ROTH, V.: 'Secure recording of itineraries through cooperating agents'. Proceedings of the ECOOP workshop on *Distributed object security* and 4th workshop on Mobile object systems: secure Internet mobile computations, INRIA, France, 1998, pp. 147–54

17 VIGNA, G.: 'Protecting mobile agents through tracing'. Proceedings of the third ECOOP workshop on *Operating system support for mobile object systems*, Finland, June 1997, pp. 137–53

18 YEE, B.: 'A sanctuary for mobile agents', *in* VITEK, J. and JENSEN, C. (Eds): 'Secure Internet programming', LNCS, vol. 1603 (Springer-Verlag, New York, 1999) pp. 261–74

19 YOUNG, A. and YUNG, M.: 'Sliding encryption: a cryptographic tool for mobile agents'. Proceedings of the 4th international workshop on *Fast software encryption*, FSE '97, LNCS, vol. 1267 (Springer-Verlag, Berlin, 1997) pp. 230–41

20 RIORDAN, J. and SCHNEIER, B.: 'Environmental key generation towards clueless agents', *in* VIGNA, G. (Ed): 'Mobile agents and security', LNCS, vol. 1419 (Springer-Verlag, Berlin, 1998) pp. 15–24

21 SANDER, T. and TSCHUDIN, C.: 'Towards mobile cryptography'. Proceedings of the IEEE symposium on *Security and privacy*, Oakland, CA, May 1998 (IEEE Computer Society Press, 1998) pp. 215–24

22 BARAK, B., GOLDREICH, O., IMPAGLIAZZO, R., *et al.* 'On the (im)possibility of obfuscating programs', *in* KILIAN, J. (Ed.): Proceedings of the *21st Annual International Cryptology Conference*, Santa Barbara, CA, USA, LNCS, vol. 2139 (Springer-Verlag, Berlin, 2001) pp. 1–18

23 HOHL, F.: 'Time limited blackbox security: protecting mobile agents from malicious hosts', *in* VIGNA, G. (Ed.): 'Mobile agents and security', LNCS, vol. 1419 (Springer-Verlag, Berlin, 1998) pp. 1419

24 KOTZANIKOLAOU, P., BURMESTER, M. and CHRISSIKOPOULOS, V.: 'Secure transactions with mobile agents in hostile environments', *in* DAWSON, E., CLARK, A. and BOYD, C. (Eds): 'Information security and privacy: Proceedings of the 5th Australian conference – ACISP 2000', LNCS, vol. 1841 (Springer-Verlag, Berlin, 2000) pp. 289–97

25 BORSELIUS, N., MITCHELL, C. J. and WILSON, A. T.: 'A pragmatic alternative to undetachable signatures', *ACM SIGOPS Operating Systems Review*, 2002, **36**, (2), pp. 6–11

26 NECULA, G. C. and LEE, P.: 'Safe, untrusted agents using proof-carrying code', *in* VIGNA, G. (Ed.): 'Mobile agents and security', LNCS, vol. 1419 (Springer-Verlag, Berlin, 1998) pp. 61–91

27 CHESS, D., GROSOF, B., HARRISON, C., LEVINE, D., PARIS, C. and TSUDIK, G.: 'Iterant agents for mobile computing', *in* HUHNS, M. N. and

SINGH, M. P. (Eds): 'Readings in agents' (Morgan Kaufmann, San Francisco, CA, 1997) pp. 267–82

28 FARMER, W., GUTTMANN, J. and SWARUP, V.: 'Security for mobile agents: authentication and state appraisal', *in* BERTINO, E., KURTH, H., MARTELLA, G. and MONTOLIVO, E. (Eds): Proceedings of the European symposium on *Research in Computer Security* (ESORICS), LNCS, vol. 1146 (Springer-Verlag, Berlin, 1996) pp. 118–30

29 BORSELIUS, N., HUR, N., KAPRYNSKI, M. and MITCHELL, J. C.: 'A security architecture for agent-based mobile systems'. Proceedings of the third international conference on *Mobile Communications Technologies* – 3G2002, London, UK, May 2002, IEE Conference Publication 489 (IEE, London, 2002) pp. 312–18

30 SCHILLO, M., FUNK, P. and ROVATSOS, M.: 'Using trust for detecting deceitful agents in artificial societies', *Applied Artificial Intelligence*, 2000, **14**, (8), pp. 825–48

Chapter 13

Security issues for downloaded code in mobile phones

*Derek Babb, Craig Bishop, Terence Dodgson
and Vaia Sdralia*

'Software defined radio' (SDR) is a technology that will appear in future generations of mobile phones, that is, following the third-generation mobile phone technology that is currently being defined and developed. Early versions of 'pragmatic' SDR will allow the terminal to be reconfigured at any level of its protocol stack. Ultimately, the 'pure' SDR technology will allow a mobile phone or terminal to have its air interface software configured or reconfigured by other software (or software parameters) that have been downloaded to the terminal, for example, over the air, or from a remote server via the Internet and one's personal computer (PC). A number of security issues arise with downloaded code that implements the air interface functions, and these may not be obvious simply from looking at the way PC software is updated online today. This chapter starts with an outline of the code that allows a mobile phone to operate over a particular air interface. This sets the baseline for a discussion of the security issues surrounding the change of this code from one that is fixed and downloaded once only, to code that is reconfigurable during the life of a product.

13.1   Introduction

Today's mobile phones are small, hand-portable, battery powered devices that can communicate using radio frequency (RF) signals on one or more of a variety of air interfaces. The Global System for Mobile communications (GSM), as specified by the European Telecommunications Standards Institute (ETSI), is currently the most widely used digital, 'second-generation', air interface standard, though other systems are in use in some regions of the world.

A new, 'third' generation (3G) of mobile phone technologies, which can be grouped under the International Mobile Telecommunications 2000 (IMT 2000) family of standards and which include (amongst others) the Universal Mobile Telecommunications System (UMTS[1]) standard, are (at the time of writing) being launched in some areas. These technologies offer higher data rates and hence a wider range of services can be supported by terminals. Some 3G systems bring with them a shift that involves the use of wideband code division multiple access (W-CDMA) technology.

Research on a future generation [also known as 4G, 'Beyond 3G', B3G and NG (Next Generation)] of mobile communication systems is in its early stages and products will probably start to appear on the market in around 10 years. Such systems will almost certainly employ software-defined radio (SDR) techniques in the mobile equipment. The air interface and communication method of a terminal employing SDR might be changed by reconfiguring both the software and the hardware, for example, the firmware (FW), digital signal processor (DSP) and multiple RF sections utilising microelectromechanical switches (MEMS). Such reconfiguration might be achieved by downloading software.

The definition of SDR, that is, what it is, can vary depending on what school of thought is being consulted. The two extremes could be regarded as the 'pure' software radio, which is able to reconfigure its air interface software, and the 'pragmatic' software radio, in which any level of the protocol stack can be reconfigured. This paper encompasses both extremes, the air interface configuration concept, together with multiple protocol stack configuration. The pragmatic SDR concept can be taken to generate a roadmap towards the pure SDR, working down from the application, presentation and session layers, through transport and network layers to, ultimately, the data link and physical layers (the latter two layers comprising most of the processing pertinent to the air interface itself). This chapter identifies software security issues associated with downloading code for reconfiguring a future generation of SDR mobile phones. Mobile phone design and architecture issues related to the download of software code are discussed from a *generic* point of view only, as there are many different approaches to the associated security issues. This chapter does not claim to represent the views of any specific research programme, standards body or manufacturer, although there is a definite focus on the European approach. Because of space constraints, the discussion is confined to basic mobile phones, as combined personal digital assistant (PDA)/mobile phones add an extra dimension to the hardware and software architecture that needs separate and careful attention.

As this chapter is about security, it is perhaps useful to state the meanings of the word 'security' as it is used here. In a general sense, 'security' can refer to:

- precautions that are taken to protect against theft, espionage or other danger;
- the state of being free from danger, damage or worry;
- something given or pledged to guarantee payment.

From the mobile telecoms network point of view the 3rd Generation Partnership Project (3GPP), which is responsible for defining the UMTS standards, defines security [1] as: 'The ability to prevent fraud as well as the protection of information availability, integrity and confidentiality'.

Furthermore, the concept of downloadable applications, and the requirement for their secure interoperability across mobile phones from different manufacturers, has also led to the standardisation of a Mobile application Execution Environment, or MExE [2] (see later). This includes the idea that security mechanisms should also: 'prevent attack from unfriendly sources or transferred applications . . .'.

This introduces some of the elements of security as they will relate to SDR, in that mechanisms are required to ensure that downloaded software is fit for purpose, and that it must neither 'attack' (whether maliciously or otherwise) the equipment nor have a detrimental impact on the environment in which that equipment operates. In particular, an SDR environment should provide the following.

- Confidentiality, it guarantees that certain information is never disclosed to unauthorised entities.
- Integrity, it guarantees that code being transferred is never corrupted.
- Availability and Reliability, it guarantees the survivability of services despite denial of service attacks.
- Authentication, it enables a node to ensure the identity of the peer node it is communicating with.
- Non-repudiation, it ensures that the originator of a message cannot deny having sent the message.

A major part of setting the scope is to detail the relevant points of the current status of code in mobile phones, whether downloaded or otherwise, so before looking into the future it is useful to look at the current situation. Many of the reasons for the current status on downloaded code for mobiles will have to be examined in order to understand what is needed to facilitate the download of code for reconfiguration in an SDR environment.

13.2 Today's second-generation digital mobile phone

Figure 13.1 shows a typical GSM product, the T100 from Samsung. Not much downloadable code can be used on phones such as this. Most are limited to using downloaded ring tones and screen graphics. The reasons for this lie in the history of the mobile phone and the rules and regulations that have been required to get these products to market.

13.2.1 Regulatory issues

There are three main types of rules.

- *Regulatory.* For a mobile phone to carry the European Community's CE mark it must conform to the radio (basic RF parameters, interference, etc.) and safety regulations in force throughout the European Union (EU). Compliance is determined by parameters of the software driving the hardware as well as by the performance of the hardware itself.

Figure 13.1   A typical GSM mobile today, the Samsung T100

- *Type approval*[2]. The phone must comply with the full rules of the protocol stack and function correctly in the RF domain. Once more, this depends on the software executing correctly and on a very tight relationship between software and hardware.
- *Operator approval.* This ensures that the phone functions in the desired manner when connected to a particular operator's network. Network configurations are very complex and the network operator has many options to choose from (e.g. very few networks rely solely on equipment from a single supplier). Operators typically have to balance the trade-offs that occur between capacity (number of users on a system), services offered (which is related to the associated bit rate), coverage, user priority, system grade of service, etc. This has led to a basic design philosophy according to which software is downloaded once only – during production – and then made secure. Some software upgrades are available at service points, but, in general, the latter tend to replicate the facilities available during production.

13.2.2   Some security issues

It may come as a surprise to learn that almost every user feature in today's digital mobile phone has an impact in terms of security, and a large part of this is driven by software execution in the phone. Table 13.1 lists some major user features of today's mobile phones and their associated security considerations.

In addition, it is expected that a phone can operate at all times; indeed, there are regulatory obligations that require the ability to make emergency calls at any time (providing the battery has enough power and there is network coverage at the location

Table 13.1   Impact of user features on phone security

Phone feature	Expected security/security impact
Making voice and data calls	Nobody should be able to listen to a call by 'eavesdropping' or recording the radio signals in any way. This security is mainly achieved through software execution. The software execution is secure as it cannot be changed in the phone – it is a one-time only secure load
TFT LCD (thin film transistor driven liquid crystal display) screen with 4096 colours	The security impact is indirect: any downloaded code in the form of pictures and wallpapers must not interfere with the operation of the phone, which must work and display correct information
Sixteen polyphonic ring tones	The security impact is indirect: any downloaded code in the form of new ring tones or tunes must not interfere with the operation of the phone. There is also a minor safety issue: the phone must not cause physical side-effects, such as harm to the ears due to high volume
Voice dial, voice command	This feature provides (sometimes unwanted) security in that only one person – the main user – can use it on any one phone, as this facility has to be 'trained'
Voice memo	This generally uses the same technology as voice dial. The security considerations relate to how and where the data are to be stored. If the memos are stored on an external memory card, such as Smart Media or Compact Flash, then the user may want to protect these data in some way when it is removed from the phone
Personal organiser	The user data must be secure: it should not be possible for an unauthorised entity to read it out of the phone in any way, for example, via the air interface
Games	Games tend to be processor and memory intensive. They must not use so much memory and processor power that they prevent the phone from maintaining its connection to the network or from receiving calls

of the mobile phone). This must be reflected in the design of both the hardware and the software.

13.2.3   Basic architecture

In this section, we discuss the basic architecture of a mobile phone, as this will serve as a reference in subsequent sections on downloaded code security issues in future phones. Figure 13.2 provides an overview of a typical mobile phone software

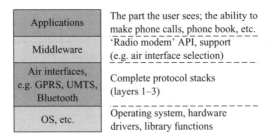

Applications	The part the user sees; the ability to make phone calls, phone book, etc.
Middleware	'Radio modem' API, support (e.g. air interface selection)
Air interfaces, e.g. GPRS, UMTS, Bluetooth	Complete protocol stacks (layers 1–3)
OS, etc.	Operating system, hardware drivers, library functions

Figure 13.2   Typical present-day mobile phone software architecture

architecture. The hardware is fixed and driven by the software (a point of discussion when looking at SDR technology), so that the performance and characteristics of the hardware and the product are defined by software. Today's assumption is that the software, and, in particular, the one related to the terminal's air interface capabilities, is supplied at the point of sale and does not change. This implies that there is no download of software code after purchase for the purpose of configuring software below the application layer (as mentioned above, it is already possible to download simple applications such as ringtones and screen graphics). The hardware has safety features (e.g. power supply control and RF power control) to make sure that the product behaves within preset limits.

The software may well reside in several components within the phone, as there will be at least two processors executing various parts of the code:

- a main microprocessor for executing radio modem software and user applications
- a digital signal processor (DSP) for executing baseband functions such as speech coding, channel coding and digital modulation.

This raises additional issues when designing the software security architecture if it is assumed that security threats can occur at run time and can affect any part of the terminal.

Additionally, if a Bluetooth device or a wireless local-area network (WLAN) device (e.g. one that conforms to the IEEE 802.11b standard) is included in the mobile equipment (and such equipment is beginning to appear on the market), this will probably introduce additional hardware and memory support, although from the overview software architecture point of view Figure 13.2 remains valid.

The physical memory for the phone may or may not be shared between the various processors, and may be a mixture of flash memory, static random-access memory (SRAM) and read-only memory (ROM). Also, it may or may not have hardware memory protection, which is another detail to look at in the overall security analysis.

Another point, which from a customer's viewpoint is very important and should therefore be considered when looking at mobile phone security architectures, is that many mobile phones are expected to run for up to a week without charging! This is achieved by fine-tuning the software and hardware architectures so that the phone operates in a highly power efficient manner, and it is also dependent on the air interface design. Complex security algorithms tend to use a lot of processor power and memory, if only for a short time, and this may have a significant influence on the life of the phone's battery if we select algorithms that are complex and run many times. One solution that reduces overall complexity is to partition the memory into several protected areas using a memory protection unit, so that each stratum runs in a single address space only. Unfortunately, there is only a limited amount of memory to achieve this, and memory is expensive.

This all sounds like bad news; however, solutions have been found that make current digital mobile phones very secure, very reliable and not too 'power hungry' (in comparison, e.g. with the average home PC, whose associated Internet protocols were not really designed with privacy in mind).

The current digital mobile phone has three basic security architectures. These are independent but all need to be present and correct for the phone to operate properly. The three basic architectures are as follows.

- *Software security*: this is achieved by virtue of the fact that the software cannot be changed by the user (except in the case of simple downloaded applications). Today, software security is tightly linked to hardware security as this also does not change.
- *Subscriber identity security*: this is typically taken care of by the subscriber identity module (SIM) card and the system itself together with the associated signalling protocols. A valid SIM card (together with valid terminal equipment identification) will allow a user access to the corresponding system. The SIM card itself can be protected through the use of a valid personal identification number (PIN), which, when activated, will require that it is entered every time the phone is switched on. The phone itself may have a user blocking code (independent of the SIM PIN) that would also, when activated, require input to operate the terminal.
- *Air interface security*: this ensures that someone who is deliberately searching for a radio signal cannot understand the radio signals that are sent and received by the phone. Moreover, if someone does try to listen with malicious intent, they cannot easily gather useful information, such as the subscriber identity. This is achieved by having a fairly complex air interface in which the digital data are encrypted. This is standard in GSM, where a robust time-varying encryption procedure makes 'listening in' impossible.

If one compares this security to that provided when logging on to the Internet or an Internet POP [3] mail server from a home PC one will realize that although the details can vary depending on the Internet service provider and software packages being used, it is likely that at least the user name and sometimes the password will be sent as plain text and, once logged in, the data will not normally be encrypted. At this point one can get into a major argument from the security point of view as to how easy it is to listen to a mobile phone or somebody's home phone, asymmetric digital subscriber line (ADSL) or cable connection, but that debate is not for this chapter. Suffice it to say that once the radio system has gone digital, with a few selected security mechanisms such as encryption on the air interface, then it really is very secure. However, no matter how complicated you make the security, someone who really wants to, will find a way around it.

It is apparent that currently downloadable code is limited to simple applications and cannot be used to configure software below the applications layer. Nevertheless, as phones and the applications that can be downloaded to run on them become more complex, an additional level of security is considered necessary to continue to ensure the integrity both of the applications and, as a consequence, of the phones on which they are running. Indeed, it is the integration of a (very secure) mobile phone technology with a (comparatively insecure) Internet technology that raises the question of how to make the resulting, integrated, system itself secure.

13.3   The protected application environment: mobile execution environment

MExE is an open specification, first created by ETSI and now maintained by 3GPP, that specifies a standardised environment for executing applications in enhanced second generation, for example, General Packet Radio Service (GPRS) and third-generation (e.g. UMTS) mobile terminals [3]. It has already been implemented on some of the current GPRS phones and certain aspects of it, that is, at least those relating to download and security, will probably be a standard feature in dual-mode GPRS/UMTS phones. The aim of MExE is to allow software to be downloaded to mobile phones, with a specific aim of downloading applications (as opposed to software for reconfiguration of software/firmware below the application layer) and providing a standardised interface between the application and the phone (or mobile telephony) air interface. The execution environment is shown in Figure 13.3 as a wrapper between the Java applets and .NET applications, and the rest of the phone. In particular, the aim here is to provide standard interfaces to phone features, so that

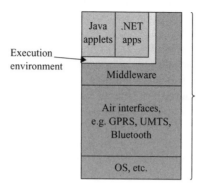

Figure 13.3   Location of the MExE in the software architecture of a mobile phone

downloaded applications can be run in a secure environment and in a standardised manner regardless of the phone on which they are being run.

Note that this is not the only software architecture approach that can be adopted. Moving the applications to a separate processor and operating system, such as Psion's EPOC operating system, can have the same effect as putting a wrapper between the applications and the phone air interfaces. Another solution is Qualcomm's BREW (Binary Runtime Environment for Wireless), which has a slightly different architecture but gives the same overall result [4].

In the MExE specification mobile devices are categorised based on their performance using classmarks. Four classmark types have been specified.

- Classmark 1 supports Wireless Application Protocol (WAP), used by all WAP browsers.
- Classmark 2 supports PersonalJava (a subset of the Java 2 Platform, Standard Edition), used for Java MIDlets[3] with no interface to phone functions.
- Classmark 3 supports J2ME Call Connected Limited Device Configuration (CLDC) and Mobile Information evice Profile (MIDP), which allows Java MIDlets to interface to the phone air interface for limited phone functions.
- Classmark 4 supports Common Language Infrastructure (CLI) based applications running on a broad range of connected devices.

Even in these environments the downloaded software will have various levels of trust, from untrusted (source unknown) to trusted by either the mobile phone manufacturer or the operator. The support framework for defining and testing trust is still being designed and discussed at the time of writing this article, but one key point will be independent testing of both the execution environment within the mobile phone and how the applications get 'trusted' status.

There are a number of important points to note about MExE as specified:

- a capability to make emergency calls is always present;
- the execution environment tends to augment the functions on a mobile phone (e.g. by addition of a WAP browser or games) rather than replace older phone software;
- applications running in the execution environment tend to have lower priority than voice or data calls using the air interface.

Although there is no reconfiguration of the air interfaces, with the march towards SDR, there has been some discussion about future MExE classmarks encompassing the air interface part of the software architecture. Current thoughts will be to manage the configurability in such a way that there is no chance of the air interface becoming 'contaminated' through the various threats mentioned earlier. The architecture would strive to make a clear demarcation/interface between the air interface software and application software.

13.4   Fixed multiple radio access technologies: third-generation phones

In the coming years, the digital mobile phone will have several new important technology features, the major one being the ability to work with multiple fixed air interfaces (reconfigurable air interfaces will be discussed in Section 13.5, on SDR), which will enable 'seamless' roaming, for example, between UMTS and GSM/GPRS. Indeed, some of today's GPRS phones are already equipped with a second air interface in the form of Bluetooth. Another feature will be the increased speed of the data link that is available via the air interface. If you have GPRS today then speeds of up to around 48 kbit/s are available, depending on the network and a number of other variables. Speeds higher than this are deemed possible and can be obtained from the appropriate system specification. With the arrival of UMTS, data rate of up to 384 kbit/s should be available in urban areas, with an upgrade route to much higher speeds under certain conditions over the next few years (e.g. using High Speed Downlink Packet Access – HSDPA). These improved air interface data rates enable improved services; for instance, fast Web browsing and quick download of medium-sized files.

To support these features the software, as well as the hardware processing power, will grow dramatically. The software architecture in Figure 13.2 is still valid, but the security aspects grow since one must consider all the possible download services that might be obtainable with a high-speed interface. Some of these features have already been considered, such as that on MExE, which will have enhanced capabilities once there is a high-speed air interface. To be secured then is the area shown inside the grey boxes in Figure 13.4, namely the core air interfaces and the basic applications for making and receiving calls.

With UMTS the subscriber identity module becomes a USIM (Universal SIM) – the basic technology is the same but there are some enhanced features. The security

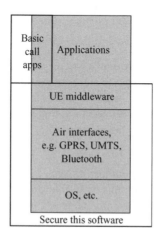

Figure 13.4   Basic areas (in grey boxes) of UMTS phone software architecture that must be secured

of the UMTS air interface is enhanced over that of GSM (though it should be noted that the GSM air interface is already extremely secure).

13.4.1 User data memory cards

These are not tied to any particular air interface. However, as they can carry quite a lot of data and as UMTS provides higher data rate services than does GSM/GPRS, it is quite likely that 3G phones will be developed to take, and be supplied with, standardised memory cards such as those already in widespread use for digital cameras, MP3 players, etc. Smart Media, Compact Flash and the Sony Memory Stick are widely used examples of such cards, and today can typically store up to 128 Mbyte of data, although 256 Mbyte versions are becoming available. This is a considerable amount of data – the total memory inside a UMTS phone will probably be in the region of 20–40 Mbyte, depending on the configuration and functionality. The memory cards in themselves do not present a threat as long as they are treated as pure data devices. At the low level they are usually configured to act like floppy disks.

There are, however, some security aspects to be considered of the data that is read from and written to these devices. First, data that are read from the card should only go to pre-assigned or controlled areas in the phone. If the data are phone-specific, such as phone books or screen pictures, that is fine; but if the data are something that runs a kind of executable code on the phone, such as a Java applet, then there is a real security threat: does the code contain a virus, for example? The approach on PCs today is to virus check the external storage media before copying files to the PC; the same approach should probably be used in the case of user data memory cards for phones.

Digital rights management (DRM) [5] is also an issue with such memory cards, and the same will apply to ring tones with the advent of the increased polyphonic capabilities that are appearing on phones. Security is required when transferring any copyright data, such as music, via any part of the mobile phone, as opposed to just playing the media. The problem here is that in a UMTS phone (or even in an earlier generation phone that has a WLAN interface), multimedia data such as music can be received, played and then sent on to many people very quickly. If the data contain marketing information, that is probably the desired effect, but if it contains copyright material that has been purchased on a limited licence, such as music for personal listening, wide distribution is not a desired result.

13.5 Software-defined radio: future generation phones [6, 7]

As mobile communication systems evolve, there will be a significant change in the architecture of mobile phones. It is expected that future-generation mobile phones will not just include all the required air interfaces, but will be able to reconfigure the air interface software and, to a more limited extent, the associated hardware. Figure 13.5 presents an architecture roadmap in which the end-to-end timescale is possibly between 10 and 15 years.

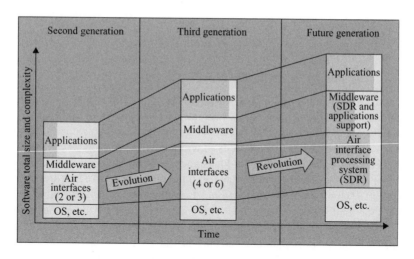

Figure 13.5   Roadmap of mobile phone architecture. Areas where download of code is permitted are highlighted

The first step in evolution will be the addition of more air interface variants (centre of Figure 13.5), probably using a technology similar to that to be used in 3G multimode phones. Further into the future (right of Figure 13.5), there will be a significant change in architecture from the fixed hardware and software architectures described in earlier sections to one in which the architecture can be reconfigured.

In Figure 13.5, the areas where and the extent to which download of software code is permitted have been highlighted. This roadmap is a view of what might happen; it can be challenged, and perhaps we should challenge it more. One of the reasons for not showing software downloads in the air interface area in the close future is that today this air interface is treated with care. Until recently the radio regulators (effectively the government), a type approval agency and the radio network operators were all involved (as outlined in Section 13.2) for every single phone, to make sure that it performed in an exact, predetermined, repeatable and approved manner, in particular following the appropriate air interface standard. The reason for all this effort is that the air interface is a shared medium. The consequence of this is that the software and hardware that are responsible for implementing the complex air interfaces have to be designed so that they do not interfere with each other or stop others from using their system. The advantages of this are also clear: a mobile phone can 'roam' and be used in other countries; it can be used on networks with equipment supplied by different manufacturers; and it will not stop other people using the network.

SDR adds to the roadmap those parts of the software architecture that can be downloaded (from any of the diverse sources listed later). In the more distant future (right of Figure 13.5), not only will software download be targeted, but also dynamic reconfiguration of the hardware will be addressed, so that the result is that best suited to the user requirements. The consequences of moving to this type of system are enormous. The security checking procedures in the downloaded software, and in the supporting software system and environment, will have to answer the following questions.

- Have I got the software I thought I wanted – does it perform the relevant function – does it reconfigure the part of my phone that I wanted?
- How can I be sure that the software is not malicious, it does not execute any unwanted functionality such as launching denial of service attacks or access my private and security information?
- Is it authenticated, authorised and accounted (paid for)? This is commonly abbreviated as AAA [8].
- Does it work in my hardware configuration?
- Does it maintain its regulatory status? (Can I still use it and not interfere with the network or other users?)
- If it turns out not to pass 'security' checks when downloaded, can the phone still perform as it was without further downloads?
- While upgrading the software or reconfiguring the software or hardware, can I still make calls?
- How do I (if I want to) make sure before performing the download that the downloaded code will be both usable and acceptable?
- How do I check for a 'good' source – is this source who it says it is – how can I be sure that I am billed properly by this source and my privacy is respected?
- Do I have any guarantees that the software cannot be altered or corrupted while being transferred from the source?
- Does the new software work with my existing software?
- Can I be certain that only I and other authorised entities can reconfigure my phone?

In a similar situation on a PC, generally during any software upgrade, there are periods when the PC is not usable, either while the machine is being restarted or while key parameters are being changed. Is this acceptable on mobile phones?

To answer these questions the software reconfiguration systems will have to run a complex database system that knows not only the limits of your hardware, but what you have paid for in terms of services and what your current software is. This raises the question of where this database should reside. It is probably too complex for the phone, which will, however, have to hold basic data. Using this database, the software reconfiguration system will have to produce new software that can be downloaded to a phone, and then reconfigure the phone's hardware to run in a totally new configuration.

A situation that could arise with SDR is that the service provider might decide to reconfigure the air interface on behalf of the user, in order to get better performance from the network (see the panel 'A scenario for SDR'). How will this take place? If the phone is being used at the time it might be disruptive. If the phone was in an 'always on' data connection downloading e-mail or performing a Web page update, etc., the reconfiguration might not be noticed, provided it happened quickly and with minimal interruption to the data. Real-time voice or video calls may not be so forgiving; so the required software download would have to take place in the background and prepare the hardware so that some sort of 'fast' changeover could occur.

Ensuring that the software is suitable to run on the phone, so that the phone remains legal and continues to function, is a daunting task in the light of today's technology (when dealing with small, mobile and portable, battery-powered devices that people rely on for basic communications).

13.5.1   Sources of downloaded software

In looking at the security of downloaded code, and in particular at the impacts for SDR, one aspect that has to be addressed first is how the code is delivered to the phone. As with an ordinary PC, there are a number of sources for downloading code. In the future mobile phone environment these may be the following.

- Over the terminal's primary air interface, for example, the GSM/GPRS or UMTS air interface, perhaps utilising a protocol such as WAP (though it is appreciated that WAP has been designed to facilitate the transfer of Web-based content to limited-capability terminals, i.e. mobile phones). This is very similar to the way you would download software via the Internet to your home PC (using the Hyper Text Transfer Protocol, HTTP). The user would be expected to have been involved in the set-up of this transfer, probably by first browsing a Web site to find the required data.
- Over the terminal's secondary air interface, perhaps using Bluetooth or WLAN technology. This is where things start to get complicated: if a phone has a Bluetooth interface today, then it has the ability to transfer files quickly from another Bluetooth-enabled device to, for example, a laptop computer. At this point the 'downloaded' file route/path is not certain, but this could be important information to have if you are using the software to reconfigure your phone.

A scenario for SDR

Consider the following simple scenario:

You are on your way from your office to a meeting and using a (W-CDMA) monile phone in an urban area in 'always on' mode to connect to your mail server, collect e-mails as they appear and automatically download them to your laptop. The meeting is in a medium-sized town, and when you arrive and settle in to the meeting, the network notices that you have stopped moving, but are still receiving calls on your e-mail. Your accounting preferences allow you to roam to other types of network.

In the location of your meeting there is a public WLAN that is known by your service provider and, as their voice network is very busy, they decide to download some new software to your phone, which is of the SDR type. Your data service then switches the e-mail download over to the WLAN, since this is not only cheaper for you but also allows the W-CDMA network to be reused for other services.

The possibilities for network management, roaming, getting the service you want on demand, etc., are extremely interesting, but only if the software security is good and makes it all work first time, every time.

- From a memory device that has been inserted into or connected to the phone in some way, for example, Smart Media, Compact Flash and Sony Memory Stick. Again, as these data could have been stored on the card from anywhere the information on the data source becomes important. In the future this could be a miniature hard disk or other mass storage device.

13.5.2 A code downloading scenario

One possible method for downloading code to a mobile phone is described in the sequence below and illustrated in Figure 13.6. This is just one of many permutations, and is probably not the preferred way, but it does show the complications that will have to be addressed in any overall security architecture, and the solutions that are eventually designed.

- *Stage 1*: A software upgrade is downloaded to a PC and the software copied to a memory card. The details do not matter for this example, but assume it is some sort of performance upgrade to the air interface for use when roaming.

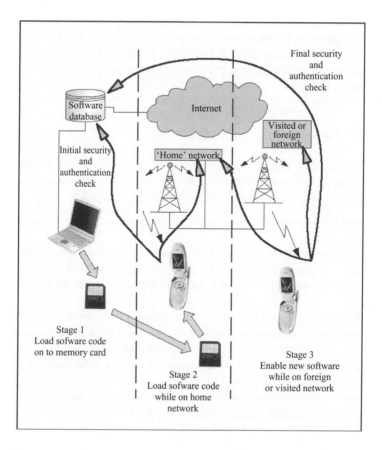

Figure 13.6   An example software download procedure

- *Stage 2*: While the phone is in use on the normal 'home' network, the memory card is inserted into the phone and the software possibly copied into the phone. Some sort of security check has to be made both with the original software database and with the home network. This check needs to show that the software is valid for the mobile phone and for the network it is currently being used on.
- *Stage 3*: Not until the user visits a foreign network is the new software actually executed, so the security and authentication should probably (this is a topic for research and discussion) be repeated on the visited network to make sure that the new software is secure for use. This may also mean accessing the home network to authenticate the mobile equipment, and accessing theoriginal software database for details of the software characteristics.

All of these imply that downloading 'authorised' code may not be simple. It is more likely to involve complex combinations of security and configuration management, and the scenarios presented here are ideal cases. What would happen if a user was downloading code while roaming and then could not find (for any reason) the original source server? Would the phone shut down? Customers would be unlikely to be happy about this; and the operators could lose revenue, so they would not be happy either. If the phone is to make its own decisions, how would this be implemented, and at what level should the phone's internal software security operate? How much intelligence would the phone need and how would this intelligence be implemented in a fail-safe way?

13.6   Conclusions

It would be difficult to derive a 'solution to security' that could be deemed all-encompassing at the outset, however from the arguments and discussions in this chapter sensible guidelines might be suggested, including the following.

1. If possible, separate the application downloadable software area from the air interface area.
2. Incorporate robust, less 'power hungry' air interface encryption techniques.
3. Adopt a 'hand-shaking' man–machine interface (question and answer type procedure verifying that download content and source are trusted by the user).
4. Networks should continue to adopt a policy of minimum transfer of non-encrypted personal information (e.g. International Mobile Subscriber Identity and equivalents thereof should not be transmitted unnecessarily).
5. Avoid the use of static ID where possible (e.g. use dynamically assigned IP addresses for Internet sessions).
6. Provide extremely secure minimum back-up (fixed) configuration software with good default security settings (to ensure that complete system lock-out cannot occur).
7. Provide (updatable) virus capture software.

8. Provide content encryption of stored data and all removable memory (this will also ensure integrity if mobile devices are lost/stolen).
9. Operate secure user authentication (PIN numbers/pass phrases, etc.).
10. Use methods for the authentication of the software source identity.
11. Employ cryptographic techniques for software integrity.
12. Develop an algorithmic and theoretical framework to ensure secrecy of the transmitted data and privacy of the user's preferences.
13. Encourage security consciousness. Provide clear/simple guidelines for using terminals in a secure manner.

There is much to research in this area and some of the scenarios currently proposed look large and complex for mobile devices. Everybody relies on mobile phones, and their reliability cannot be compromised (simply for the sake of technical flexibility), especially if prices to the consumer are likely to increase. A balance will, therefore, have to be struck.

Notes

1 The term UMTS is no longer in favour within 3GPP (3rd Generation Partnership Project), which is now developing the standard, as it is thought to be too Euro-centric (having been defined by ETSI). The preferred term is '3GPP System'. However, for the sake of clarity, and given that GSM/GPRS is also a '3GPP System', the term UMTS is used in this chapter as the default third-generation system name.
2 Formal type approval itself no longer exists, as it has been replaced under the Radio and Telecommunications Terminal Equipment (R&TTE) directive by a regime based on manufacturer's declaration of conformance. However, the term is still in widespread use to describe tested compliance to the requirements of the core specifications.
3 A reduced version of Java called Java 2 Microedition (J2ME) has been developed for mobile information devices. Applications conforming to this standard are called 'MIDlets'.

References

1 3GPP TS 21.905 v3.3.0: 'Vocabulary for 3GPP specifications'
2 3GPP TS 23.057 v3.4.0: 'Mobile station application execution environment functional description, Stage 2'
3 MExE Forum Web site: http://www.mexeforum.org and http://www.mobilemexe.com/whatis.asp?link=1
4 See http://www.qualcomm.com/brew/
5 WALLER, A. O. *et al.*: 'Securing the delivery of digital content over the Internet', *Electronic Communications and Engineering Journal*, 2002, **14**, (5), pp. 239–48

6   TUTTLEBEE, W. (Ed.): 'Software defined radio: enabling technologies' (Wiley, New York, 2002)

7   WWRF: 'Book of visions 2001'. See http://www.wirelessworld-research.org

8   HASAN, H. *et al.*: 'The design of an extended AAAC architecture'. IST Mobile & Wireless Telecommunications Summit 2002, 17–19 June 2002, Thessaloniki, Greece, pp. 36–40

Part IV

Application security

This part contains three papers addressing security issues for major applications of mobile systems. Both Chapter 14, *Secure mobile commerce* and Chapter 15, *Securing the delivery of digital content over the Internet*, are concerned with applications of mobile technology. M-commerce is an already widely discussed application domain, and one with enormous practical potential. However, for this to become a practical reality, the security issues considered in Chapter 14 need to be addressed.

Another major application domain for mobile technology is the delivery of digital content. Again, for this to become a commercial reality means that security issues associated with content protection need to be addressed – this is the topic of Chapter 15. The third chapter, Chapter 16, also relates to security issues for the delivery of proprietary content over mobile channels, and questions some of the prevailing views about how this might be achieved. This is certainly an area where a consensus about the 'correct' approach is yet to emerge.

Chapter 14

Secure mobile commerce[1]

Heiko Knospe and Scarlet Schwiderski-Grosche

M-commerce, or mobile commerce, is a major application domain for mobile devices, enabling users to perform commercial transactions wherever they go. However, these applications require a high level of security. In this chapter, we identify the special characteristics of m-commerce and reflect on some important security issues.

14.1 Introduction

The term e-commerce (electronic commerce) denotes business processes on the Internet, such as the buying and selling of goods. There is a distinction between B2B (business-to-business) and B2C (business-to-consumer) markets. In the first case, the business processes are carried out between businesses; in the latter case, they are carried out between businesses and end consumers. This general definition of e-commerce does not say anything about the kind of device that the end-user employs to gain access to the Internet. The underlying technology could be wireline (e.g. using a home PC as end-user device) or wireless (e.g. using a mobile phone as end-user device).

The term m-commerce (mobile commerce) is all about wireless e-commerce, that is, where mobile devices are used to do business on the Internet, either in the B2B or B2C market. As such, m-commerce is a subset of e-commerce.

With the omnipresent availability of mobile phones (and other mobile devices), m-commerce services have a promising future, especially in the B2C market. Future applications include buying over the phone, purchase and redemption of ticket and reward schemes, travel and weather information and writing contracts on the move. However, the success of m-commerce very much depends on the security of the underlying technologies. For example, today the chargeback rate for credit card transactions on the Internet is 15 per cent, versus 1 per cent for point-of-sales (POS) credit card transactions. Chargeback rates grow to 30 per cent when digital products are sold.

For m-commerce to take off, fraud rates have to be reduced to an acceptable level. As such, security can be regarded as an enabling factor for the success of m-commerce applications. In this chapter, we discuss three main areas of m-commerce that are relevant to security.

- *Device security* – mobile devices are more prone to theft and damage than fixed devices. Hence, the security of the mobile device is much more critical. Here, we highlight major issues of device security.
- *Network technology* – in m-commerce, all data are transmitted via a mobile telecommunication network. Here, we consider existing network and service technologies for 2G (2nd generation), 3G (3rd generation) and other wireless systems.
- *M-payment* (mobile payment) – doing business on the Internet requires the payment of goods and services. M-payment systems have different requirements and characteristics than e-payment systems. Here, we give an overview of current payment technology.

In the following section, we provide background information on m-commerce. Section 14.3 looks at the security of the mobile device, whereas Section 14.4 is devoted to network technologies for m-commerce. Different approaches to m-payment will be discussed in Section 14.5.

14.2   M-commerce and its security challenges

14.2.1   Definition of m-commerce

There are many definitions of the term m-commerce [1, 2]. Common to all definitions is that a terminal or mobile device is employed to communicate over a mobile telecommunication network. There are different views on the purpose of this communication. Some definitions restrict m-commerce to transactions involving a monetary value, whereas other definitions generalise the term to services that involve communication, information, transaction and entertainment. Summarising, we define m-commerce as using a mobile device for business transactions performed over a mobile telecommunication network, possibly involving the transfer of monetary values.

14.2.2   Mobile devices

M-commerce is not just about using mobile phones as end-user devices. The following list gives an overview of different kinds of mobile devices:

- laptop,
- sub-notebook/laptop or mini-notebook,
- personal digital assistant (PDA),
- smartphone – the smartphone combines mobile phone and PDA technology into one device,
- mobile phone,
- earpiece (as part of a Personal Area Network).

Each mobile device has certain characteristics that influence its usability, such as

- size and colour of display,
- input device, availability of keyboard and mouse,
- memory and CPU processing power,
- network connectivity, bandwidth capacity,
- supported operating systems (e.g. PalmOS, Microsoft Pocket PC),
- availability of internal smartcard reader (e.g. for a SIM card in mobile phones).

Depending on these factors, the services that the end-user can receive differ considerably. Moreover, depending on the network technology used for transmission, the bandwidth capacity varies and influences the kind of services that the end-user is able to receive.

In mobile phones, there exist three solutions to internal smartcards: single SIM, dual chip and dual slot. Single SIM is the solution that is most widely available today, where all confidential user information is stored on one smartcard. Dual chip means that there are two smartcards in the mobile phone, one for user authentication to the network operator and one for value-added services like m-payment or digital signature. A dual slot mobile phone has a SIM card and a card slot for a full-sized external smartcard. With this solution, different cards can be used one after the other. Moreover, the cards can also be used in traditional POS and ATM terminals.

14.2.3 Differences between m- and e-commerce

In comparison to e-commerce, m-commerce offers both advantages and disadvantages. The following list summarises the advantages of m-commerce [1].

- *Ubiquity* – the end-user device is mobile, that is, the user can access m-commerce applications in real time at any place.
- *Accessibility* – accessibility is related to ubiquity and means that the end-user is accessible anywhere at any time. Accessibility is probably the major advantage in comparison to e-commerce applications involving a wired end-user device.
- *Security* – depending on the specific end-user device, the device offers a certain level of inherent security. For example, the SIM card commonly employed in mobile phones is a smartcard that stores confidential user information, such as the user's secret authentication key. As such, the mobile phone can be regarded as a smartcard reader with a smartcard.
- *Localisation* – a network operator can localise registered users by using a positioning systems, such as GPS, or via GSM or UMTS network technology, and offer location-dependent services. Those services include local information services about hotels, restaurants, and amenities, travel information, emergency calls and mobile office facilities.
- *Convenience* – the size and weight of mobile devices and their ubiquity and accessibility makes them an ideal tool for performing personal tasks.
- *Personalisation* – mobile devices are usually not shared between users. This makes it possible to adjust a mobile device to the user's needs and wishes (starting with the mobile phone housing and ringtones). On the other hand, a mobile

operator can offer personalised services to its users, depending on specified user characteristics (e.g. a user may prefer Italian food) and the user's location (described earlier).

The following list summarises the main disadvantages of m-commerce.

- Mobile devices offer limited capabilities (such as limited display, see Section 14.2.2). Between mobile devices these capabilities vary so much that end-user services will need to be customised accordingly.
- The heterogeneity of devices, operating systems and network technologies is a challenge for a uniform end-user platform. For this reason, standardisation bodies consisting of telecommunication companies, device manufacturers and value-added service providers integrate their work (see Section 14.5.5). For example, many current mobile devices implement an IP stack to provide standard network connectivity. At the application level, the Java 2 Micro Edition (J2ME) offers a standardised application platform for heterogeneous devices.
- Mobile devices are more prone to theft and destruction. According to a government report, more than 700 000 mobile phones are stolen in the United Kingdom each year [3]. Since mobile phones are highly personalised and contain confidential user information, they need to be protected according to the highest security standards.
- The communication over the air interface between mobile device and network introduces additional security threats (e.g. eavesdropping, see Section 14.4).

14.2.4   Security challenges

As mentioned earlier, m-commerce is not possible without a secure environment, especially for those transactions involving monetary value. Depending on the point of views of the different participants in an m-commerce scenario, there are different security challenges. These security challenges relate to the following.

- The mobile device – confidential user data on the mobile device as well as the device itself should be protected from unauthorised use. The security mechanisms employed here include user authentication (e.g. PIN or password authentication), secure storage of confidential data (e.g. SIM card in mobile phones) and security of the operating system. The different types of mobile devices and their security characteristics will be discussed in Section 14.3.
- The radio interface – access to a telecommunication network requires the protection of transmitted data in terms of confidentiality, integrity and authenticity. In particular, the user's personal data should be protected from eavesdropping. Different security mechanisms for different mobile network technologies (i.e. in 2G, 3G and other systems) will be presented in Section 14.4.
- The network operator infrastructure – security mechanisms for the end-user often terminate in the access network. This raises questions regarding the security of the user's data within and beyond the access network. Moreover, the user receives certain services for which he/she has to pay. This often involves the network operator and he/she will want to be assured about correct charging and billing.

- The kind of m-commerce application – m-commerce applications, especially those involving payment, need to be secured to assure customers, merchants and network operators. For example, in a payment scenario both sides will want to authenticate each other before committing to a payment. Also, the customer will want assurance about the delivery of goods or services. In addition to the authenticity, confidentiality and integrity of sent payment information, non-repudiation is important. Section 14.5 will review security mechanisms for m-payment.

14.3   Security of the mobile device

The security of the mobile device is mainly influenced by its operating system (OS). The following list summarises the most relevant current OS for mobile devices:

- Symbian OS on smartphones [4],
- Palm OS on PDAs [5],
- Microsoft Pocket PC and Windows CE OS on PDAs [6],
- Microsoft Windows OS (98, ME, 2000, XP) on (sub-)laptops,
- Linux OS on PDAs and (sub-)laptops [7, 8].

In the following, we discuss different aspects of the security of mobile devices.

14.3.1   User authentication and authorisation

Mobile telecommunication networks (GSM, UMTS) require user authentication with a smartcard (SIM and USIM, respectively) and a personal PIN; smartphones usually adopt this mechanism to protect general access to the phone. Most PDA devices (e.g. Palm or Pocket PC PDAs) do not require mandatory authentication but the user can optionally configure an access code. Brute force or other attacks on this authentication mechanism cannot be excluded, which underlines the need for additional data security. Mobile devices may also be a target for unauthorised synchronisations, even over the air. Authorisation mechanisms for mobile devices are very limited: smartphones and PDA devices are personalised and an 'all or nothing' policy regarding the user permissions applies.

User authentication and authorisation are part of current desktop OS, which are also used for (sub-)laptops. But, these mechanisms are sometimes switched off for user comfort. Furthermore, an attacker can circumvent the authentication procedure with special boot media. BIOS and boot passwords or encrypted file containers or disks can then provide additional security.

14.3.2   Data security

The mobile device often contains important and confidential documents. Although this imposes important risks, most mobile OS do not provide native data encryption or integrity protection capabilities. However, there exists additional cryptographic software from third party vendors offering strong file encryption and data integrity protection. It is crucial that the protection does not only cover special file systems but also Personal Information Management (PIM) data (e.g. calendar information) and mail messages [9].

The storage capacity of many mobile devices can be extended by removable memory media, for example, Compact Flash (CF) cards, Multimedia cards (MMC) or Secure Digital (SD) cards. Only the latter incorporates security functions (mainly for Digital Rights Management) and data protection mechanisms should hence be extended to these removable devices.

14.3.3   Secure networking

The different network security technologies are discussed in the next chapter. Here we briefly discuss their availability on the different mobile devices.

Smartphones support the generic security mechanisms of the respective mobile technology (GSM, GPRS, UMTS). Similarly, devices with WLAN or Bluetooth radio support their native security mechanisms (which may have specific weaknesses as discussed in Section 14.4). It should be noted that a mobile device supporting GPRS, UMTS or WLAN can be permanently connected to the Internet and hence exposed to all kinds of network attacks. A personal firewall then protects the device but the network traffic may still be at the user's expense.

Secure IP networking with IPsec is available from third-party vendors for mobile OS, whereas built-in and user-friendly support for IPsec is still quite rare. SSL/TLS is a very popular transport layer security technology and most mobile browsers support it. But sometimes certificate management functionalities are missing (e.g. for the Pocket PC 2002 OS), so that the authenticity of a particular secure web server cannot be checked.

14.3.4   Software security

Additional software from external sources can pose another security risk since it may contain viruses, trojan horses or do any kind of harm.

Mobile phones from the 1990s did not provide execution environments for external software, but this has changed with current smartphones [37]. J2ME or Microsoft's .NET applications can be installed but the downloadable code runs in a special environment where access is restricted to specific areas. Future generations of smartphones may have extended access to core middleware or even radio processing functionalities. Then, additional security mechanisms are needed.

PDAs and, in particular, (sub-)laptops offer more capabilities for downloadable software or even allow exchanging of OS components. This increases the risks of harmful software or drivers. Digitally signed software can then provide the assurance that a piece of code comes from a trusted source. Additionally, an anti-virus software (e.g. [9]), which scans the mobile device for harmful content, should be installed and regularly updated.

14.3.5   Backup

The backup and restore capabilities of a personal device contribute to the system's availability. Mobile devices are particularly exposed to damage or theft and important data or information may be lost if the data can not be restored from a backup device. Limited mobile devices can backup their data to a PC via the synchronisation process (e.g. with a SyncML compliant protocol [11]), which may result in new security threats:

- synchronisation data is stored unencrypted on the PC (but the files can, e.g. be put on an encrypted file system);
- synchronisation data can become corrupted or get lost on the PC (hence the PC should be incorporated with a backup strategy);
- personal data is copied to other devices by unauthorised synchronisations (the mobile device should require authentication before synchronisation).

14.4   Security of the radio interface

In this section, we give an overview of the technologies that are relevant to secure m-commerce transactions. We focus on those network and service technologies that are specific to mobile and limited devices. The security architecture of current and potential future mobile systems has been studied in the IST SHAMAN project.[2]

14.4.1   Security of network technologies

We first discuss the security of network technologies used for mobile commerce.

14.4.1.1   Global System for Mobile Communication (GSM)

GSM is the current European standard for mobile communications. Since GSM handsets are popular and widespread, they have to be considered as the major device for mobile commerce at the moment. In the first years of GSM (beginning of the 1990s), the devices were very limited with respect to their capabilities other than telephony. Dial-in data sessions over circuit switched connections were possible but relatively slow (9.6 kbit/s) and required a separate device (computer), which reduces mobility. As the GSM core network was extended with more and more data service elements, the cellular phones also became more powerful. A number of data services were established.

- Short Message Service (SMS) allows the exchange of 160 character short messages over the signalling channel.
- Wireless Application Protocol (WAP) permits access to Internet content and applications formatted in Wireless Mark-up Language (WML). At first, WAP was based on circuit switched connections.
- High Speed Circuit Switched Data (HSCSD) provides higher data rates by channel bundling.
- General Packet Radio Service (GPRS) extends GSM with packet oriented services. With GPRS, the mobile node can stay 'always on' without blocking a connection timeslot with the base station. GPRS can also be used as a bearer service for WAP and SMS.

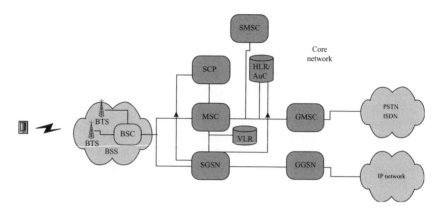

Figure 14.1   GSM architecture

The basic architecture of GSM including GPRS, intelligent network (IN) and SMS components is depicted in Figure 14.1.

The mobile station communicates over the wireless interface with a base transceiver station (BTS) which is part of a base station subsystem (BSS). The base station controller (BSC) is connected with a mobile switching centre (MSC) and a Serving GPRS Support Node (SGSN). The latter two are the central switching components for circuit and packet switched data.

When a customer subscribes, the GSM home network assigns the mobile station a unique identifier, the international mobile subscriber identity (IMSI), and an authentication key K_i. The IMSI and the secret authentication key K_i of the mobile station (MS) are stored in the subscriber identity module (SIM), which is assumed to be tamper proof. On the network side, the IMSI, K_i and other information are stored in the home location register (HLR) and the authentication centre (AC).

GSM provides the following security features for the link between the mobile station and the network, as described in References 12 and 13:

- IMSI confidentiality,
- IMSI authentication,
- user data confidentiality on physical connections,
- connectionless user data confidentiality,
- signalling information element confidentiality.

GSM provides the basic security mechanisms for m-commerce transactions. In particular, the *mobile customer authenticates* towards the network with a challenge/response protocol based on the secret key K_i. Furthermore, the wireless link between the mobile station and the BTS is encrypted with a symmetric key which is also derived from K_i. The secret key K_i is never sent over the network. But there are

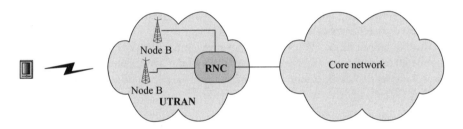

Figure 14.2   UTRAN system

weaknesses ([13]): since the network is not authenticated, a false base station can perform a 'man-in-the-middle' attack. The base station can suppress IMSI confidentiality and encryption and this is not even visible to the mobile station.

14.4.1.2   Universal mobile telecommunication system (UMTS)

UMTS is the next generation (3G) mobile telecommunication system and a further development of GSM. The major difference to GSM is the radio network (UTRAN) with its transition to the Wideband Code Division Multiple Access (WCDMA) radio technology. Two new network components, Radio Network Controller (RNC) and Node B, are introduced in UTRAN. Furthermore, the security protocols have been modified and now the RNC is responsible for (de)ciphering.

The main components of the GSM/GPRS core network with MSC, SGSN, etc., can be reused or evolved to UMTS. This is depicted in Figure 14.2.

In general, the security architecture of UMTS is carefully designed to fix the security weaknesses of GSM [13, 14]. As is described above, the main problems of GSM originate from two facts: authentication is one way (the mobile station does not authenticate the network), and encryption is optional. In UMTS, authentication is mutual, and encryption is mandatory unless the mobile station and the network agree on an unciphered connection. In addition, integrity protection is always mandatory and protects against replay or modification of signalling messages. Sequence numbers in authentication vectors protect against reuse of authentication vectors by network impersonators. UMTS introduces new cipher algorithms and longer encryption keys. Thus, UMTS does not seem to have any obvious security holes. UMTS security is discussed in detail in Reference 15, Chapter 6.

14.4.1.3   Wireless local area network (WLAN)

The IEEE standard 802.11 specifies families of WLAN that operate in the unlicensed 2.4 and 5 GHz band. The standards specify the physical layer (PHY) and the medium access control (MAC) layer. For the network layer and above, WLAN employs a classical IP stack. A number of commercial products (even for PDAs) are available, and IEEE 802.11b, offering 11 Mbit/s raw bandwidth, is popular and widespread in 2002/2003. When operated in the infrastructure mode, the mobile station attaches to an access point (AP), which provides connectivity to fixed net IP networks (e.g. the Internet) or to other mobile stations.

In the default mode, WLAN does not provide any security. This means that a mobile attacker can eavesdrop and manipulate all the wireless traffic with standard tools. In order to provide a certain level of security, the IEEE defined wired equivalent privacy (WEP), which was designed to provide:

- authentication to protect the association to an AP,
- integrity protection of MAC frames,
- confidentiality of MAC frames.

The protection is based on secret WEP keys of either 40 or 104 bits. Concatenated with a clear text initialisation vector, the secret key serves as input for the RC4 stream cipher. But, it has been shown [16] that authentication and integrity protection is completely insecure and encryption is at least partly insecure. It suffices for an attacker to intercept a single successful authentication exchange between a mobile station and the AP to be able to authenticate without knowing the secret keys. Furthermore, since a CRC checksum is used for integrity protection, an attacker can modify the data and adapt the checksum accordingly. For example, if the position of commercially sensitive information (e.g. an amount) within a datagram is known, the corresponding bits can be EXORed with any value. With a large number of intercepted frames, the WEP keys can even be recovered, breaking the encryption. Furthermore, since the WEP keys are network keys, preserving their secrecy is difficult for private networks and impossible for public WLAN hotspots.

As a reaction to WEP's cryptographic weaknesses, the Wi-Fi alliance has announced Wi-Fi Protected Access (WPA) [10], which shall be available as a software update for existing WLAN equipment in 2003. WPA shall provide an improved encryption and integrity protection protocol (TKIP, Temporal Key Integrity Protocol) and user authentication via the IEEE standard 802.1X [17] and the Extensible Authentication Protocol (EAP) [18]. WPA may be regarded as an intermediate step towards the new security standard IEEE 802.11i, which is expected to be published around 2003/2004.

In recent work of the IEEE Task Group on Security (TgI), the new security standard IEEE 802.1X has been adopted. 802.1X defines mechanisms for access control and employs the EAP framework for a variety of authentication protocols, for example, certificate based TLS. The known deficiencies of plain EAP can be remedied with Protected EAP (PEAP) [19] or EAP Transport Layer Security (EAP-TTLS) [20] (see Chapter 8, Section 8.4.2 for public key based EAP methods). These protocols negotiate (after server authentication with certificates) a protected channel which is then used for authentication and key agreement with EAP methods or attribute-value-pairs.

Another approach is to employ virtual private network (VPN) technologies and in particular IPsec in order to establish *network layer* security. The IPsec protocol (or more specifically the ESP Tunnel protocol) is an Internet standard [21] for the protection of IP packets between two nodes (e.g. a mobile station and a security gateway). This architecture is depicted in Figure 14.3. Note that link layer specific information (e.g. MAC addresses) is still unprotected.

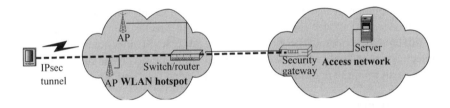

Figure 14.3   WLAN security with IPsec

14.4.1.4   Bluetooth

Bluetooth is a wireless technology developed by the Bluetooth Special Interest group[3] and is mainly aiming at ad hoc piconets and connections to peripheral devices. Bluetooth is also operating in the unlicensed 2.4 GHz band and can be considered as a de facto standard. The Bluetooth specification defines a complete OSI stack, so, unlike WLAN, it is not restricted to IP connectivity. Although raw bandwidth is limited to 1 Mbit/s, the Bluetooth technology will probably often be used in the future to connect devices in the personal environment, which makes it relevant for m-commerce.

Bluetooth specifies three security modes, including 'no security'. Bluetooth provides link layer security with a challenge–response protocol for authentication and a stream cipher encryption of user and signalling data [22]. When the connecting devices do not share a key in advance, they have to establish initialisation and link keys in a pairing procedure. This is based on a PIN, which must be entered into both devices (or imported from some application).

Bluetooth can currently be considered secure for small ad hoc networks, provided the pairing happens in a safe environment and the PIN is strong enough. The existing attacks are still theoretical in nature. However, privacy requirements may not be met since the Bluetooth device address (unique MAC address) allows the tracing of personal devices and hence their owner.

14.4.2   Transport layer security

The technologies mentioned on the previous sections provide security for the wireless link between mobile customer and access network or access device. If the access network is considered secure and the m-commerce transaction is completely handled within the access network, this may be sufficient. But often, an m-commerce transaction involves parties outside the access network (merchant, payment service provider, etc.). In this section, we discuss end-to-end security for mobile devices. This protects applications which communicate over an IP port. More information, in particular on certificates and PKI, can be found in Reference 23, Chapter 2.

14.4.2.1   SSL/TLS

The SSL/TLS (Internet Secure Socket Layer) [24] protocol is by far the most widely used Internet security protocol. Its main application is the HTTPS protocol (HTTP over SSL), but it may also be used as a standalone protocol. SSL requires a bidirectional byte stream service (i.e. TCP). SUN has implemented a client side version of SSL for limited devices, called KSSL (kilobyte SSL). KSSL does not offer client side authentication and only implements certain commonly used cipher suites, but it has a very small footprint and runs on small devices using the J2ME platform.

14.4.2.2 WTLS

The WAP forum has standardised a transport layer security protocol (WTLS) as part of the WAP 1 stack [25]. WTLS provides transport security between a WAP device (e.g. a mobile phone) and a WAP gateway, which performs the protocol transformation to SSL/TLS. Hence, no real end-to-end security is provided and the WAP gateway needs to be trusted. Note that the WAP Forum now proposes a WAP 2 stack [26], which is a classical TCP/IP stack on a wireless bearer medium. This permits end-to-end SSL/TLS sessions.

14.4.3 Service security

Here, we discuss the security of network services which can be used for m-commerce transactions.

14.4.3.1 Intelligent network

With the introduction of the IN technology to GSM networks, additional services could be realised. The IN architecture for GSM (called CAMEL, Customised Application for Mobile Enhanced network Logic) [27] was adapted from the fixed network standard ETSI Core INAP, and was originally designed for circuit switched calls (CAMEL phase 1 and 2). The IN is triggered during call handling at the MSC if the HLR entry indicates subscription to an IN service. With CAMEL phase 3 [28], the IN services can also be applied to SMS and to packet data services. The IN component Service Control Point (SCP) controls the call or data service via the CAMEL Application Part (CAP) protocol that runs on top of the SS7 (Signalling System Number 7) protocol.

Prominent examples of IN services are the transformation of dialled numbers (e.g. to realise Virtual Private Nets) and prepaid services. The IN platform provides some flexibility for the generation of m-commerce services. IN handling can, for example, be triggered by a specific called party, a calling party, an USSD string (requiring CAMEL phase 2), mobile originating SMS (requiring CAMEL phase 3) or mobile terminating SMS (requiring CAMEL phase 4). The security of an IN service depends on the underlying GSM or UMTS network security (see preceding text) and on the specific characteristics of the service application.

14.4.3.2 Parlay/open service access (OSA)

Parlay/OSA is an initiative of the industry (Parlay group), ETSI and 3GPP and aims at introducing standard interfaces to network services [29]. The IN platform and their SS7 based protocols like INAP and CAP are relatively complex and generation of services is reserved to operators and manufactures. Now Parlay offers standard application programming interfaces which allows service provisioning on IT platforms using standard middleware (e.g. CORBA). The Parlay/OSA framework then provides gateway functionality between applications and Service Capability Features (SCFs) of the IN. M-commerce applications can then access core network functionality, for example, inquire status and location of a mobile user, send messages or place calls. Parlay/OSA applications are portable among networks which is usually not possible with IN services.

Security is an important issue, since Parlay/OSA potentially opens the core network to intruders. Parlay/OSA specifies authentication and encryption on the application layer [30]. But the security also depends on the underlying network architecture, e.g. firewalls and strict policies should protect core network components.

14.4.3.3 Short message service

SMS is a very popular data service for GSM networks. Although SMS messages are limited to 160 characters, a considerable number of m-commerce scenarios are based on this service. The sender and receiver of an SMS is identified by its IMSI, which an attacker cannot forge without breaking the GSM/UMTS security mechanisms (e.g. by cloning a SIM card). Hence, SMS messages can be used for authentication (at least towards the network). Furthermore, SMS data is transmitted in the GSM (UMTS) signalling plane, which ensures the confidentiality of messages. However, the protection ends in the GSM or UMTS network, there is no end-to-end security, and the network operator and its infrastructure (e.g. SMSC, Short Message Service Centre) must be trusted (when no other security mechanisms are applied to the SMS message, refer to Section 14.4.3.6).

14.4.3.4 Multimedia message service (MMS)

MMS can be regarded as the next step in mobile messaging. With MMS capable handsets, the user can send and receive text messages, pictures, audio and short video streams. The messages are relayed by the network operator's Multimedia Message Service Centre (MMSC). Multimedia messaging uses the WAP protocol stack and GPRS or circuit switched data as bearer medium. The security mechanisms of the wireless bearer medium apply, but within the operator's network, the messages are unprotected.

14.4.3.5 Unstructured supplementary service data (USSD)

The GSM Unstructured Supplementary Service Data (USSD) service allows data communication between a mobile station and either the HLR, VLR, MSC or SCP in a way transparent to the other network entities. Unlike the asynchronous SMS service, an USSD request opens a session which may induce other network operations or an USSD response before releasing the connection. Mobile originated USSD may be thought as a trigger for a network operation. USSD works with any mobile phone since the coded commands are entered in the same way as a phone number (e.g. *123#1234567890#).

With USSD, roaming can be offered for prepaid GSM customers before IN services (CAMEL) are implemented in a network. Another USSD application (requiring CAMEL phase 2) is replenishing a prepaid account by incorporating the voucher number in an USSD string. In principle, any transaction, for example, a payment operation, could be triggered by USSD data.

USSD possesses no separate security properties; instead it relies on the GSM/UMTS signalling plane security mechanisms.

14.4.3.6   SIM/USIM application toolkit

The SIM and USIM application toolkits (SAT and USAT, respectively) allow operators and other providers to create applications which reside in the SIM/USIM. These applications can, for example, send, receive and interpret SMS or USSD strings. Currently, there exist banking applications using SAT. In Reference 31, a format for secured packets is defined. This permits the sending application (e.g. the one residing on the SIM card) to send protected messages to the receiving application (which, e.g. runs at a payment service operator). The required security mechanisms are:

- authentication,
- message integrity,
- replay detection and sequence integrity,
- proof of receipt and proof of execution,
- message confidentiality,
- indication of the security mechanisms used.

However, it depends on the SIM applications whether these security mechanisms are implemented and whether their cryptographic strength is sufficient.

14.5   Security of m-commerce

This section gives an introduction to payment mechanisms for m-commerce. First, we will discuss the heterogeneity of payment system solutions, before categorising e- and m-payment systems. Section 14.5.4 states examples of m-payment systems.

14.5.1   Background on payment systems

E-payment systems provide means for payment of goods or services over the Internet. In contrast to conventional payment systems, the customer sends all payment-related data to the merchant over the Internet; no further external interaction between customer and merchant is required (e.g. sending an invoice by mail or confirmation by fax). To date, there exist more than 100 different e-payment systems [32–34].

14.5.1.1   Distinctive features of payment systems

There exist a large number of payment systems for e- and m-payments. The following distinctive features of payment systems motivate this diversity:

- time of payment,
- payment amount,

- anonymity issues,
- security requirements,
- online or offline validation.

Time of payment denotes the relation between the initiation of a payment transaction and the actual payment. In prepaid payment systems, the customer's account is debited before the payment and the amount is stored, for example, on smartcards, in specific customer accounts or as electronic cash. In pay-now payment systems, the customer's account is debited at the time of payment (e.g. ATM card or debit card with PIN) and in post-payment systems, payment can be regarded as a 'payment promise' where the merchant's account is credited before the customer's account is debited (e.g. credit card systems).

The payment amount has an influence on the design of e-payment protocols. For example, payments in the order of 1 € are only viable if the incurred computational and communications overhead is kept small. Accordingly, there is a distinction between

- micropayments (up to about 1 €),
- small payments (about 1–10 €),
- macropayments (more than about 10 €).

E-payment systems often originate with conventional payment systems. As such, cash-like payment systems should provide *anonymity* to the customer. There are different degrees of anonymity: complete anonymity means that the customer remains anonymous to the merchant and the bank. However, in many payment systems, only partial (or no) anonymity can be provided.

The *security requirements* of electronic payment systems differ. Generally, integrity, authentication, authorisation, confidentiality, availability and reliability issues need to be considered, depending on the specific requirements of an e-payment system.

Offline payment validation means that no third party (e.g. a bank or credit card institution) is involved during the payment procedure, whereas online payment validation involves some kind of background payment server as a trusted third party. The latter causes an additional communication overhead, but reduces certain risks, for example, double spending.

The earlier discussion summarises some distinctive features of payment systems. There are other issues such as

- overhead imposed on customers and merchants (e.g. installation of software, registration);
- performance (e.g. response times);
- cost incurred per payment transaction;
- fulfilment of the ACID (atomicity, consistency, isolation, durability) principle for payment transactions (i.e. transactions have to be executed all or nothing, leaving the system in a consistent state, and their effect should be durable);
- national or international deployment.

This list of distinctive features gives an idea of the complexity and variety of payment systems.

14.5.2   Categorisation of e-payment systems

E-payment systems are typically modelled on conventional payment systems. As such, there are the following categories [34]:

- direct cash,
- cheque,
- credit card,
- bank transfer,
- debit advice.

Direct cash. In direct cash-like payment systems, the customer withdraws money from the issuer, that is, the third party interacting with the customer (e.g. a bank or service provider), and hands payment tokens for the payment amount to the merchant. The merchant deposits the payment tokens with its acquirer, that is, the third party interacting with the merchant (e.g. a bank or service provider). The issuer and acquirer then settle the payment. This payment scenario is sketched in Figure 14.4.

Since digital cash is trivial to copy, direct cash-like payment systems involve either tamper proof hardware (i.e. smartcards) or online validation by the issuer (i.e. double spending test).

Cheque. In this scenario, the customer hands a cheque (a payment authorisation) to the merchant. The cheque is presented to the acquirer who redeems it from the issuer. Cheque-like payments are sketched in Figure 14.5.

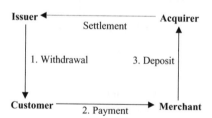

Figure 14.4   Direct cash-like payment system

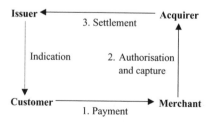

Figure 14.5   Cheque-like payment system

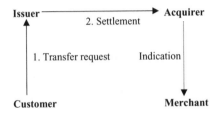

Figure 14.6   Bank transfer payment system

Credit card. In terms of the information flow, credit card based payment systems are similar to cheque-like payment systems, with the difference that credit card based payment systems use the existing credit card infrastructure for settling the payment.

Bank transfer. The bank transfer model is sketched in Figure 14.6. Here, the customer instructs the issuer to transfer money to the merchant's account at the acquirer. The merchant is notified of the incoming payment.

Debit advice. This model describes the opposite case to the bank transfer model. The merchant instructs the acquirer to charge the account at the issuer. The customer is notified of the outgoing payment.

14.5.3   Categorisation of m-payment systems

Most e-payment systems are not suitable for use in a mobile context, that is, using a mobile device and communicating over a mobile telecommunication network. This is due to the special characteristics of mobile devices and mobile telecommunications (see Sections 14.2.2 and 14.2.3). In the following, we categorise m-payment systems according to the whereabouts of the customer's money.

(1)   Software electronic coins – electronic money stored on the mobile device in file format.
(2)   Hardware electronic coins – electronic money stored on the mobile device on a smartcard.
(3)   Background account – electronic money stored in a remote account at a trusted third party.

Software electronic coins. In this case, monetary value is stored on the mobile device and the customer has full control of his/her money wherever he/she goes and whatever he/she does. An electronic coin is represented as a file containing, among other information, a value, a serial number, a validity period, and the signature of the issuing bank. Since software electronic coins are easy to copy, the validity of an electronic coin depends on its uniqueness in terms of its serial number. The customer transfers electronic coins to the merchant, who forwards them to the issuing bank for the 'double spending test'. In this test, it is checked whether the electronic coin has been spent beforehand. If yes, it is rejected. Otherwise, its serial number is entered into the double spending database and the money is credited to the merchant's account.[4] The generation and storage of electronic coins is an orthogonal problem. Due to the limitations of mobile devices, electronic coins may have to be generated and stored externally, until they are downloaded onto the mobile device.

Hardware electronic coins. In this case, monetary value is stored on a secure hardware token, typically a smartcard, in the mobile device. The presentation of electronic money is not important, as long as it is stored securely on the smartcard. Electronic money could be represented as a simple numeric counter. In order to get to the money, the customer's smartcard and the merchant's payment server authenticate each other and a secure channel is set up between them. Then, electronic money can be transferred from one to the other. This approach is quite attractive because smartcards provide an additional level of mobility. That means the payment smartcard can also be used in POS transactions.

Background account. Here, the money is stored remotely in an account at a trusted third party. Depending on the specific payment system, the account could be a credit card account, a bank account, or an account held at the network operator. Common to all scenarios is that, on receipt of an invoice, the customer sends an authentication and authorisation message to the merchant that allows the trusted third party (that holds the account) to identify the customer and to verify the payment authorisation. The accounts can then be settled. There are numerous payment systems that fall into this category. The differences are regarding the nature of the trusted third party and the procedure to send authentication and authorisation data. For example, in some cases this data is sent in the clear (e.g. a credit card authorisation) not providing any security against eavesdropping, and in some cases this information is encrypted and digitally signed, providing anonymity to the customer (e.g. SET – Secure Electronic Transactions).

14.5.4   Examples of m-payment systems

In the previous section, we gave a categorisation of m-payment systems. In this section, we survey existing payment solutions for m-payment, as well as e-payment solutions that are suitable for mobile use.

Software electronic coins. There are several e-payment systems that are based on electronic coins. As for cash, one main advantage is that the customer can potentially remain completely anonymous to the merchant as well as to the bank, while staying in full control of the money. E-commerce solutions of this type include eCash, NetCash, and MilliCent. Due to the storage and processing constraints of mobile devices, an adaptation of the software is necessary. Moreover, storing electronic coins on the mobile device is problematic. One option is to run the full-fledged payment system on a home PC and download electronic coins when needed.

Hardware electronic coins. There are various e-payment systems that implement an e-purse, that is, electronic cash on a smartcard, for example, GeldKarte and Mondex. In both cases, electronic money is stored on the card and can be transferred directly from the customer to the merchant. Shadow accounts are held at the bank to log transactions. Currently, these payment schemes are being adapted for m-commerce, where the GeldKarte and Mondex cards can be used in dual slot mobile phones.

One smartcard based system for m-commerce is already in operation; the system by BarclayCard and Cellnet uses a dual-slot Motorola mobile phone for payments of up to £50.

Background account. Depending on the type of trusted third party, there are various different approaches to a background account solution.

If the background account is held at a network operator, the charged amount is transferred to the existing billing solution and included in the customer bill. Customers pay their bills using traditional systems, such as direct debit, cheque or cash. Examples in this category are the M-Pay Bill service from Vodafone and Mobilepay by Sonera. These systems work for micropayments only; accepting higher payments would imply that mobile operators become subject to a host of banking regulations.

If the background account is held at a credit card institution, the payment mechanism is all about secure transmission of credit card data to the credit card company. The first possibility uses a dual-slot mobile phone, whereas the second possibility employs a dual chip mobile phone. The dual-slot solution (e.g. ItiAchat by Mastercard, Oberthur smartcards, France Télécom, Europay, and Motorola) has the advantage that it works with 'normal' credit cards that can also be used in traditional POS terminals. On the other hand, dual-slot mobile phones are quite bulky. This solution is favoured by credit card companies and banks, since it allows them to stay in control of the payment functionality. The dual chip solution (e.g. EMPS – Electronic Mobile Payment System by MeritaNordbanken, Nokia and Visa) has the advantage that handsets can be kept small. This solution is favoured by mobile operators, because they are in control of the dual chips.

Finally, if the background account is held at a bank, the existing banking infrastructure and technology can be reused. Examples are Paybox and MobiPay by BBVA and Telefónica. Typically, the merchant receives the customer's mobile phone number (or a pseudonym) and passes it on to the payment server, together with the payment details. The customer authorises the payment by providing a PIN.

In References 35 and 36, we have studied the use of software electronic coins and hardware electronic coins for a specific application in m-commerce, namely for ad hoc payment of the access of a mobile device to an access network with which no former relationship exists (i.e. the access network is not the home network and does not have a roaming agreement with the home network). Hence, the service for which the user of the mobile device pays is the current use of this device.

14.5.5 Standardisation and forums

One important aspect of m-commerce is standardisation. Due to the heterogeneity of technologies for mobile devices, and the need for transmission and payment over the air interface, it is essential to find common approaches, both at a national and an international level. The following list summarises standardisation bodies and forums dealing with issues relating to m-commerce.

- PayCircle® (www.paycircle.org) – is a vendor-independent non-profit organisation. Its main focus is to accelerate the use of payment technology and develop or adopt open payment APIs (uniform Application Programming Interfaces) based on XML, SOAP, Java and other Internet languages.

- MoSign (www.mosign.de) – banks, technology partners and end-device manufacturers have joined forces in the MoSign (Mobile Signature) project to create a platform for secure, legally binding mobile transactions based on existing standards.

- Mobile Payment Forum (www.mobilepaymentforum.org) – is a global, cross-industry organisation dedicated to developing a framework for standardised, secure and authenticated mobile commerce using payment card accounts.

- mSign (www.msign.org) – the Mobile Electronic Signature Consortium is an association of companies and organisations from the mobile phone and Internet sectors. The objective is to establish and develop a secure cross-application infrastructure for the deployment of mobile digital signatures.

- MWIF (www.mwif.org) – the Mobile Wireless Internet Forum (MWIF) is an international non-profit industry association. Its mission is to drive acceptance and adoption of a single open mobile wireless and internet architecture that is independent of the access technology.

- Radicchio (www.radicchio.org) – as a non-profit organisation, Radicchio brings together market leaders to establish a common foundation for secure m-commerce by reaching a consensus on important inter-operability issues.

- Encorus (www.encorus.com) – Encorus Technologies is focused on building a flexible and open infrastructure and efficient payment processing services to drive the acceptance and usage of mobile payments worldwide.

14.6   Conclusions

There will be no m-commerce without security of the underlying technologies. In this chapter, we discussed security issues relating to the mobile device (see Section 14.3), the system, network and service technologies (see Section 14.4), and m-payment (see Section 14.5). Regarding m-payment, some systems are under development or already operational. One of the main future challenges will be to unify payment solutions, providing the highest possible level of security.

Notes

1   The work described in this chapter has been supported by the European Commission through the IST Programme under Contract IST-2000-25350. The information in this document is provided as is, and no guarantee or warranty is given or implied that the information is fit for any particular purpose. The user thereof uses the information at his/her sole risk and liability.
2   IST-2000-25350 SHAMAN, http://www.ist-shaman.org
3   Bluetooth Special Interest Group: http://www.bluetooth.com
4   This description is based on the eCash electronic payment system.

References

1 WEITZEL, T. and KÖNIG, W.: 'Vom E- zum M-Payment' (in German), http://much-magic.wiwi.uni-frankfurt.de/profs/mobile/infos.htm

2 Datamonitor, Mobile Payment Systems, Whitepaper, November 2000, http://www.sun.co.uk/wireless/resources/pdf/datamonitor.pdf

3 BBC News, Tough penalties for mobile phone theft, 3 May 2002, http://news.bbc.co.uk/hi/english/uk/newsid_1966000/1966247.stm

4 Symbian OS technology, http://www.symbian.com/technology/technology.html

5 Palm OS, http://www.palmsource.com/palmos/

6 Microsoft Mobile Devices, http://www.microsoft.com/mobile/pocketpc/default.asp

7 Sharp Zaurus, http://www.zaurus.com

8 Yopy Mobile Community, http://www.yopy.com/

9 F-Secure Handheld Solutions, http://www.f-secure.com/wireless/

10 Wi-Fi Alliance, Wi-Fi protected access, http://www.wi-fi.com/OpenSection/secure.asp#resources

11 SyncML Sync Protocol, http://www.syncml.org/technology.html

12 GSM 02.09 version 7.0.1 Release 1998. Digital cellular telecommunication system (Phase 2+); security aspects

13 WALKER, M. and WRIGHT, T.: 'Security aspects', *in* HILLEBRAND, F. (Ed.): 'GSM and UMTS: the creation of global mobile communication' (John Wiley and Sons Ltd, Chichester, UK, 2001)

14 3GPP TS 33.102 3.9.0 Release 1999, 3rd generation partnership project; Technical Specification Group services and system aspects; 3G Security; security architecture

15 UMTS security. This book (Chapter 6)

16 BORISOV, N., GOLDBERG, I. and WAGNER, D.: 'Intercepting mobile communications: the insecurity of 802.11'. Proceedings of *MOBICOM 2001*, http://citeseer.nj.nec.com/article/borisov01intercepting.html

17 IEEE Standard 802.1X-2001. Port-Based Network Access Control

18 BLUNK, L. and VOLLBRECHT, J.: 'PPP extensible authentication protocol (EAP)', RFC 2284, March 1998

19 JOSEFSSON, S. *et al.*: 'Protected EAP protocol (PEAP)', Internet draft, September 2002

20 FUNK, P. and BLAKE-WILSON, S.: 'EAP tunneled TLS authentication protocol (EAP-TTLS)', Internet draft, November 2002

21 KENT, S. and ATKINSON, R.: 'Security architecture for the Internet protocol', RFC 2401, November 1998

22 LÖHLEIN, B.: 'Bluetooth – technology, security and weaknesses'. 6th World Multiconference on *Systemics, Cybernetics, and Informatics* (Orlando, FL, USA July 2002)

23 PKI in mobile systems. This book (Chapter 2)

24 DIERKS, T. and ALLEN, C.: 'The TLS protocol', Version 1.0, RFC 2246, January 1999

25 WAP Forum, Wireless transport layer security, Version 06-Apr-2001, http://www1.wapforum.org/tech/documents/WAP-261-WTLS-20010406-a.pdf

26 WAP Forum, Wireless application protocol, WAP 2.0, Technical white paper, http://www.wapforum.org/what/WAPWhite_Paper1.pdf

27 3GPP TS 03.78 version 7.7.0 Release 1998. Digital cellular telecommunications system (Phase 2+); Customized applications for mobile network enhanced logic (CAMEL) Phase 2 – Stage 2

28 3GPP TS 23.078 version 4.4.0 Release 4. Digital cellular telecommunications system (Phase 2+); Customized applications for mobile network enhanced logic (CAMEL) Phase 3 – Stage 2

29 ETSI ES 201 915-1 V1.1.1 (2001-12) Open service access; application programming interface, Part 1: Overview. http://www.parlay.org/specs/index.asp

30 ETSI ES 201 915-3 V1.1.1 (2001-12) Open service access; application programming Interface, Part 3: Framework. http://www.parlay.org/specs/index.asp

31 GSM 03.48 version 8.3.0 release 1999. Digital cellular telecommunication system (Phase 2+); Security mechanisms for the SIM application toolkit

32 O'MAHONEY, D., PIERCE, M. and TEWARI, H.: 'Electronic payment systems for e-commerce' (Artech House Computer Security Series, 2nd edn., 2002)

33 ePayment Systems Observatory, Database on E-payment Systems, http://epso.jrc.es/

34 WEBER, R.: 'Chablis – market analysis of digital payment systems', Technical Report TUM-I9819, TU Munich, 1998, http://chablis.informatik.tu-muenchen.de/Mstudy/

35 KNOSPE, H. and SCHWIDERSKI-GROSCHE, S.: 'Future mobile networks: ad-hoc access based on online payment with smartcards'. 13th IEEE International Symposium on *Personal, Indoor and Mobile Radio Communications* (PIMRC 2002), Lisboa, Portugal, September 2002, pp. 197–200

36 KNOSPE, H. and SCHWIDERSKI-GROSCHE, S.: 'Online payment for access to heterogeneous mobile networks', *IST Mobile & Wireless Telecommunications Summit* 2002, Thessaloniki, Greece, June 2002, pp. 748–752

37 BABB, D., BISHOP, C. and DODGSON, T.E.: 'Security Issues for downloaded code in mobile phones', *IEE Electronics & Communication Engineering Journal*, **14**, October 2002, pp. 219–27

Chapter 15

Securing the delivery of digital content over the Internet

Adrian Waller, Glyn Jones, Toby Whitley,
James Edwards, Dritan Kaleshi, Alistair Munro,
Bruce MacFarlane and Angus Wood

In this chapter, we will look at the security issues that arise for the browsing, selection and delivery of digital content over the Internet. Particular emphasis will be placed on the problems of finding security solutions for microtransactions (small items of content) and micropayments (low-value content), and the digital rights management issues concerning the protection of content after it has been delivered to consumers. The chapter will conclude with a description of the secure content delivery system developed during the Secure Interactive Broadcast Infotainment Services (SIBIS) project, which addresses many of the issues raised.

15.1   Introduction

Currently, the purchasing of digital content over the Internet is in its infancy. The most widespread use of the Internet for purchasing is for items of relatively high-value content using the credit card system, mirroring the use of well-established telephone mail order systems. However, the delivery of large items of high-value digital content represents only a small fraction of the potential services that could be supported by the Internet. An ideal content delivery system should be able to support efficient and secure browsing, selection and delivery of content from large sizes down to just a few bytes (microtransactions) and of high value down to potentially a fraction of a penny (micropayments). Such a system would allow, for example, individual charging for the delivery of stock market quotes, newspaper articles and web pages, or even pay per use charging for video streams (such as a penny a minute). Being able to support the payment on delivery for such low-value items of content as a single web page potentially allows a far greater range of people and organisations to become content providers. To make this possible however, the content delivery system must have low technical, financial and legal barriers to becoming a content provider.

In this chapter we will give an overview of the current state-of-the-art in secure content delivery systems for the Internet, focusing on the issues identified above. Section 15.2 gives an overview of the potential requirements for a general-purpose content delivery system for the Internet. In Section 15.3, we look at systems designed to protect the content during delivery, and analyse how well they meet the requirements of Section 15.2. These systems, however, do not protect the content after it has been delivered to the consumer. Recent high-profile cases, such as Napster [1], have highlighted the need to protect content against illegal copying and other licence infringements. Digital Rights Management (DRM) systems that aim to protect content against such licence rule infringements are looked at in Section 15.4. In Section 15.5, we briefly cover micropayment systems, and finally in Section 15.6 we describe the SIBIS secure content delivery system, which has been developed as part of a 2-year DTI/EPSRC Link Broadcast project.

15.2   Requirements

The participants in a content delivery system, and the high-level interactions between them, are shown in Figure 15.1. In general, consumers will perform some form of browsing followed by content selection and possibly payment to the content provider. The content provider will then deliver the content to the consumer (note that the producers and distributors of content are, in general, different but have been combined here into one entity for clarity). To handle the payments, there is typically a third party payment broker that may interact online or offline with the content provider and consumer. For example, with the credit card system the payment broker interacts with the consumer offline to set-up and manage his/her account and obtain monthly payments, and interacts with the content provider online to authorise the payment for each purchase.

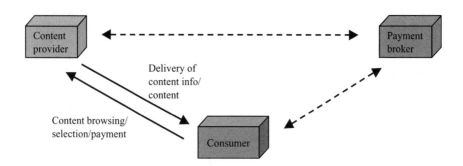

Figure 15.1   Overview of a content delivery system

Some of the potential requirements of each of the participants are listed below.

- *The Consumer*. Seeks guarantees of the integrity of content, ease of use and low cost of the system, protection from payment fraud, privacy protection for their personal and payment details and anonymity in their dealings with content providers (which can help in achieving privacy).
- *The Content Provider*. Content providers are primarily concerned with receiving payment for use of the content and hence preventing those who have not paid for it from using it. This implies requirements for the confidentiality of content during content delivery to prevent eavesdroppers from obtaining content, authentication of consumers and receipt of payment from them before they can use the content and preferably protection of the content after it has been delivered to the consumer (DRM).
- *The Payment Broker*. The payment broker does not have requirements from the system as such, other than the ability to charge for the services that it provides.

Ideally, the content delivery system should be suitable for a wide range of scenarios to make it as flexible as possible. The following three requirements are crucial to achieve this flexibility.

- *Scalability*. The system should be able to support thousands of content providers and millions of consumers. Ideally, almost anyone should be able to be a content provider, and for this to be possible the barriers in terms of costs of licences and hardware and the necessary trust required to be a content provider should be as low as possible.
- *Granularity of content size*. This refers to the delivery of content from just a few bytes (microtransactions) up to potentially gigabytes of data. An example of a microtransaction could be the purchase of a stock market quote or foreign exchange rate. Being able to support large volumes of microtransactions in an efficient way is a particular challenge.
- *Granularity of payment*. Support for payment from a fraction of a penny (micropayments) up to tens of pounds or even more should be possible. The ability to support micropayments would allow charging, for example, for Internet search results, individual Web pages or for individual articles in an online magazine.

In the remainder of this chapter we will look at existing security systems to determine the degree to which they meet these requirements.

15.3 Securing the content delivery

Technologies designed to protect the delivery of content can be divided into two types, which are those that set-up a secure connection to send the content down ('Secure the pipe') and those that apply security to the content directly ('Secure the content').

15.3.1 'Secure the pipe'

As mentioned earlier, this type of content delivery security is characterised by the setting up of a secure session between the content provider and the consumer, which can then be used to transmit the content securely. The most commonly used technology of this type is Secure Sockets Layer/Transport Layer Security (SSL/TLS) [2] which is widely supported in Web browsers and Web servers. We will use TLS to illustrate the potential advantages and disadvantages of this type of content delivery.

The process involved in setting up a secure TLS session, the 'handshake', is illustrated in Figure 15.2. This figure shows the flow of messages between a client and server in the handshake, as well as the computationally intensive public key type operations (in this case RSA sign/verify or encrypt/decrypt operations) that the client and server have to do. Other operations have been omitted for clarity. As can be seen, this handshake involves several round-trip communications and public key operations. It may also involve the client and server interacting with a Public Key Infrastructure (PKI) to obtain certificate status information in order to validate each other's certificate. However, once the session has been set up, TLS uses symmetric key encryption and integrity protection techniques and can, therefore, provide protection to data in a relatively efficient way.

In terms of achieving the potential requirements identified in Section 15.2, TLS has many advantages. For the content delivery itself, it will protect the confidentiality and integrity of the content and will do this in an efficient way once the session has been set up. The secure session can also be used to protect the privacy of consumers by protecting the content browsing, selection and payment communications from potential eavesdropping. From the content providers' point of view the secure session can also provide them with authentication of the consumer's identity.

The major drawback of TLS is a potential lack of scalability, particularly for the scenario of microtransactions where the session set up is a significant overhead. Handling a large number of connections on a TLS server within a short period of time is a problem that can require special hardware. In the case of microtransactions, all of this effort has to take place just to transfer a small amount of content, which is not very efficient and may even be uneconomic for very low-value content. Another potential drawback is that the protection has to be applied at the same time as the content delivery. If the content provider is simultaneously handling a very large number of consumers, or if high data rate content delivery is required (e.g. real-time protocols), then this represents a significant computational load that again can require special hardware at the server to enable it to cope. Finally, as the key used to decrypt the content is essentially the key of the secure session, the payment and authorisation decisions have to be part of the content delivery decision and, therefore, the content provider has to be involved in this. This has implications for consumer privacy, as consumers will, in general, have to provide sensitive information, such as payment details (e.g. credit card numbers), to the content provider. It can also place a barrier on becoming a content provider, as content providers may need to be relatively highly trusted.

In general, therefore, this type of content delivery security is most suited to the scenario of a consumer obtaining content in a session, where they are either obtaining

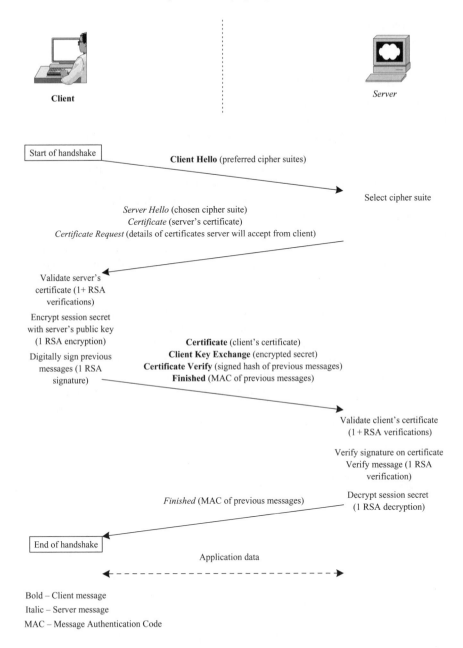

Figure 15.2 A typical TLS handshake with client authentication

a large volume of content or are selecting a number of different items of content from the same content provider.

15.3.2   'Secure the content'

With this type of content delivery, protection is applied directly to content and, therefore, a secure delivery channel between the content provider and the consumer is not required. Typically, the content is encrypted and wrapped into some kind of container, with appropriate header information added, and integrity protection is then applied by, for example, digitally signing the container. This container can then be delivered to any number of consumers without further processing being required by the content provider. The remaining problem is how to distribute the decryption keys to the consumers who are authorised to use the content (i.e. to those who have paid for it).

One way to solve this problem is to make use of a third party transaction broker, and this is the approach used by digital container systems. Commercial systems usually integrate this approach with DRM, which we will look at in more detail in Section 15.4.2.

The basic digital container's architecture is illustrated in Figure 15.3. As mentioned earlier, digital content (in this case represented by an arbitrary file) is wrapped into a protected container using a secret key, and this container can then be delivered to any number of consumers over any, possibly insecure, distribution network without compromising the security of the content. When a consumer wishes to use the content, they have to contact the broker with the appropriate authorisation details, such as a valid payment. The broker will then return the decryption key allowing the consumer to use the content, and will also handle the clearing of the payment with the

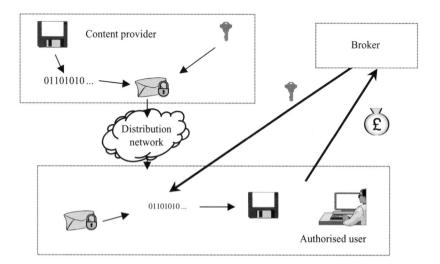

Figure 15.3   Digital container's architecture

content provider. The fine details of how the broker can generate a decryption key, and the interaction between content providers and the brokers differ from system to system.

This approach has many advantages, particularly in terms of scalability. Protection can be applied once, offline, for any number of consumers, which means that potentially expensive cryptographic operations do not need to take place in real-time at the content provider's end. This also protects the content while it is in storage. For these reasons, digital containers are particularly suited to store and forward content delivery systems. The content may also be delivered through any distribution channel, which means that even though usually it would be distributed directly from the content provider via the Internet, it could also be distributed on CDs for example. Another advantage is that it separates the content payment and authorisation from the content delivery, which frees content providers from having to handle payment and authorisation decisions and allows them to concentrate on content delivery. This freedom significantly reduces the amount of trust required between the broker and the content provider, thus reducing the content providers' costs and at the same time allowing a far larger number of people and organisations to be content providers. Furthermore, this offers an advantage from the point of view of the consumer, in that the content provider is not responsible for handling potentially sensitive payment and authorisation details of the consumers, and indeed consumers can be completely anonymous to the content providers with this approach.

For these reasons, digital containers are particularly suited to store-and-forward delivery of discrete items of content. Real-time content delivery, such as video streaming from point-to-point or even multicasting applications, could also be supported by dividing up the stream into suitably sized chunks for placing into containers. However, this 'chunking' approach could place a significant load on the broker in terms of the number of key requests it has to handle.

15.4 Digital rights management (DRM)

The technologies discussed in Section 15.3 are concerned with protecting the content while it is being delivered, but not after it has been delivered. DRM aims to address the protection of content throughout its lifetime. This leads to requirements such as copyright protection, prevention/detection of illegal copying and enforcing of other licensing rules (e.g. expiry of content, limited number of viewings, etc.). Developing a secure content delivery system that incorporates DRM is attracting a lot of interest at the present time, and there are a large number of companies and organisations working on developing such systems. Major companies include Microsoft, IBM, RealNetworks and Sony, to name just a few, who are working on both the component technologies of the systems as well as providing completely integrated secure content delivery systems. Organisations include the Motion Picture Association of America (MPAA) and the Secure Digital Music Initiative (SDMI) who have been looking at developing systems for particular applications such as DVD copy protection and the secure delivery of digital music. In terms of the DRM technologies used in these proposed systems, there are two major types at present. These are digital watermarking and digital containers, and each of these is discussed below.

15.4.1   Digital watermarking

Digital watermarking is a technique that enables information to be embedded within digital content. This information could be, for example, the copyright holder's identity or licence rules that apply to the content. For a good introduction to digital watermarking techniques and how they can be applied, see the paper by Podilchuk and Delp [3]. Example applications include:

- keeping track of content to detect illegal copying;
- controlling, in conjunction with special hardware, the number of copies that can be made by embedding copy control information;
- detecting modifications to digital content by the use of 'fragile' watermarks, which any such modifications will destroy;
- monitoring broadcast services to, for example, track the number of uses of adverts or a song for royalty and other payment purposes.

For our scenario of thousands of content providers wishing to protect digital content after it has been delivered to consumers, the main application of watermarking would be in embedding information within content to allow illegal copying and distribution to be detected. For this application, watermarks need to be 'robust' in that it should not be possible for consumers to remove the watermark without harming the quality of the content so much that it becomes worthless. This implies that the watermark must be embedded throughout the content, otherwise just the part that contained the watermark could be removed. There are two kinds of robust watermarks, visible and invisible.

Visible watermarks can be used to embed copyright information within images. Visible watermarks are only of limited use due to their effect on the content.

Invisible watermarks aim to embed information within redundancy in the content in such a way that the watermarked content is perceptually the same as the original. As the watermarks are invisible, this helps with robustness as attackers will not know where the watermark is embedded. It does mean, however, that a secret key must be used to embed the watermark and allow its later extraction by the content provider. Techniques to do this exist for most content types, including pictures, audio and video.

An example of a potential use of invisible watermarking is fingerprinting, which is illustrated in Figure 15.4. With fingerprinting, when a consumer purchases some digital content, the content provider embeds a watermark containing the consumer's identity. If at a later stage that consumer decides to illegally distribute the content, and if the content provider obtains a copy, then they are able to extract the consumer's identity from the watermark and know who was responsible for the illegal distribution. As the watermark is invisible, it does not affect in any way the consumer's use of the content. In fact, they will, in general, not be aware that there is a watermark in

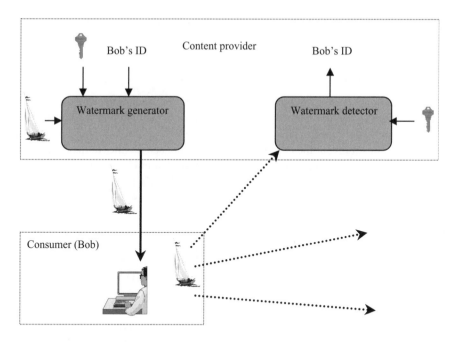

Figure 15.4   Fingerprinting using robust, invisible watermarks

the content, although for this application the fact that their identity is embedded as a watermark would be likely to be made aware to them to act as a deterrent against illegal copying and distribution.

An important advantage of using robust, invisible watermarking for DRM is that the watermarks are transparent to the consumers and to the software in existing content delivery systems. They can, therefore, be used with minimal impact on any existing system. However, there are doubts about how robust they actually are when subject to a deliberate attack. A good example of this is the SDMI challenge. The SDMI is a consortium of music industry companies who are working to develop and standardise technologies to protect digital music. In September 2000, they issued a challenge to try and remove the robust, invisible watermarks on audio content they had created using four different watermarking schemes (including the Verance scheme used to protect DVD audio). Despite the conditions of the challenge being very restrictive, with only three weeks allowed and no details of the algorithms used being provided, a team from Princeton and Rice universities managed to defeat all four watermarking schemes [4]. Work on trying to improve the robustness of watermarking schemes is, however, taking place. An important initiative in this area may be the production of benchmarking tools, such as Checkmark [5] and the work of the Certimark [6] project, which combine a large variety of the previously known and new attacks. The aim of these tools is to assess the robustness of watermarking schemes and their suitability for particular applications, and hopefully to lead to improvements in new watermarking schemes.

15.4.2   Digital containers

DRM technologies based on digital containers combine 'secure the content' content delivery with the use of 'tamper-resistant' software or hardware at the consumer's end. This software is responsible for handling the decryption and rendering of the content and must prevent the consumer from gaining access either to the decryption key or to the decrypted digital content. This is illustrated in Figure 15.5.

Commercial examples of such systems include IBM's Electronic Media Management System [7] and Microsoft's Windows Media Rights Manager [8], and for a comparison of some of the more prominent schemes see the White Paper by Sonera Plaza Ltd Medialab [9].

This approach inherits all of the advantages of the digital containers' content delivery architecture in terms of scalability, and for DRM it should be able to prevent copyright or licence fraud, not just detect it. It also enables the enforcement of new licensing rules such as pay per use or expiry of content as these can be enforced by the tamper-resistant software or hardware. A final advantage is that the so-called 'superdistribution' (the idea that consumers themselves, or indeed anyone, can become distributors of content without any loss of security) is supported, which can reduce the burden on the original content provider and enhance scalability. As an example, consumers who like a particular item of content may wish to distribute this to their friends. They are prevented from accessing the digital content itself by

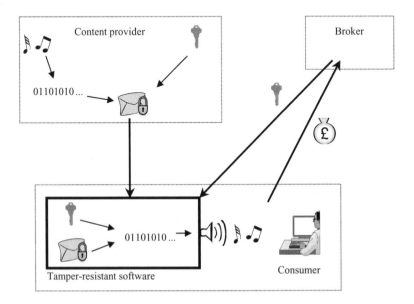

Figure 15.5   Digital containers with DRM

the tamper resistant software or hardware, but they can still distribute the protected digital containers. Anyone who receives these will still have to pay for the content to be able to use it.

There are some potential problems however. Perhaps the most serious issue, as with digital watermarking, are doubts about the security of this technique, particularly if implemented in software. How 'tamper resistant' is the software? An example of digital containers' software that has recently been circumvented, and which was widely reported in the press [10], is version 2 of Microsoft's DRM software. This particular attack also illustrated the problem that it only takes one person (in this case 'Beale Screamer') to work out how to get around the system, and they can then distribute software to automate the attack which allows anyone else to get around the system. Another potential problem is that from a practical point of view, this technique requires special software or hardware for each type of digital content as the content decryption has to be tightly coupled to the rendering application. This is to prevent consumers gaining access to the decrypted digital content. Existing rendering applications would, therefore, have to be modified to support digital container systems and this could also lead to consumers having to have multiple renderers for the same type of content. This situation could be eased by standardisation of digital container technologies, or perhaps more likely by the consolidation of the market to only a handful of competing systems. In any case, consumers already have to have several different renderers for the same type of content at present, such as for digital audio, so this may not be a significant problem.

15.5 Micropayments

For micropayments to be viable, the intrinsic costs of the content delivery system must be evaluated in terms of computation, communications and equipment required. As mentioned in Section 15.3, 'secure the pipe' delivery can suffer from significant overheads at session set-up, which mean that these system costs are likely to be relatively high. Therefore, 'secure the content' delivery is likely to be more suited to the support of micropayments.

In terms of micropayment systems themselves, current support is dominated by indirect payment systems where content is paid for by means other than direct purchase at the time of the content retrieval. This covers advertising and subscription. Advertising can meet many of the requirements; however, it is not suitable for the support of a large number of small content providers and would need to be supplemented by another payment system for high-value content. It also provides a barrier to starting up as a content provider. Subscription can also meet many of the requirements; however, it is not suited to the support of infrequent microtransactions or where consumers may wish to use a large number of different content providers.

A more general solution would, therefore, need a direct micropayment system. In recent years, there has been a considerable amount of research into developing such systems, although few if any have gained widespread support. It is beyond the scope of this paper to cover these systems in detail; however, a good source of information on proposed micropayment systems is the 'ePayment Systems Observatory' [11], which has a comprehensive list, as well as a description, of a large number of them.

Micropayment systems can be divided into two types. Token systems use some kind of marker representing a value as the medium of exchange. A physical example of this would be cash. Notational systems exchange a value by authorisation, such as with a typical bank account. While it is true that either type of system could meet the requirements for a secure content delivery system, the notational systems can provide a particularly good fit with 'secure the content' delivery systems. This is because a broker has to be contacted by the consumer to obtain the decryption key. If at the same time the broker handles the billing of the consumer for the content, such as by transferring money from the consumer's to the content provider's account, micropayments can be handled in a particularly efficient way. This adds relatively little in the way of overhead to the content delivery system in order to support payments.

15.6   The SIBIS system

The SIBIS project is a DTI Link Broadcast project with the following aim:

> To develop the protocols, methods and functionality to enable scalable, secure, and reliable electronic microtransactions to take place between consumers and content providers.

The original project partners were Thales Research and Technology (UK), University of Bristol (Department of Electrical and Electronic Engineering and Department of Computer Science), Pedagog and Zygo Communications. During the project we have produced a specification of a content delivery system. To achieve the aim of being able to support microtransactions and of being able to scale to a large number of content providers and consumers, we decided to base the system on the digital containers architecture described in Section 15.3.2. This architecture is particularly suited to microtransactions due to the relatively low overhead of content delivery. It is also particularly suited to micropayments as payment occurs within the brokers and can, therefore, use efficient account based billing. Due to the separation of payment and authorisation from content delivery, the brokers in the SIBIS system can also treat content providers as little more than special consumers, thus opening up the possibility of almost anyone being a content provider.

The SIBIS system does not, however, include provisions for DRM after the content has been paid for. There were a couple of reasons for this. First, the system is specifically designed for microtransactions and micropayments where the value of content is relatively low and hence the copying and redistribution of content is likely to be a relatively minor concern. Having said that, the system may be suitable for delivery of high-value content that is only short lived, such as, for example, stock market quotes or certain kinds of live event (e.g. a football match). Second, we decided that the maturity of the rights management technologies was not sufficient to provide enough guarantee of the security of the content to make their incorporation worthwhile. This may change in the future, and if so either digital watermarking or the rights management facilities of digital container technologies would be relatively easy to integrate into the SIBIS system.

Some details of the components in the SIBIS system, as well as the protocols and data formats specified by SIBIS, are given in the following sections. Note that the final part of the project will involve the production of a demonstrator and the execution of medium-scale trials with it at the University of Bristol in order to validate the performance of our design.

15.6.1 The SIBIS components

A component view of the SIBIS system, including the SIBIS API interface, is shown in Figure 15.6. These APIs provide the interface to SIBIS components that handle the packaging of content into digital containers at the content provider and the extraction of content from the containers at the consumers' end and are designed to allow seamless integration with existing content delivery systems. Note that this approach means that the SIBIS system is independent of the content delivery channel, and can, therefore, support content delivery channels other than the Internet.

The content pre-preparation stage at the content provider divides content up into suitably sized chunks to be placed in containers. The chunk sizes are determined based on the type of content being requested, the characteristics of the delivery channel, and the payment method. In many cases this will simply provide the content as it is to the SIBIS API, but could be used for example to divide up a stream of data, such as a video stream, into chunks to allow consumers to be charged only for the particular

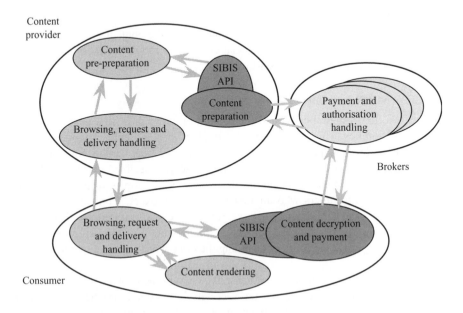

Figure 15.6   Component view of the SIBIS system

parts of the stream that they actually use. In either case, the rest of the content delivery should be unaffected.

At the consumer, the content delivery handling software (e.g. a Web browser) will invoke the consumer SIBIS API to handle the SIBIS content (i.e. exchange the payment with the broker for the keys to the digital containers), and deliver the decrypted content to the rendering software.

15.6.2   The SIBIS specification

The SIBIS specification defines the interaction between a content provider and its broker, a consumer and its broker and the construction and format of the digital containers. The main considerations when designing this specification were to make sure that the system would be secure and scalable. In terms of security, the SIBIS participants need to be protected not only from the usual external attackers, who may wish to eavesdrop on communications for example, but also from the other SIBIS participants. Measures were, therefore, included within the SIBIS specification to prevent the participants gaining from acting maliciously within the system.

It is not within the scope of this chapter to provide a detailed description of the SIBIS specification; however, a summary of some of the important features is given below.

15.6.2.1   Content provider–broker interactions

The interaction between a content provider and its broker is limited to two areas. First, in order to receive payments for content the content provider needs to open and maintain an account with its broker. Second, to allow its broker to regenerate content encryption keys (to be passed to consumers) the content provider needs to share a 'master' secret with its broker that will be used to derive these keys. These interactions mostly take place at set-up time and can be performed offline.

15.6.2.2   SIBIS containers

To create a SIBIS digital container, the content provider must first generate a content encryption key. This key is derived using the 'master' secret, a unique content ID (to make the key specific to that item of content) plus certain other information, including the price of the content, whose integrity needs to be protected from consumers. The purpose of including the latter information is to protect content providers from consumers who attempt to mislead brokers. Consumers may, for example, attempt to tamper with the price of the content so that they are charged less for it. Including the price of the content in the derivation of the key prevents this from happening since the result of changing the price would be the generation of the wrong key.

Once the key has been derived it is used to encrypt the content. The details used to derive the key, excluding the 'master' secret itself, of course, are then placed in the container together with the encrypted content. Other information that may be useful to consumers, such as a description of the content, are also included and, finally, the

Table 15.1   Simplified view of a SIBIS container

Header	Version information
	Content description – *includes the MIME type of the encapsulated digital content as well as a user readable description*
	Key derivation details – *includes information needed by the broker to regenerate the decryption key, including the content ID, price, content provider ID, etc.*
	Sequence number (optional) – *this can be included to identify a container if it is part of a sequence, such as a video stream*
Content	Encrypted content – *the encrypted digital content*
Integrity protection	Digital signature – *this is calculated on the previous fields and protects both the header and the content*
	Certificates (optional) – *this contains the content provider's digital certificate chain, which is needed by consumers to validate the signature on the container. Due to the potentially large size of these, they may be distributed by out of band means instead*

container is digitally signed by the content provider. This digital signature enables consumers to check the integrity of the content. In the case that the content provider accidentally or maliciously makes errors in containers, the signature also provides evidence, which consumers can use to prove wrongdoing by the content provider.

The precise format of a SIBIS container is quite complicated; however, a simplified view of its contents is given in Table 15.1.

15.6.2.3   Consumer–broker interactions

When a container is received by a consumer, the consumer needs to obtain the decryption key from its broker. In the SIBIS system, a secure connection is established using SSL/TLS between the consumer and its broker for this purpose. The secure connection provides authentication of the consumer and broker to each other, as well as privacy and integrity protection for their communications. As mentioned in Section 15.3.1, there is a significant overhead when this connection is established. However, it need only be established once at the start of the consumer's content purchasing session and can remain open until the end of this session. Therefore, the overhead can, in general, be spread over key requests for many items of content.

To request a decryption key for the content, the consumer sends to its broker the key derivation details, which include the price of the content, obtained from the container. Note that it is also possible for the consumer to batch requests for keys and therefore obtain the keys for several containers at the same time to enhance scalability. It is also important to note that the key derivation details only contain a content ID and no other description of the content. If this ID, as would be the case in general, is simply a random number, then the consumer's privacy is protected as the broker will not know what content the consumer is purchasing.

On receipt of the consumer's request, the broker regenerates the key and returns it to the consumer. This step will, in general, require the consumer's broker to forward the request details to the content provider's broker and obtain the key from it. On receipt of the key, it is possible that the consumer will fail to decrypt the content for some reason, such as the key being derived incorrectly. In order to protect consumers from being charged for content which they cannot use, charging will not take place until a confirmation message is sent from the consumer to its broker that the decryption was successful. Clearly, this provides an opportunity for consumers to claim that they have not successfully decrypted content and thus enable them to obtain content for free. This will need to be dealt with by means outside of the SIBIS specification, such as monitoring consumers for abnormal behaviour and investigating persistent offenders in more detail.

15.7   Conclusions

We have seen that in order to support microtransactions and micropayments the digital containers' architecture offers many advantages. By combining this architecture with the use of tamper resistant software or hardware at the consumer or with digital watermarking, digital rights management for the lifetime of the content can also be achieved. However, further work is likely to be required on improving and demonstrating the security of these rights management systems, as well as investigating the applicability of this architecture to real-time content delivery.

Acknowledgements

The authors gratefully acknowledge the financial support provided by the DTI/EPSRC LINK Broadcast project 'Secure Interactive Broadcast Infotainment Services (SIBIS)' (No. TJBF/C/002/00034) for the work on which this chapter is based, and Thales Research and Technology (UK) for allowing its publication.

References

1 http://news.bbc.co.uk/hi/english/entertainment/new_media/newsid_1586000/1586226.stm

2 DIERKS, D. and ALLEN, C.: 'The TLS Protocol', RFC 2246, IETF Network Working Group, January 1999

3 PODILCHUK, C. and DELP, E.: 'Digital watermarking: algorithms and applications', *IEEE Signal Processing Magazine*, July 2001

4 CRAVER, S., DEAN, D., FELTEN, E. W., *et al.*: 'Reading between the lines: lessons from the SDMI challenge'. Proceedings of the *10th USENIX Security Symposium*, 13–17 August 2001

5 PEREIRA, S., VOLOSHYNOVSKIY, S., MADUEÑO, M., MARCHAND-MAILLET, S. and PUN, T.: 'Second generation benchmarking and application oriented evaluation', Information Hiding Workshop III, Pittsburgh, PA, USA, April 2001

6 CERTIMARK: 'Certification for watermarking techniques', IST-1999-10987, http://vision.unige.ch/certimark

7 http://www.ibm.com/software/emms

8 Microsoft Digital Media Division: 'Security overview of Windows Media Rights Manager', September 2001, http://www.microsoft.com/windows/windowsmedia/WM7/DRM/whitepapers.asp

9 Sonera Plaza Ltd Medialab: 'Digital rights management white paper', 3 February 2002, http:www.medialab.sonera.fi/workspace/DRMWhitePaper.pdf

10 http://www.cnn.com/2001/TECH/internet/10/25/ms.hacked.idg/

11 http://epso.jrc.es

Chapter 16

Security for future standardised DRM

Timothy Wright

Digital Rights Management (DRM) is a generic term used to cover the protection of proprietary digital content against misuse, including unauthorised distribution and use. This chapter seeks to re-examine possible solutions to the provision of DRM by first examining the threats and resulting security requirements, and then classifying approaches to DRM depending on the nature of the operating system on which the proprietary content will be used. This then leads to a detailed analysis of security requirements for a DRM solution running on an 'open' operating system platform.

16.1   Introduction

This chapter was originally written as in input to the OMA DL/DRM group meeting in Rome, 27–28 August, 2002, to contribute to the discussion they will be having on future standardisation of DRM within the Open Mobile Alliance (OMA) [1]. The OMA phase 1 standards [2] contained minimal security features in one option (content encryption with delivery of the Content Encryption Key (CEK) in plaintext) and two other options with no security features. This chapter was therefore written to discuss the additional security features that could be added to a future phase of the OMA DRM standards.

   This chapter is written with the assumption that future OMA DRM standards will contain 'strong' security. Further, corresponding to the fact that although the existing OMA DRM solution does not provide 'strong' security, it will be reasonably secure on 'closed' Operating System (OS) devices, it is assumed that future DRM solutions will be designed so that they are secure on 'open' OS devices.

   This chapter is not solely focused on DRM standardisation, as it contains many recommendations that cannot be standardised but are put into the public domain to encourage best practice by implementors and to encourage a high level of debate within the OMA.

The aims of this chapter are to:

- examine the likely attacks on mobile DRM devices;
- in particular, examine the characteristics of open OS devices that make them particularly susceptible to attacks on their DRM functionality;
- examine the effectiveness of potential additional security mechanisms that might be added on top of the existing OMA DRM specification in the context of the identified attacks and the identified characteristics of open OS devices;
- make appropriate recommendations for future DRM standardisation and, where recommendations cannot sensibly be converted into standardisation requirements, to make recommendations for best practice in device implementation.

Only security features that can be added on top of the existing DRM specification will be recommended.

This chapter does not deal with:

- whether or not additional security features are required. This much is assumed, as the OMA DRM specification states [2] that the existing solution does not provide strong security and is only designed for low value content;
- the use of (U)SIM from a non-security perspective (e.g. as a license store for portability reasons);
- any non-security related functionality that might be added to future OMA DRM standards.

16.1.1   Terminology and system model

The term 'device' is defined as the terminal (ME) plus USIM, though in most cases it will refer to the terminal only.

The term 'rights' is defined as in [2] as an 'XML document specifying permissions and constraints defining under which circumstances access is granted to DRM content'. The rights also contains the Content Encryption Key (CEK) if the content is encrypted by the content provider.

The term 'content distributor' is defined as the source of both content and rights. The content distributor would possess a 'DRM Packager', as defined in [3].

The term 'executable' is used to cover any instance of executable code that can be downloaded to a terminal after that terminal has been manufactured and is in service and executed on that terminal. Executable can refer to executable code that runs within a defined environment on the terminal such as Java, Visual Basic or C-sharp. It can also refer to scripting languages such as WMLScript, JavaScript and ECMAScript and to native applications compiled from a high level language and to shell languages such as Perl.

16.2   Threat model

16.2.1   Types of attack

16.2.1.1   DRM specific attacks

We distinguish three broad types of attack that are specific to DRM: content extraction, content injection and on device rights manipulation.

Content extraction is the removal or manipulation of DRM protection from protected content so that the content exists either in a plaintext form and without attached usage conditions or in a protected form but with usage conditions that are not those of the original content provider.

The content has therefore been extracted from the protected/premium content value chain/system where the use of the content can be restricted and charged for according to the conditions of the original content distributor. Content extraction includes the process of moving content off a device but not the reception of the content by another device.

Content extraction can take place when the content is in transit between the content distributor and the device or can take place on the device.

Content injection is the unauthorised distribution of content extracted from one device to other devices and its reception by these devices.

On device rights manipulation is where the usage conditions in the rights associated with the protected content are changed so that the content can be used with fewer conditions that was the intention of the content distributor. For example, the rights might be changed so that it gives an unlimited number of plays instead of a finite number of plays.

These classifications can be analysed and rationalised.

We assume that no measure can be taken in terms of device functionality to prevent content injection. This assumption is justified as follows. It is assumed that devices, in order to play user-generated or otherwise free, unprotected content will allow the rendering of unprotected content. Further, it is assumed that once protected content has been extracted, it cannot be distinguished from content that was always free.[1] As the device is able to play unprotected content and cannot distinguish between extracted and now unprotected content and content that was always unprotected, content injection cannot be prevented by device functionality. However, although there are no means to prevent content injection, it can be made harder by limiting the means by which unprotected content can be distributed from one device to another. For instance, if unprotected content can only be distributed using network based means (e.g. MMS or http download) then fraudsters must pay for content injection, which will be a discouragement to some potential fraudsters, and also allows for network based detection of injection and preventative measures such as the closing down of web sites hosting extracted content.[2] However, there are also issues of device functionality and usefulness that need to be considered here.

A (device functionality) measure that might be thought to prevent content injection is the definition of new content types and the programming of devices to only accept content in this new type unless it is protected. If content of this type is also to be user-generated, then there has to be some way either for users or entities working on their behalf to put protection around the user generated content. However, this measure is likely to be ineffective for the following reasons.

- It would take some time for a large number of devices and content generators to support this new content type. Until such a time, and only with the active suppression of the generation of content in older formats, content would exist in older content types, which admit content injection, in significant quantities.
- Transcoding applications, to convert content in the new format into an existing format which admits content injection, are likely to be available and extracted content in the new format can therefore be converted into an older content type and then injected into the system.

In addition to these points, content watermarking is also not considered within this chapter. Many watermarking schemes have been broken, but to be effective, they must be implemented by all parts of the system. It does not therefore seem an effective use of time and effort for all parts of the system to implement something of dubious effectiveness.

With respect to device rights, manipulation and content extraction can be differentiated as follows.

If protected content is attacked whilst present on a particular device so that it can be used with more freedom (but still with some conditions, as device rights manipulation is not rights *deletion*) on that particular device, but no effort is made to make a plaintext version of the content available for export to other devices, then this is an instance of device rights manipulation. If the content is made available in plaintext form, or encrypted but with manipulated rights for *export to other devices*, that this is an instance of content extraction.

There is a question of whether content extraction will happen but with the content still being distributed in a protected format, but with restrictions imposed by the content extractor and not the original content provider. If we assume that this is never done, and that an extractor will only make extracted content freely available, then there would not be value in rights integrity measures and/or in providing authentication and authorisation of the content provider to the client, as it is assumed that anyone who makes content available with rights is an authorised content distributor. However, apparently, Napster was used in many cases by students who were compiling CDs for other students for a charge. It seems reasonable in any case to guard against the danger of organised crime being involved in content extraction and not to assume that all content extractors will altruistically make the extracted content available for free to all.

It also seems reasonable to guard against content extraction with rights manipulation because it may, for some devices, be easier to manipulate the rights than to extract the content entirely. For instance, the rights may be stored in insecure memory, so can be changed easily, but the API to be used for content decryption is actually only available to authorised executables so cannot be used by some 'ripping' executable. In such a case, a fraudster may just make the content available still in protected form but with no restrictions, and this is functionally equivalent on the device to making the content available in plaintext and without restrictions.

16.2.1.2   Non-DRM specific attacks

In addition to the DRM specific attacks mentioned, there are generic communications attacks that can be applied to DRM systems. These are as follows.

Eavesdropping. This is where an attacker observes the content that is being consumed by a particular device but makes no effort to extract the content. We will assume that this is not a threat we wish to explicitly guard against. If certain content distributors wish to provide end to end eavesdropping protection, this can be done using transport level means, for instance using transport layer security (TLS) [4].

Subscription masquerade or spoofing. This is where an attacker uses the identity of another subscriber instead of their own, to purchase content so that the other subscriber pays for the cost of the content.

Device type masquerade or spoofing. This is where an attacker uses the device type identification of another device, not their own, in requesting content in order to evade any measures that the content distributor may be taking not to distribute content to certain device types. Device type authentication might be performed by the content distributor so that content is not distributed to particular device types that are known to have a poor DRM implementation that is susceptible to content extraction. Though the limited goals of the original OMA DRM specifications are admitted, it is clearly a drawback of those specifications that they provide no means to reliably discern if a device requesting content is a compliant device with a secure internal implementation of the OMA DRM specifications, or if it is PC running extraction programmes and using a mobile phone merely as a modem.

16.2.2   Security requirements

At this point, one might traditionally list security requirements/features for the solution, such as 'Authentication of the client by the content provider' and the like. However, many of the attacks on DRM systems are actually conducted by the owner of the device on the device itself. In such a case, the device will contain both the protected content and the CEK and we therefore need security requirement and features that are effective on the device itself and not in the device to content provider link. This point is particularly true when the device in question has an open OS. We will therefore examine attacks on an open OS device in detail before listing security requirement and features.

Note, however, that there is definite value in security requirements and features that apply to the device to content provider link, such as authentication of the device type to the content provider, and these will not be ignored.

16.3   Threats to open OS devices

16.3.1   Differences between an open OS and a closed OS terminal

We assume that the difference between a closed OS terminal and an open OS terminal is that an open OS terminal may have *some or all* of the following features and a closed OS terminal has *none of them*.

1. A generic file system that can be used by the user[3] (using a tool such as Windows File Manager) or by executables downloadable by the user to find and examine/open any file on the terminal that are not specifically hidden from such tools or applications.
2. A method of downloading applications to the terminal after manufacture, where these applications have *direct* access to functionality and APIs (such as a file management tool or to device drivers) of the terminal OS that are not specifically and robustly protected from access from any entity other than the core, provisioned, unchangeable elements of the terminal OS.

    It might be thought that Java midlets, which execute within a Java Virtual Machine (JVM), would *not* fall into this category as they should only have access to a filtered subset of the whole terminal functionality and should only have access to their own data. However, this subset of terminal functionality can be sufficient either to perform some part of the DRM evasion or to masquerade as a DRM UA, as all the functions necessary to build a DRM UA that could request, rip and re-post (but not use on the ripping device) content protected using the forward lock or combined delivery mechanisms, is provided by MIDP 1.0 [5]. Thus Java midlets do fall into this category.
3. A programming language ('scripting language') for writing programs on the terminal that can be used to directly access functionality and APIs of the terminal OS that are not specifically and robustly protected from access from any entity other than the core, provisioned, unchangeable elements of the terminal OS. An example of such a scripting language would be Unix PERL.
4. A method for downloading a DRM client to the terminal and a method for resetting the terminal so that this DRM client receives all protected content. (This DRM client may be specifically written to ignore the rules attached to protected content.)
5. A method for downloading other components, for example, device drivers that have been compromised or have been designed to assist in the DRM evasion process.

In summary, we might say that we could characterise an open OS by saying there is a file system which allows open access to post-factory installed executables (closed OS terminals have a file system and executables – but they are not open to post-factory executables). However, as we want in many cases to have a file system open to post-factory executables, we need a way to differentiate between elements in the file store and between different post-factory provisioned executables. We also require protection against unauthorised reprogramming of devices in general.

A general question when examining attacks on devices, particularly open OS devices is, to what extent can the user change the behaviour of the terminal?

16.3.2   Generic attacks on open OS terminals

This section will examine the types of attack which can be performed on DRM clients on an open OS terminal.

Unless suitable protective mechanisms are employed in the terminal OS, the following attacks could be performed.

(a) *Extract protected but un-encrypted content.* If the content is not encrypted, the file management tool could be used (either by the user but more probably by a rogue executable) to retrieve the un-encrypted content and make it available for transport off the terminal.

(b) *Extract protected content when decrypted.* As the content has to be decrypted at some point, so that it can be rendered, it could potentially be obtained at this time. If the decryption process left temporary copies then the file management tool could be used to retrieve these. If the internal buses of the device can be monitored, again this might be a way to obtain the decrypted content.

Another option is to install a modified device driver which will copy the decrypted data to memory as well as, or instead of, writing it to the correct output device.

(c) *Decrypt and extract protected content.* The file management tool could be used (either by the user but more probably by a rogue executable) to retrieve the CEK (the key used to encrypt and decrypt the content) for the encrypted content and make it both available to an executable which decrypts the content (either using the encryption functionality on the terminal or its own[4]) and makes it available for transport off the terminal.

A variant of this attack is where one user extracts the CEK from their device and then posts it on a web site. Another user may then, with more ease than the initial CEK extraction, use the downloaded CEK to decrypt encrypted content on their device.

This implies that there should also be access control on encrypted files and that it should not be assumed that they can be generally accessible just because they are encrypted. The protection of the CEK on the device in question may be good, but if CEK protection on other device types is not so good, then this will affect the strength of the content protection on the first device, if its encrypted files do not have adequate access control applied.

(d) *Change rules attached to content.* The file management tool could be used to retrieve the usage rules associated with the content and make them available to an executable which changes the rules, for instance, to give the user unlimited access to the content.

If the rules are protected with symmetric integrity protection mechanisms, such as Message Authentication Codes (MACs), where the same key is used to verify the integrity as is used to generate the integrity protection data, then this key could be retrieved by the file management tool and made available to the rogue executable which generates compliant MACs for the modified rules.

Only if the usage rules are digitally signed, an asymmetric technique where the key used to verify the integrity (the 'public key') is not the same as that used to generate the integrity protection (the 'private key') is this attack impossible (assuming the rule parsing software in the device is valid).

(e) *Replace the compliant DRM client.* This could be done on the terminal with a non-compliant DRM client, either partly or wholly.

The executable download mechanism of the terminal OS could be used to download a DRM client that was non-compliant. The terminal could then be configured so that the (new) non-compliant DRM client received all the data (protected content and usage rules) that was destined for a compliant DRM client.

Alternatively, if the complaint DRM client existed on the terminal in modular form[5] or in an interpretable and non-compiled form (e.g. Java) then an attacker could perform the above replacement attack but need only replace individual modules of the DRM client or indeed need only modify certain lines within the interpretable code of the DRM client. The module that parses the rights would be an obvious module to compromise, as would the audio or video device drivers.

(f) *Use of debug tools.* Debug tools could be used to perform either content or CEK extraction. Terminal implementors need to make sure that boot into debug mode by the terminal is not allowed.

There may be tools that can be installed on a device that can look at 'random' memory locations, but they may need to know precisely where to look and when.

(g) *Use of other system components.* The OS also needs to protect against the use of other system devices, such as DMA engines, to ensure that they can not be used to bypass the protection on areas of memory.

(h) *Replace the OS.* If the device OS is in EEPROM, then the authorised OS can be replaced with a compromised OS by 'reflashing' the OS. Clearly re-writing an OS is a highly non-trivial task but in some cases critical functionality of the OS can be compromised by just manipulating selected bytes within the OS code. In such a case, reflashing and OS compromise are possible if the location of these critical bytes within the OS is known.

It therefore seems very sensible that implementations should not include, perhaps for test purposes, some internal constant having an ON or OFF value that the DRM UA examines run time to see if DRM restrictions should be observed or not.

(i) *Exploitation of security flaws in the OS to give to access protected areas.* As a particular OS becomes heavily used on mobile terminals, it may become subject to considerably more attack by the hacking community. This may lead to weaknesses in the OS being discovered. It may be possible to exploit these to compromise the security of the system.

16.4 Insertion of non-compliant DRM UAs

As the capabilities and prevalence of executables environments on mobile devices is increasing rapidly, it is appropriate to consider separately the threat of replacement of a complaint DRM UA with a non-compliant DRM UA.

As had already been stated, Java MIDP 1.0 [5] contains enough functionality to build a DRM UA that could request content by the forward lock and combined delivery methods and then strip off the protection on the content. A compliant MIDP

1.0 implementation should only allow a midlet to share its data with other midlets in the midlet suite in which it was delivered. An extraction midlet should not therefore be able to make extracted content generally available to the phone functionality for instance, to another media player on the device. However, an extraction midlet could post the extracted content to a web site using its http capability. Every MIDP 1.0 device could therefore be a potential content extractor. Devices that support the Messaging JSR 120 [6] or MIDP 2.0 [7] should strictly enforce the defined restrictions on the SMSs that midlets can have access to and must only allow midlets to have access to SMSs sent with the ID of the midlet and must not allow midlets to have access to SMSs sent with the tag prescribed for the rights in [2]. There are JSRs for multimedia capabilities [8] and the use of midlets using APIs from this JSR in combination with midlets using APIs from MIDP 1.0 and 2.0 should be investigated to see if it is possible for a midlet suite to be designed that could rip and render content on the device.

Non-compliant DRM UAs can probably also be written to run within other executable environments and implementors of these environments should investigate this.

It is therefore important that in designing future DRM standard solutions, the present and future capabilities of executable environments are taken note of. If all the security mechanisms that are specified for a DRM UA can be performed within an executable environment, then there must be some robust method within the OS or executable environment of only allowing suitably authorised executables to access critical parts of the functionality that could be used to build a DRM UA.

In designing the capabilities of future executable environments, the possibility of these environments being used to build deliberately non-compliant DRM UAs should be considered. Standardisation and design of such executable environments should not extend the functionality to include critical functionality within a DRM UA unless there is also some way to only allow this to suitably authorised entities as described above.

With regard to authorisation of particular entities to give them access to critical DRM UA functionality, the complexity of the design and operation of this authorisation framework should not be underestimated. Delegates involved in the design of MExE [9] and MIDP2.0 and the Recommended Practice document for MIDP2.0 (RP MIDP2.0) [10], specifications will know the considerable discussion that took place regarding the relative and absolute capabilities of (authorisation) domains (e.g. 'operator domain', 'manufacturer domain', 'third party domain'). If such authorisation frameworks can be used to designate entities that are authorised to produce DRM UAs (that may, if the authorised entity does not behave correctly and so produces non-compliant DRM UAs that can be used to extract premium content) the liability issues alone are extremely complex. It seems that in order for the device supplier to be seen in any meaningful way to be 'responsible' for their device, that the only entities that are authorised to produce DRM UAs should be those explicitly authorised by the device supplier (that is to say, in terms from [9] and [10], that the ability to construct a DRM UA should be within the manufacturer domain only).

It seems better in some ways, and certainly easier, not to allow executable environments to have the capabilities required for the construction of DRM UAs in the first place.

16.5   Security features for DRM on open OS devices

The following security requirements and features are believed to be necessary to meet the threats listed in earlier sections. Many of the recommendations below cannot be standardised. Where the measure can be standardised, this is mentioned and Section 16.6 will provide a summary of recommendations for additional security features for future standardised DRM.

There are, in addition, some non-cryptographic measures that can be taken to reduce DRM evasion. Two significant such measures are as follows.

- The device UI should make it obvious to the user that they are doing something illegal – this may deter many users that have downloaded a hack.
- The DRM standard should make it really easy to get hold of legal content – one of the reasons Napster took off was that it was really easy to use and find music.

Content confidentiality whilst in transit. This can be provided either at the application layer, in that the content itself is encrypted, or by mandating use of end to end encryption (e.g. TLS) at the transport layer. Note that if symmetric application layer content encryption is used, as for the separate delivery case in Reference 2, the CEK must also be encrypted in transit and not sent in the clear as for the separate delivery case.

Content encryption can be standardised.

Content confidentiality on the device with respect to unauthorised executables. This can in theory be achieved by storing the content in plaintext but only allowing access to authorised executables (the authorised DRM UA) but this leaves open the way to access to the content by physical attack on the device, leaves the content exposed if faults are found in the OS. Therefore, it seems sensible to provide content confidentiality on the device using mandatory encryption of the content itself at the application layer, when in transit and whilst stored on the device. This defeats attack (a).

However, it should be noted that although content encryption would allow the content, when encrypted, to be 'accessible' to any executable on the device or even exported off the device, access to the CEK and the content when in plaintext form, must still be highly controlled.

This measure cannot be standardised as such, as it depends on device implementation. However, standardisation can aid good implementation. Mandatory content encryption by the content distributor and mandatory secure delivery of the CEK to the device by the content provider will allow easier implementation of secure access control to the content whilst on the device, in that access to decryption functions and APIs for the encrypted content can only be given to authorised executables on the device.

CEK confidentiality on the device. This can be provided, in theory, if the CEK is in a secure location on the terminal where it cannot be accessed by file management tools accessible by the user, by downloaded executables or by any executable save the authorised DRM UA. This defeats attack (c). As the CEK is delivered to the terminal it must not be possible for executables on the terminal to obtain the CEK as it arrives at the terminal and prior to its storage in a secure location. This implies that the CEK should not be delivered in a manner whereby it might reside in some location accessible by a generic file management tool at any time.

However, the same remarks apply to the provision of content confidentiality on the device by merely restricting access to the content to the authorised DRM UA only. This leaves the CEK open to discovery using physical means or through weaknesses in the device OS. Therefore, to avoid the threat of physical attack or OS faults, the CEK can itself be encrypted in transit and whilst in storage on the device. Of course, the same remarks apply to the storage on the device of the key that is used to decrypt the CEK and, at some point, the decryption key must exist in plaintext. It is therefore recommended that the CEK decryption key is stored securely using physical means, that is, in a Hardware Security Module (HSM).

As it is desirable that a number of content distributors can encrypt a CEK for each device, it is appropriate that public key cryptographic techniques are used for CEK encryption. That is, the device possesses a private–public key pair. The private key is stored in a HSM on the device and the public key is presented to the content distributor in an integrity protected form, that is, in the form of a certificate issued by an appropriate entity that the content distributor trusts. The content distributor then encrypts the CEK using the public key of the device.

It must not be possible to extract the private key from the device HSM. Further, the HSM must only accept requests for CEK decryption from the authorised device DRM UA.

In addition, CEK encryption using public key methods means that the method of transporting the CEK to the device does not need security features itself. This removes the need for transit encryption at the transport layer or for the use of transport mechanisms, for example, SMS, as used for separate delivery, that are perceived as harder to intercept in transit or on entry to the device than transport mechanisms involving the device browser.

CEK encryption by public key methods can be standardised, including the content of the device public key certificate. The resulting flexibility in the (encrypted) CEK delivery methods can also be standardised.

Temporary copies of the decrypted content. These must not be left after decryption and software must not be allowed to monitor the signals on internal buses if these are carrying content in plaintext form. This defeats attack (b).

Temporary copies must not be held in files. However, it is also important to understand how much protection a particular OS provides between different executables. If inter-process-communication is allowed by reading and writing into the memory space of other executables then there is still a risk even without the creation of any temporary files.

Integrity of the usage rules whilst in transit and on the device. It must not be possible for the rights and its usage rules to be manipulated whilst in transit or on the device.

Integrity of the rules whilst in transit can be provided using transport layer security mechanisms such as TLS [4]. Alternatively, application layer integrity mechanisms could be used on the rights themselves. There are a number of options here.

Symmetric methods, such as MACs, could be used but this would require the establishment of a shared secret key between the content distributor and the device. As it is desirable that the device be able to use a number of content providers, it would not be possible to use a provisioned secret key (e.g. one held in the SIM) on the device. Use of public key cryptography would provide this flexibility.

One option for provision of transit integrity would be for the content provider to digitally sign the rights. This would require the content distributor to acquire a key pair and certificate and for the device to possess the root key that could be used to verify the content distributor certificate. This also provides for content provider authentication (and if certain conditions are imposed on the issuance of a certificate to a content provider, content provider authorisation).

An alternative means would be for the rights to be encrypted with the device public key, along with the CEK, as recommended earlier. However, it is always good security practice to integrate protection with explicit integrity protection mechanisms such as MACs, and not to provide integrity protection with encryption.

Therefore, if CEK encryption using the public key of the device is being performed in any case for CEK confidentiality, it seems sensible to use the CEK to also calculate a MAC over the rights to provide rights integrity whilst in transit. Alternatively, a separate Content *Integrity* Key, could be used to generate the MAC and this sent to the device encrypted, along with the CEK, with the device public key.

Integrity of the rights on the device could be provided if the device stored the usage rules in a secure location on the device so that the rights are not accessible by generic file management tools, which could be used to make the rules available to rogue executables that would change the rules. However, the same comments apply to this method as to other methods of providing security features using access control on the terminal.

If the rights are being encrypted with the device public key, for integrity in transit, it might be thought that this method could also used to provide integrity whilst *on* the device. However, it would be useful for the user to be able to see at any time the rights that he/she has acquired to some content (whether for five plays, or unlimited). The process of providing this information would be made more difficult if the rights had to be decrypted using the device private key each time the user wished to see them. The use of a symmetric integrity (but not confidentiality) means in combination with the use of the device public key for transport of the rights integrity key, as mentioned above, would be one method or using the device public key to allow the device to have integrity relationships with multiple content providers whilst allowing easier visibility of the rights to the user when on the device. However, this would require secure storage of the integrity protection key in the clear on the device (if the integrity protection key were to be stored encrypted with the device public key, this would still leave the problem of easy user visibility of their license rights), and would therefore be dependent on access control mechanisms on the device.

In all cases mentioned, rights integrity protection whilst in transit from the content provider to the device can be standardised, as can content provider authentication.

Device type authentication. One of the conclusions of Section 16.2 was that content injection could not be defeated and that we should concentrate our efforts on preventing content extraction and on device rights manipulation. One clearly effective method to prevent both of these is to prevent transport of protected content to devices of a type that is known to be susceptible to these attacks. This method requires that the content provider be able to identify and authenticate the device type before content is distributed to the device.

As it is desirable that the device be able to authenticate itself to many content providers, it is appropriate to use public key methods. The device could therefore possess a public–private key pair, with the private key held securely in an HSM on the device and the public key available to be provided to the content provider in an integrity protected form, that is, as a certificate that the content distributor can verify. As it is the device manufacturer who has provisioned this private key, it seems appropriate that the device manufacturer should issue the certificate for the device public key.

Note that the use of a private–public key pair assigned to the SIM is not sufficient in this case, as this would provide authentication of the SIM but not of the device. It is the device type that has to be identified and authenticated, as it is on the device that DRM evasion will take place.

It might be argued that if a particular device type possessed robust reprogramming protection, so we could always be sure that devices actually of that type would perform in a compliant manner, we could simply use device *identification*, and ask the device to provide its IMEI (say) and use some sort of transport level integrity protection to prevent manipulation whilst in transit. However, an IMEI could be extracted from such a compliant device and used by a reprogrammed and non-compliant device to masquerade as a compliant device.

Equally, device type authentication does not guarantee that a device will be acceptable to a content provider unless it is accompanied by reprogramming protection. That is, device type authentication just does that – it does not of itself guarantee that the device behaves or will continue to behave in compliance to rights. The device supplier must ensure that their devices cannot be reprogrammed in an unauthorised manner, and communicate this to content providers, as well as providing secure device type authentication.

The above text has all related to device *type* authentication as it is assumed that DRM vulnerabilities in a particular device type will exist for all instances of that device type. If device type authentication is provided using device public–private key pairs with a public key certificate indicating the device type, the key pair can also be used for authentication of the device instance itself. It may be useful to do this to spot certain frauds, so the method of device type authentication should also allow device authentication. This amounts to requiring that the device certificate contains a unique device identifier in addition to its device type identifier.

Device type and device authentication can be standardised.

Subscription authentication. This is required to prevent an attacker posing as a legitimate subscriber when requesting and purchasing content so that the cost of the content goes on the legitimate subscriber's account.

It is for discussion whether subscriber authentication should be provided within future DRM standardisation or whether existing means, such as UMTS/GSM authentication and derivation of the device MSISDN are sufficient. Needless to say, a DRM standard is not of use unless there are robust billing mechanisms available for content distributors.

Also, there may be some value and convenience in performing both subscription authentication/identification with device (type) authentication to the content provider. This would allow the content provider to spot a single device identity being used with a large number of subscription identities (evidence of possible device private key compromise and distribution) or the reverse (evidence of a single subscription being used for multiple devices, 'SIM stuffing').

Protection of the integrity of the authorised DRM UA. This covers protection of this and related critical components, and prevention of the replacement of these by unauthorised, potentially non-compliant UA or other critical components.

One method to achieve this is the provision of the DRM UA and other critical components on the device at manufacture, in a fully compiled and non-modular manner and for them to be integrated with the terminal OS, or the device OS be configured such that the DRM UA cannot be replaced.

Alternatively, if the download of the DRM UA or other critical components is allowed after manufacture, this method should ensure that only authorised and integrity protected components can be downloaded and run on the device.

Per device CEKs. In such a case, content is not encrypted with a single CEK for all users but is encrypted with a different key for all users, or there are a number of CEKs for a particular piece of content. This is useful if it is much easier to write and/or use a downloaded executable to decrypt encrypted content with a CEK downloaded from a web site than it is to obtain the CEK from the device itself and then decrypt the content. If the former is easier, then someone will just hack their device and then post the CEK (and the ripping executable) on a web site for everyone else to use to rip their own instance of the content. If we make the assumption that encrypted content will be stored in the normal device store, and that use of APIs to decryption functionality on the device does not require authorisation, then it may well be the case that content decryption using a downloaded CEK will be easier than hacking one's own device to obtain the CEK.

It is thought that per device CEKs stop superdistribution, but if a CEK identifier (CEKID) is generated at the same time as the CEK and delivered with the DCF, a device that has received the DCF by superdistribution can submit the CEKID with the rights request and be given the correct CEK. This assumes however that the content provider can:

- perform on the fly CEK generation and content encryption, and
- keep a database of CEKs and CEKIDs.

These tasks are non-trivial. Instead a content provider might generate a number, say 50, CEK and CEKID pairs, store them on her server and randomly or cyclically choose one version of the encrypted content to distribute when requested. This removes the need for on the fly encryption and CEK generation but does give some division of the device population with the content and some reduction in the usefulness of a CEK posted on a web site.

Protection of the integrity of the device OS. Clearly, in all the cases mentioned, it is imperative that the device OS itself is not compromised.

If the OS is stored in ROM on the device then it has some level of intrinsic hardware security. However, it is more likely that much of the device OS, apart from a small 'boot' component, will be in EEPROM or flash memory. In such a case, the device must have mechanisms to prevent the 'reflashing' of the device OS with a version of the OS that allows DRM evasion. One method to prevent this reflashing would be the requirement for the device OS to be digitally signed and for the hardware boot OS to verify the signature on the flashed OS before it is accepted and run.

16.6   Standardisation recommendations

The following security features are regarded both as useful and as functionalities that can be standardised.

- Mandatory content encryption (as opposed to the optional content encryption in [2]).
- Secure CEK delivery to the device. As it is recommended that device type authentication is provided using a device public key, this device public key can also be used to provide secure CEK delivery to the device.
- Integrity protection of rights from content provider to device.
- Device type and device authentication using device public key techniques.

The following are for further discussion.

- Content provider authentication/authorisation. Under some fraud scenarios (commercial content injection by organised crime), it is useful to be able to authenticate/authorise content providers. However, this does impose some conditions on content providers that might be seen by some as unacceptable.
- Per device content encryption keys.

Acknowledgements

The following made valuable contributions to discussions vital in the production of this document: Oliver Bremer and Markku Kontio from Nokia; Craig Heath and Will Bamberg from Symbian; Nick Bone from Vodafone. The final document, however, represents the views of the author alone.

Notes

1   There will be some identity associated with the content that could perhaps, be put on some sort of content revocation list that is propagated to devices. However, the content identity will have no cryptographic association with the content once the

content has been ripped and can therefore be changed by the device performing the content extraction.

2   However, web site closure is no good against true peer to peer file sharing methods.

3   We assume there is only one class of user with regard to the terminal which has maximum privileges with respect to functionality that the terminal makes available to the user (editing of the OS kernel for instance, is not made available to the user). We do not assume the existence of users with extra privileges, an example of which would be the Windows Administrator.

4   The rogue executable may need to do its own decryption if the APIs to do this are not exposed. It is perfectly possible for the executable to have its own decryption – an encryption algorithm can be implemented in about 20 Kbytes.

5   A DRM UA might be modular so that only those elements of the UA that need high privileges on the device are given those privileges.

References

1   http://www.openmobilealliance.org

2   *Digital Rights Management*, Candidate Version 1.0, OMA-Download-DRM-v1_0-20020905-L. At http://www.openmobilealliance.org

3   *Download Architecture*, Candidate Version 1.0, OMA-Download-Arch-v1_0-20020610-L. At http://www.openmobilealliance.org

4   *Transport Layer Security*, RFC 2246, www.ietf.org/rfc.html

5   http://java.sun.com/products/midp

6   http://www.jcp.org/jsr/detail/120.jsp

7   http://www.jcp.org/jsr/detail/118.jsp

8   http://www.jcp.org/jsr/detail/135.jsp

9   Mobile Execution Environment (MExE); Functional description; Stage 2, 23.057

10 Available to members of JSR 118, soon to be public

Part V

The future

This final part contains a single chapter (*Pioneering advanced mobile privacy and security*) devoted to a roadmap for the future development of mobile privacy and security. This chapter covers a wide range of topics where future research is needed to develop security solutions for future mobile systems, and has been developed as part of a series of European research framework projects.

Chapter 17

Pioneering advanced mobile privacy and security

Robert Hulsebosch, Christian Günther, Günther Horn,
Silke Holtmanns, Keith Howker, Kenneth Paterson,
Joris Claessens and Marko Schuba

To achieve the goal of security and privacy in future mobile communication networks, further research and technology development will be required. The roadmap presented in this chapter pioneers the boundaries of mobile privacy and security from a broad perspective. It registers the mobile privacy and security requirements of the actors on whom the success of future mobile communications, systems and services depends, it gives an overview of the current state-of-the-art in mobile security and privacy, addresses non-technical aspects, conducts a SWOT analysis from a European perspective and ultimately identifies the areas where research, standardisation and development will be most needed and beneficial in the coming years, so that we can progress rapidly and efficiently towards a trusted mobile environment.

17.1 Introduction and overview

Although the basic needs for privacy and security in wireless communication are widely known and accepted, there are still gaps between these needs and the state-of-the-art. The existence of such gaps immediately leads to the question: how wide are the gaps? Is only a bit further effort required to close them, can they be closed in the next few years, or is it rather unlikely that they will be closed at all within the next 10 years?

But the width of the gaps between existing and the desired solutions is not the only important issue. Since there are so many gaps and only limited resources available, it is essential to partition and prioritise the work so that there is no waste of valuable effort. Therefore, a second question has to be answered: what should the priorities be when closing gaps? What challenges are the most urgent to solve?

This roadmap answers both questions for future research challenges in the area of mobile privacy and security.

17.1.1   Future mobile security and privacy work

The fundamental requirement, or rather assumption, that is identified here is the need to establish foundations for security provision at the outset of the design and development of the next generation of mobile communications technology.

The goal is secure, trusted communication between people, and access to information and services at our finger tips (probably hands-free, actually) – pervasive, ubiquitous, ambient – wherever and whenever we want it.

There are no real surprises in the requirements that are identified: the headline topics are the same as ever. What is new is a change of emphasis and changes to the complexities and difficulties of providing security. The history of IT and communications security runs alongside the dissemination of technology and the provision of ease of access.

The scope of this next generation is as yet undefined, but it will paradoxically reach out in one direction towards greater integration with the fixed networks and their established procedures and in the almost opposite direction towards ad hoc networking with its opportunities (or threats) for free-for-all quasi-anarchy. The roles of network operator and service provider will become increasingly fuzzy as responsibilities move from well-established commercial entities to less well defined niche operations or even individuals. None of this can really be for free, especially in a society where information and other digital assets are seen as high-value commodities, so there remains an increased need for recording, collecting and distributing costs, charges and payments.

Ad hoc networking is not the only wireless development that raises the security stakes for mobile networks. Packet-switched data, seamless roaming with heterogeneous access networks, reconfigurable terminals and software-defined radio all add to the complexities of delivering an uninterrupted, secure, high-quality service to users. There are implications for all the usual security primitives such as authentication, confidentiality, integrity, availability, etc. The problem as ever is to devise protocols to do the job and then to deliver the right keys to the right places at the right time – simple!

A parallel need that emerges is for some persistent control over owned information. This may be personal or enterprise data that is to be protected in some specifiable way, or it may be system information concerning identity or location relating to the user. In both cases the user is looking for rights over its use, propagation and disclosure.

This is well illustrated by the scenario of the roadside medical emergency where the victim/patient is unconscious, her mobile terminal is active and possesses, or has access to, personal identity, medical and location information vital to her expeditious treatment. On the one hand is the recognised need to maintain the privacy and confidentiality of location, identity and personal data; on the other hand is the life and death interest of the user in ensuring that in certain circumstances such information is not only released but may even be multi-cast to appropriate recipients. A collateral need then is for those recipients to allow visibility of their location and profile information to certain (authorised) services whilst maintaining vigilance against junk/spam traffic.

The key need here and in many other situations that has been identified in the course of our study is the protection and control of information once it has left the shelter of trusted elements in the server or the user's terminal configuration. How can we ensure that information – including software, active agents, copyright material, etc. – is used only for the purposes intended by its owner? What 'owner' means may need to be defined in the appropriate legal framework, but sidestepping that issue for the moment, the questions concern the technical means of maintaining control over sensitive or valuable digital assets in a hostile environment with the potential for accidental or malicious loss and for abuse of resources.

A further need is to establish priority communication channels to appropriate recipients – in our scenario, designated emergency services, say, or even anyone in the immediate vicinity who might be able to assist in some way. This is common to many if not all safety-related applications.

The mobile terminal or device must be able to protect its information and control the processes that can access and use it. It may be assumed that the powerful, static server can fairly easily look after itself using established technologies and procedures, but the designer of a trusted mobile device must be very conscious of the practical concerns of costs and power consumption, portability and usability in addition to basic technical feasibility.

These privacy and confidentiality issues point in the direction of a need for common approaches to the provision and use of trusted execution and storage environments. Cryptography traditionally protects our information during transmission, but what ensures that it is handled according to rules or expectations at the end points, particularly in the highly dynamic picture that is starting to appear?

These and more aspects (see Figure 17.1) form the boundary conditions of the playground for future mobile security and privacy research and development.

17.1.2   Current situation and trends

The Global System for Mobile Communications (GSM) is the most popular mobile phone system in the world, accounting for 70 per cent of the world's digital mobile phones. One of the key features of GSM is its relatively high level of security. The security methods used for authentication, anonymity and confidentiality are adequate for normal usage, though a number of weaknesses remain (weak encryption and revealing of user identity and location during the authentication process). The success of GSM security is largely determined by the fact that security was taken into account from the start of its development and was defined and embedded in standards. The development of GSM was mainly carried out by European enterprises under the supervision of the European Telecommunication Standards Institute (ETSI). Together with a well-coordinated introduction of GSM, which was supported by the European Union, this proved to be a highly successful approach.

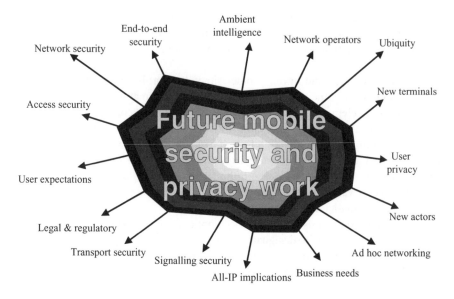

Figure 17.1   Boundary conditions for future mobile security and privacy work

The explosive growth in wireless networks and services over the last few years, in contrast, can be characterised by a notorious lack of effective security. For instance, deployments of the IEEE 802.11b based Wireless Local Area Network (WLAN) standard, one of the most prominent WLAN technologies available today, often feature very little security control and potentially present a 'backdoor' gateway to gain unauthorized access between mobile devices and data, systems and networks. Also, the Wireless Application Protocol (WAP) showed serious security gaps after its introduction and Bluetooth, again, has been widely deployed without adequate security measures. Vulnerabilities in these mobile technologies leave them open to eavesdropping, session hijacking, data alteration and manipulation, and an overall lack of privacy.

Security measures are gradually being added, but these additions do not offer a sufficient level of service for mobile users. Current solutions lack the flexibility and interoperability that is needed for securing mobile data communications.

Limitations and inherent vulnerabilities of wireless security are:

- access points are easily accessed by users outside the physical location;
- eavesdropping is relatively easy, difficult to detect and can be done from afar;
- many wireless/mobile technologies do not have built-in authentication measures;
- data encryption and key management procedures have been found to be flawed many times.

Despite the existence of mobile security solutions, often they are not used to their full extent: security may be switched off or a sub-optimal level of security is used. Why is this? The reasons for omitting or neglecting security vary. Most enterprises still do not apply security measures because they inconvenience workers and managers alike, and are widely seen as both acting as a brake on progress and as being counter-productive. Moreover, most information technology security breaches and incidents take advantage of known, patchable flaws that exist because of poor enterprise security practices, bad corporate culture and a lack of investment in system protection. Other reasons for poor security are:

- rapid development of mobile technology to meet high user expectations;
- implementing good security is seen as too much work in a compressed time line;
- building secure software is a complicated endeavour, particularly because software technology is constantly evolving and many companies have yet to focus on software as a critical piece in the security puzzle;
- many industries just did not emphasise security in the past;
- despite affected sales, many industries have deferred the decision for mobile security until a proper solution is found;
- most consumers are not concerned about security today;
- careless software installation and maintenance (e.g. security patches are not installed, WLAN encryption protocols are not turned on, etc.);
- managing passwords is seen as burdensome; consequently, many passwords are shared and rarely refreshed;
- laws in certain countries restrict the length of the encryption keys used for export, import or use;
- limited processing power and storage capabilities of the mobile device;
- people are not aware of the potential threats and their consequences when security is not in place and need to be educated about the risks;
- current solutions lack the flexibility that is needed for securing mobile data communications and are, therefore, switched off.

From these examples, it is obvious that often a combination of technical, human, social, economic and legal factors lie behind incorrect use of security.

With regard to privacy, the mobile industry has not been particularly active. There are many reasons for this. The primary reasons are that the technologies that cause privacy threats are still in their infancy and there is a general lack of understanding of privacy issues within the industry. This has resulted in privacy aspects not being considered in the design phase of many new systems. Furthermore, commercial incentives to protect the user's privacy are small. Instead, many companies have been driven by the promise of big profits related to the use and release of customer information.

Fortunately, a long-term cultural shift seems to be underway. Digital security has been growing in importance in recent years, as more and more aspects of business and personal life have come to depend on fixed and mobile terminals. As the mobile terminal will increasingly be the de facto personal communication and computing device, new technologies, facilities and capabilities will generate new requirements for security of user and service information, protection of services and other digital assets, safety and integrity of management and control of underlying systems and infrastructure. New trends include general ubiquity, new context-aware applications and services, new network and terminal technologies, flexible spectrum management and dynamic reconfiguration of terminals and networks in response to user mobility, user behaviour and capacity optimisation. Most of the trends may have an impact on the user's privacy because additional user profile attributes in a mobile context include location, context and terminal capability. Users are more and more aware of the impact of these developments on their personal privacy.

In short, computing is in the midst of a transition from an optional tool to a ubiquitous utility. And people expect utilities to be reliable and secure.

17.1.3   Integration

Privacy and security of mobile information systems have been approached in the past mainly from the perspective of technological dependability (availability, security, confidentiality, etc.). However, in order to satisfy the growing demand for mobile communications and mobile multimedia more is needed than the development of enhanced technologies alone. Recent developments in information and communication technologies, users' expectations and business models make it imperative that we take a broader perspective to privacy and security embracing the technical, socio-cultural, business and legal aspects. Effective technology cannot be divorced from socio-economic and legal factors. Whatever requirements and mechanisms are developed and deployed, they must either leave users with confidence that their security needs are safeguarded, or (perhaps equivalently) they must engender sufficient traceability to support litigation. Therefore, these three essential elements need to be integrated (see Figure 17.2).

In the light of the increasingly important role played in the economy by mobile services, the security of mobile networks and information systems is of growing public interest. Networks and communication systems have become a key factor in economic and social development and their availability and integrity is crucial to essential infrastructures, as well as to most public and private services and the economy as a whole. Secure mobile communications opens the way to applications such as mobile payment and finance, mobile ticketing, mobile voting and location-based services with increased convenience and confidence. This will lead to further

Figure 17.2   Aspects of mobile security and privacy

development of an electronic marketplace with higher efficiency and lower prices for consumers. It is also clear that in the applications of mobile communications to public safety and healthcare, security is an essential prerequisite. Socio-economic factors for mobile security are:

- adequate security is prerequisite for new ways of working and new types of business;
- security as a value creator (business perspective);
- encouraging and securing competition;
- acceptance;
- benefits for society.

Ubiquitous and secure mobile communications can bring enormous benefits to the citizen. Technologies that enable the protection of the privacy of citizens will benefit them by reducing misuse and improving the protection of privacy-sensitive information. Moreover, effective research to ensure the necessary adoption of these technologies will require concerted effort from a large segment of the mobile industry and its clients. Technology factors for mobile security are:

- need for privacy preserving technologies;
- authentication and access control;
- standardisation of mobile security;
- infrastructure protection.

We shall not only address technological and standardisation issues, but also assess possible impacts on regulation and international legislation. These issues should not be solely left for industry self-regulation, especially since the area is currently dominated by very few and mostly US-based companies from the IT/Internet industry. Moreover, technology alone will not provide the full solutions to all challenges in wirelessly connected (ad hoc) networks. Legislation, together with enforcing mechanisms, will always play an important role in such solutions. Legislative/regulatory requirements for mobile security are:

- foster conditions that will enable future digital services to thrive;
- protection of human rights.

The strengths of a regulatory body are often characterised by a methodical and integrated approach to the development of competition, regulatory practice which has managed to bring industry together to find solutions to complex regulatory and technical problems in a consensus seeking framework, and recognition of the importance of the telecommunication service industry in the fabric of the new economy.

Ultimately, the legal/social/technological mix needs to be carefully balanced. For instance, the amount of privacy must be in harmony with the overarching culture and must keep pace with technical innovations. One of the key strengths of the mobile industry in Europe, and the main reason for its worldwide success and dominance, has been its approach of 'co-opetition', that is, the well-balanced mix of cooperation and standardisation at an early stage, with competition in the market once products and solutions have matured. To continue this success story, a basis for an early, pre-competitive agreement on key areas for research and standardisation in a field that will be crucial for beyond-3G systems and applications must be created.

Integration of all these aspects is essential for the success of future mobile communications. As such there is a need to mobilise all parties with a stake in mobile communications to ensure the desired impact in the beyond-3G environment. Parallel and incompatible developments in different sectors must be avoided. After all, security is not an end in itself, but is security for something/someone.

17.1.4   PAMPAS – the European perspective on mobile security and privacy

Mobile communications has been one of the great success stories of European technology and industry over the past decade. Hundreds of millions of users around the globe are using mobile technology developed in Europe. Mobile data applications such as Internet access, mobile messaging, mobile banking, mobile payment, mobile health and location-based services are gaining in importance and becoming the driving force for the transition towards third-generation mobile systems and beyond.

Europe is in an excellent position to continue to play a leading role in mobile communications. However, the current lead of European industry over their competitors in North America and the Far East must not be taken for granted. It must be ensured that the key factors that were a pre-condition of the existing success are also in place as a basis for future success. One of the steps towards this goal was made by the PAMPAS project. The main objective of the PAMPAS project was to produce a roadmap for research and standardisation for security and privacy of beyond-3G mobile systems and applications. PAMPAS was partially funded by the European Union. The PAMPAS consortium encompassed a broad range of competencies and experience from different industrial and academic organisations in Europe. The PAMPAS roadmap presents recommendations for further research in the area of mobile privacy and security that are interesting and valuable from a European perspective. Nevertheless, we believe that the results are also of general interest. They should be seen as a contribution to the ongoing debate about security and privacy. We hope that this debate will result in a successful approach to tackle the knotty mobile security and privacy matters in future projects.

17.1.5   Roadmapping mobile privacy and security

The process for constructing this roadmap can be divided into three major phases. In the first phase, we identified the mobile security and privacy requirements of the actors on whom the success of future mobile communications, systems and services will depend. Stakeholder position papers and ambitious and imaginative scenarios for mobile security and privacy in the near future (<2010) that challenge both applications and enabling technologies were used for identification of the requirements. The results are outlined in Section 17.2. A comprehensive state-of-the-art review, taking into account ongoing research activities as well as developments in standards bodies and user forums, was performed in the second phase. This resulted in a list of existing and emerging security and privacy-enhancing technologies. A summary of this review is presented in Section 17.3. The gaps between the needs and the state-of-the-art resulted in an extensive list of preliminary recommendations for further research. To further refine and balance the priority of the research issues, non-technical aspects (socio-cultural, economic, legal) as well as a SWOT analysis from a European perspective were also taken into account (Sections 17.4 and 17.5, respectively). Section 17.6 concludes with the main topics of needed and beneficial research and standardisation in the area of mobile privacy and security.

17.2 Requirements for mobile security and privacy

17.2.1 Need for security and privacy

Security is a basic feature of any public communication infrastructure whether fixed or mobile; it must provide user confidence and economic opportunity and must protect the values of society. Security of information and communications plays a funda-mental role in ensuring that citizens realise benefits from these services. Security establishes a proper environment for the protection of privacy and confidentiality for the conduct of all aspects of personal, economic and administrative activity carried out over networks.

The potential for mobile commerce to develop as a new and additional economic sector emphasises the need for security and privacy. Free flow of trusted commercial data over mobile communication networks provides new opportunities for economic activity. Furthermore, commercial users need to be able to establish trust between themselves and the entities – human or automated – in organisations with which they are doing business.

Already most communication of sensitive data on fixed networks is subject to some protection. However, because of the dynamic topology and connectivity, secu-rity is fundamental to the successful operation of all aspects of wireless systems and must be given consideration in all new research and development undertakings in this field. It must be an essential part of the architecture in terms of placement of functionality, protocols and mechanisms – what goes on where and how.

The following standard security issues need to be addressed in many new contexts.

- *Privacy*: keeping information confidential; preventing disclosure to unauthorised users.
- *Access control*: permitting only authorised users to access specified information and services.
- *Integrity*: providing assurance that information has not been tampered with during handling.
- *Authentication*: providing proof of the credentials[1] of the originator of information or a participant in a service.
- *Non-repudiation*: preventing a participant in a service or transaction from denying having taken some specific action.

The process of identifying requirements was mainly an analysis of extensive sources including position papers submitted to our first workshop, presentations to the workshop, results from the parallel working groups, material produced by the authors themselves, and externally available material (e.g. from the IRG, ISTAG, WWRF, etc.). We also constructed a number of application scenarios, based on diverse realistic assumptions about future technologies, applications and business situations using available input material to describe actors, technologies and communication models of the scenarios. Requirements for privacy, security and data protection were described from each actor's point of view. Requirements were then grouped, structured and prioritised. The requirements list created during this process was subjected to ongoing review through additional inputs. Of particular interest was the work of related roadmaps, including AMSD (www.am-sd.org), RAPID (www.ra-pid.org), RESET (www.ercim.org/reset) and STORK (www.stork.eu.org). In the next step, the research challenges were extracted from the requirements, and a first tentative recommendation of priorities was given.

An outline of requirements for security is given below. The extended catalogue of requirements is given in the PAMPAS deliverable D02, *Preliminary Roadmap: Register of Requirements* [2].

17.2.2   Overview of security requirements

All security features and facilities that have been successful in previous generations of communications need to be carried forward, but now will need to be provided taking into account new dynamics, technologies and topologies. Some requirements are already partially addressed by fragmentary solutions, which may need to be included as legacy technologies, but these should not be allowed to distract from the goal of providing a comprehensive, integrated solution. Some other requirements we identify are not mobile-specific at all, but are long-standing goals (e.g. single-sign-on) with increased importance in the context of always-connected.

The successful evolution of new wireless into the default communications medium will depend very much on user acceptance. Critical considerations include the following.

- What benefits are provided in terms of cost, convenience and saved time, performance, availability, breadth and richness of services? From e/m-commerce to personal welfare to entertainment.
- Can the user trust services to do only what is intended, to do it correctly and to charge the right amount?
- Will services be continuously available? Whatever, wherever, whenever?
- Consistency – although precisely the same feature may not be available everywhere, can the system manage the differences to act and to inform appropriately?
- Is it sufficiently easy to use?

There are some very basic user uncertainties.

- What control can the user maintain over what happens to personal or commercially sensitive data?

- Will the user be under constant attack from latter-day mobile viruses and unsolicited intrusion into privacy?
- Can charging and billing be trusted and how can costs be kept under control?

Security lies at the heart of all these matters. A number of issues are identified as probably near the top of the league table:

- privacy – protection of identity and location information; this recurs throughout the requirements;
- control of personal/sensitive/valuable information once outside the area of the user's immediate environment in the server or terminal;
- misuse, fabrication or distortion of sensitive or compromising information;
- security for generalised roaming and heterogeneous access and the challenges introduced by IP and ad hoc networking including denial of service threats;
- security in the generalised terminal and the blurring of roles and boundaries;
- protection of the terminal against accidental or malicious use of flawed or aggressive software, and detection and management of intrusion;
- charging and payment schemes and systems;
- safety-critical application areas including health and monitoring of plant;
- telematics applications – particularly in-vehicle systems.

Security and privacy requirements grow from a large number of perspectives:

- stakeholders and their needs;
- networks and terminals;
- information and rights protection;
- applications;
- supporting technologies;
- general considerations.

As may be expected this multi-perspective can result in conflicting requirements. For example, between protection of an owner's asset and a legitimate or beneficial need of another user to access or use it.

Several key areas of requirements for future research are identified:

- a universal architectural approach to security facilities and mechanisms for use in network entities and terminals that can be utilised and exploited by users and services, and that protects the networks and services themselves;
- protection of identity, information and associated rights;
- services and applications that offer security and safety to users;
- supporting security and privacy technologies;
- a regulatory framework that supports the provision of services offering security and safety on a well understood legal and commercial basis.

The list of topics in Table 17.1 provides an overview of the full set of requirements given in the PAMPAS register of requirements [2]. This list is structured according to four domains of requirement: user-related, terminal, network, and applications and services.

Table 17.1   Research requirement topics and their relevance

Research requirement	User-related	Terminal	Network	Applications and services
Protection and management of identity, profile, role and attributes	■	■	■	■
Unified, flexible authentication mechanisms for different forms of access	■	■	■	■
Identity and location privacy	■	■	■	■
Trusted identity tokens	■	■	■	□
Pseudonymity and anonymity of appropriate access and payment	■	■	■	■
Privacy confidentiality and integrity of communication	■	■	■	■
Protection of personal information; compliance with privacy legislation	■	■	■	■
End-to-end confidentiality and integrity services	■	■	■	■
Single sign-on, user control, ease of use, security status 'visibility'	■	■	■	■
Trustworthy unified mobile charging, billing and payment mechanisms	■	■	■	■
Non-repudiation mechanisms, traceability of actions	■	■	■	■
Accountability of responsible entities, charging and billing	■	■	■	■
Biometric support for identification	■	■	□	□
Cryptographic support: high performance, constrained environment	■	■	■	■
Key management for new contexts	□	■	■	□
Protection of information held in terminal	■	■	□	□
Secure, trusted execution environment	□	■	■	■
Authentication and integrity of downloaded software	■	■	■	■
Controlled legal disclosure of information and of user identity	■	■	■	□
Protection of signalling and control information	□	□	■	□
Protection of network services	□	□	■	□
Protection of user traffic and stored information	■	■	■	□
Protection of user identity and location information	■	■	■	□
Mobility management	□	□	■	□

Table 17.1   Continued

Research requirement	User-related	Terminal	Network	Applications and services
Authenticity, integrity and freshness of authentication data and keys	☐	☐	■	☐
Control of quality of service of delivered service	☐	■	■	☐
Unified access mechanisms	☐	■	■	☐
Defence against attack on availability (e.g. denial of service attack)	☐	■	■	☐
New issues arising from ad hoc networks, software-defined-radio (SDR), etc.	☐	■	■	☐
Business models and commercial considerations	■	■	■	■
Protection of rights of information owners and service providers	■	■	■	■
Conditional access, copy protection, control of user operations	■	■	■	■
Supporting critical technologies, e.g. power supplies/consumption	☐	■	■	☐

■ Relevant
☐ Not relevant

17.3   State-of-the-art mobile security and privacy technologies

This section gives a concise survey of the state-of-the-art security technologies relevant for future mobile communication systems. Significantly more detailed information on these technologies can be found in Section 2 of PAMPAS deliverable D03 [3].

17.3.1   Network security

GSM provides security functions for checking the subscriber identity authenticity (challenge – response method), for protecting the subscriber anonymity (use of temporary identities) and for encrypting user and signalling data (use of a stream cipher). The Subscriber Identity Module (SIM) contains all the security-related data the subscriber needs to make or receive a call. The security functionality of the General Packet Radio Service (GPRS) system is similar to GSM. The Universal Mobile Telecommunication System (UMTS) security features are divided into network access security, network domain security, user domain security and application domain security features, plus an additional group of features concerning the visibility and configurability of security.

The most salient security issue for the Mobile Internet Protocol (Mobile IP) is the problem of how to authenticate the registration messages (binding updates) that inform a home agent about a mobile node's current IP address. In Mobile IP, version 4, this issue is solved by a protocol-specific authentication extension, which is based on a secret key shared between the mobile node and the home agent. In contrast to that, Mobile IP, version 6, reuses the IP Security (IPSec) protocol to secure the binding updates. The required IPSec security association can be based either on manual configuration or results from a prior execution of the Internet Key Exchange (IKE) protocol.

The radio wave based, short-range communication technology Bluetooth provides an authentication mechanism for peer device authentication and an encryption mechanism that can be applied to the payload (user and signalling data) of baseband packets, but it does not support link layer integrity protection. The WLAN communication technology, as specified in the IEEE 802.11 standard (1999), was intended to provide data confidentiality, control of access to a wireless network infrastructure and data integrity, the Wireless Equivalent Protocol (WEP) being the most crucial part of the WLAN security architecture. However, the WLAN security design has revealed significant shortcomings. A Counter mode Cipher block chaining Message authentication code Protocol (CCMP) is currently under development at the IEEE as a long-term improvement making new hardware necessary. The international association Wireless Fidelity (Wi-Fi) has specified the Wireless Protected Access (WPA), which requires a software update only.

The lack of infrastructure in ad hoc networks offers opportunities to attackers. This is due to the fact that the supported routing protocols within an ad hoc network are more vulnerable to attacks as each device acts as a relay. There are two main types of ad hoc network routing protocols, proactive and reactive, which use different techniques for finding and maintaining routes. There are also hybrid routing protocols, which use both proactive and reactive techniques to control a hierarchical architecture.

17.3.2 Transport security

The Secure Sockets Layer (SSL) and the Transport Layer Security (TLS) protocols are situated underneath the application layer and on top of the transport layer. Both protocols provide entity authentication, data authentication and data confidentiality. They can be used to secure the communication of any application, and not only between a web browser and server. They are, in principle, completely transparent to the application. However, in practice, SSL and TLS are often integrated into the application to a large extent. The IETF TLS working group is working on a second, enhanced version of the TLS 1.0 specification.

The WAP is a protocol stack for wireless communication networks, specified by the WAP Forum, which has become part of the Open Mobile Alliance (OMA). WAP is essentially a wireless equivalent of the Internet protocol stack. The communication between the mobile phone and the WAP gateway is secured with WTLS, a wireless variant of the SSL and TLS protocols. At the application layer, WAP provides digital signature functionality through the WMLScript Crypto Library.

I-mode is the proprietary protocol of NTT DoCoMo of Japan, providing Internet services based on the Personal Digital Cellular-Packet (PDC-P) standard and on compact HTML (cHTML), a subset of HTML 3.0. I-mode allows application and content providers to distribute software (Java applets) to cellular phones and also allows users to download applets (e.g. games). Information about the security of I-mode is hard to obtain. What is known is that the I-mode protocols are based on Internet protocols. Since March 2001, HTTP and SSL/TLS protocols are used end-to-end by I-mode. Lower-level protocols are proprietary NTT DoCoMo protocols.

17.3.3   Application and software management

The Trusted Computing Platform Alliance (TCPA) is an industry grouping founded by Compaq, Hewlett-Packard, IBM, Intel and Microsoft. The Alliance now includes over 190 members and has being developing a technical specification to allow the addition of trusted functionality to computing platforms. The main component of a TCPA-enabled system is a Trusted Platform Module (TPM). This module can store cryptographic keys and other information in a tamper-resistant way, and can perform a limited set of cryptographic operations. The TPM can provide a cryptographic proof of the integrity/identity of some aspects of a platform, particularly at the operating system and system software level.

Palladium can be thought of as Microsoft's attempt to provide trusted functionality for computing platforms. While sharing many common features with TCPA (such as protected storage, attestation), Palladium is much more application-oriented. It is intended to 'provide a parallel execution environment to the "traditional" Windows kernel-and user-mode stacks'. This appears to imply that (in the future) there will be Palladium-enabled and-protected applications. Presumably, TCPA and/or Palladium may lead to secure Digital Rights Management (DRM) and content protection on PC platforms.

The process of upgrading or adapting mobile equipment to user preferences or local conditions is called reconfiguration. The increased flexibility and openness of reconfigurable terminals introduce novel security threats. Reconfiguration issues are addressed within the User Equipment Management (UEM, by 3GPP), the Mobile Execution Environment (MExE, by 3GPP), SyncML Device Management (by SyncML Initiative), the OMA Download specifications for generic content download, the WAP ServiceLoad and ServiceIndication messages (by WAP Forum) and the Mobile Information Device Profile (MIDP, by Java Standard Request 118 Expert Group).

Sun Microsystems's Java 2 Micro Edition (J2ME) and Qualcomm's Binary Runtime Environment for Wireless (BREW) are two emerging technologies that provide a new model for online access by allowing applications to be downloaded from the Web. The Common Object Request Broker Architecture (CORBA) is an open, vendor-independent architecture and infrastructure that computer applications can use to work together over networks. The .NET Framework is Microsoft's new cross-language development environment for building client applications and Web Services based on the eXtensible Mark-up Language (XML).

The main driving force for developing DRM solutions is the content industry, and the ultimate goal of any DRM system is to protect digital assets from unauthorised use, thus preserving the business value of the digital content. Consequently, the security requirements and trust models are quite different than those for other data protection and security applications. Several proprietary DRM systems, mainly designed for a networked PC environment, are available on the market. Some DRM solutions do also support mobile phones, consumer electronic appliances and low-end portable computing devices. However, most proprietary mobile DRM solutions are not fully mature and secure yet, and some critical features still have to be added. The OMA has defined the first version of a mobile DRM standard, which includes a rights expression language (REL) and a container format for DRM-protected content, which has been implemented in mobile phones from several vendors. At the same time, OMA is working on a next, extended and more secure version of the mobile DRM standard.

17.3.4   Application security

S/MIME (Secure/Multipurpose Internet Mail Extensions) and OpenPGP (Open Pretty Good Privacy) are standards providing security services for electronic mail. However, the currently available implementations of PGP and S/MIME impose undesirably heavy computational demands on CPU-limited mobile devices, and also impose undesirable message transmission overhead on bandwidth-limited wireless networks. Therefore, the long-term solution to the need for security in the wireless domain is to provide mainstream security mechanisms such as PGP and S/MIME, but based on more efficient public key algorithms and mechanisms.

XML allows for representing data in a platform and programming language independent way supporting the separation of plain data from meta-information on that data. Numerous specifications have been or are being developed which provide security for XML documents: XML Signature, XML Advanced Electronic Signatures (XAdES), XML encryption, XML Key Management Specification (XKMS) and Extensible Access Control Mark-up Language (XACML).

A Web Service (WS) is an application program offering its methods to remote programs via the Web. In terms of technical realisation, Web Services are composed of several XML-based technologies: XML itself and its family of related specifications for representing data and defining how to process it, WSDL (Web Service Description Language) for describing Web Services by means of XML, SOAP (Simple Object Access Protocol) for exchanging XML data in a platform and programming language independent way, as well as UDDI (Universal Description, Discovery and Integration) for publishing and discovering Web Services. Using SAML (Security Assertion Markup Language), authentication and authorisation assertions can be encoded in XML format. Web Services can request and receive SAML assertions from a SAML compliant authority. The current technology of wireless Web Services is not yet mature, and security is among the remaining issues yet to be resolved.

Application-level Web security addresses vulnerabilities inherent in the code of a Web application itself. A major cause of these vulnerabilities is a general lack of language-level support in popular untyped scripting languages (e.g. PHP, VB-Script). Application-level firewalls are required to protect against external attacks, as well as the effects of unintentionally malicious software being transferred between business partners. A large number of companies provide application-level firewalls as commercial products.

The Session Initiation Protocol (SIP) is a signalling protocol for Internet conferencing, telephony, presence, events notification and instant messaging, standardised by the Internet Engineering Task Force (IETF). SIP is not an easy protocol to secure. Its security mechanisms address threats like registration hijacking, impersonating a server, tampering with message bodies, tearing down sessions, denial of service attacks, as well as privacy and replay protection issues. The upcoming set of wireless specifications defined by Third-Generation Partnership Project (3GPP) will include SIP as the session control protocol for IP-based voice and multimedia.

Personal firewalls are software programs installed on a mobile terminal to protect it by keeping malicious intruders outside, turning away their unwanted probes and preventing bad programs that have already staked a claim on the mobile terminal from doing further damage. Not all available personal firewalls are suited for mobile devices like Personal Digital Assistants (PDAs) and smart mobile phones, but are better suited for PCs or laptops.

The current activities of industry forums concerned with mobile secure payments include the development of payment systems based on existing technologies (e.g. WAP), the definition of the requirements for a personal trusted device, proposing different scenarios for electronic payment [e.g. Secure Electronic Transaction (SET) wallet within a chip or on a trusted server], proposing various user authentication and identity solutions [e.g. Liberty Alliance, Radicchio's Trusted Transaction Roaming (t^2r) initiative] and the development of the business relationships required to support payment (e.g. Mobile Payment Forum).

Apart from a few available commercial products, the area of mobile healthcare services is still at the stage of research and development. Numerous EU-funded research projects addressed or are addressing mobile healthcare topics such as remote monitoring and advice, as well as remote record and information access. These projects treat security and privacy aspects at different degrees of detail and granularity.

The current m-entertainment market should be regarded as experimental. To move to the next level will require new systems, alliances, phones, networks and substantial improvements in mobile phone software and security. Security technologies used in and for m-entertainment services include: Subscriber Identity Module (SIM), Secure Game Contract (by Quixotic Solutions), Java 2 Micro Edition (J2ME), and public key infrastructures (used in Multipass by Blueice Research).

17.3.5 Privacy and identity management

A number of tools have been developed to help Internet users surf the Web anonymously: anonymising proxies for submitting requests to Web sites on behalf of users (e.g. the Anonymizer, Anonymity 4 Proxy), anonymity tools around the concept of mix networks and similar concepts (e.g. Onion Routing, Freedom, Crowds), services for sending anonymous e-mail (e.g. Cypherpunk and Mixmaster remailers) and systems for anonymous data storage and publication (e.g. Freenet, Free Haven).

Single sign-on (SSO) is a technique by which a user needs to go through an authentication process only once, and will be authenticated automatically to all the web sites he/she subsequently uses. The security challenge is to ensure that the resulting system is not less secure than the traditional situation in which a user remembers different passwords for every site he uses. Furthermore, the privacy and anonymity of the users have to be protected. Existing SSO systems are Microsoft's Passport and Liberty Alliance's federated network identities.

Management of network security services may need to be centrally provided through authentication servers or Trusted Third Parties (TTPs). The Kerberos TTP system is based on a Kerberos authentication server (KAS) and a Ticket-Granting Server (TGS). The ANSI X9.17 standard on key management incorporates the use of a Key Translation Centre (KTC), which provides authentication services for key exchange. The International Standards Organisation (ISO) has published a technical report covering TTP services.

An information service that is dependent on geographical location data is called a Location-Based Service (LBS). Examples are location-based traffic jam information or breakdown services for car drivers. The IETF Geopriv Working Group will recommend a format for representing location information to ensure that the security and privacy methods are available to diverse location-based applications. Several 3GPP technical reports and specifications are related to GSM and UMTS location technology, radio and service aspects.

Devices that can be used to supplement or replace password authentication are known generically as security tokens. Important types of security tokens are: smartcards (advanced versions of magnetic strip cards, incorporating a microprocessor, memory and program storage), SIMs (the smartcards used in GSM phones), and Universal Serial Bus (USB) tokens, also known as dongles (containing a cryptographic chip for securely storing security-related data). Smartcards are not the optimal solution for comfortable and unhindered e-commerce activities. The easy to use USB security token may become an alternative.

A major drawback to the use of public-key cryptography is the need to make available authentic copies of entities' public keys. The basic idea of identity-based public-key cryptography (ID-PKC) is to derive public keys directly from public information whose authenticity is not in doubt (e.g. IP addresses or SIM identities). There are many issues standing in the way of adoption of ID-PKC. However, in the mobile arena, one can envisage future applications in the area of ad hoc networks and Personal Area Networks (PANs).

Public Key Infrastructure (PKI) is used in mobile environments by a number of security protocols and/or security schemes, in the same way as it is used in the fixed network, although it is adapted to cope with the limitations of mobile environments. These protocols and/or security schemes may be used by applications (e.g. by WTLS in WAP) to provide end-to-end security.

17.3.6 Basic security technologies

Cryptography is one of the most important means of satisfying security and privacy requirements on communication systems. Currently, there are numerous activities not only with respect to research in the various areas of cryptography (such as block and stream ciphers, hash functions, elliptic curves, provable security, key management), but also with regard to national and international standardisation. Section 2.6.1 in the PAMPAS deliverable D03 gives a more detailed survey on the state-of-the-art of cryptography [3].

Biometric techniques are beginning to provide alternative tools for authentication of persons, especially when combined with other methods of identification. Important biometric methods are the measurement of individual characteristics of a person's voice, fingerprint, face, iris or signature. For mobile telephony, voice obviously offers a convenient biometric method, while fingerprint, face, iris and signature measurements are less suitable, at least with the current level of mobile phone equipment capability.

17.4 Non-technical aspects

Secure mobile communication over open networks cannot be achieved by technological innovation alone. Its success will also be determined by the ability of end-users (business, government or the private consumer) to both appreciate the significance of security and make intelligent use of it. Further, the willingness of the end-user to conduct business in an advanced, networked environment will be determined not only by the performance of the enabling technology, but also by the legal, financial and regulatory issues surrounding this activity. In this section, we will address the non-technical aspects of mobile security.

17.4.1 Social–cultural drivers for mobile privacy and security

Mobile communication is not just about technology, nor are its considerations solely commercial. It must have a human dimension, so that its use, operation and impact respect and reflect the values that users hold, both as individuals and as a society. Even as mobile communication permeates almost every facet of daily life, it must create opportunities to improve the quality of life. Security and privacy are two extremely important aspects for socio-cultural enrichment and acceptance of mobile communications.

An increasing number of systems, for instance, use contextual information about their users. The use of wireless and mobile communication facilitates this. Context information can be used by applications for the benefit of the user and the society.

Mining of mobile networks for data on, for example, the current location of millions of users, in which direction they are moving and at what speeds could create numerous opportunities for the benefit of the society. The resulting intelligence might be used to redesign pedestrian walkways, control traffic signals or react to natural disasters. An additional benefit of the user's location could be derived in a protective system, where the user could hit an alarm when in distress so that help could be sent to his location. Combined with information about the user, direct and accurate medical assistance can be rendered. In the long term, this fast and decisive medical assistance might save medical costs. In such situations, however, the right persons, that is, medical emergency personnel, must have access to the right information, that is, the user's location and medical dossier.

Mobiles can also be used to monitor children. Parents will soon be able to keep a much closer eye on what children are up to on their way to and from school thanks to mobile monitoring systems. Text alerts will be sent to their mobile phones if the child deviates too far from their route or takes too long getting there.

The user's location and other context information can also be used for less admirable purposes. Someone could, for instance, trace the habits of a person such as his movement patterns and thereby deduce information that is highly personal.

It is, therefore, vital to consider privacy issues when designing distributed real-time systems where context information may be distributed to a number of requesters for the benefit of the society. Service creation platforms are required to enable the creation of context-aware services while enforcing user privacy and integrity. More-over, the way(s) in which the awareness of users of various mobile applications can be effectively raised with regard to privacy and identity management issues should be analysed, in the contexts of both public and private sectors.

17.4.1.1   User expectations

There are many different user expectations and needs when it comes to applications, be it in a fixed or mobile environment. What is common to most personal users is their simple expectation that the application works and is easy to use. Privacy and security need to be in place as well, but typically hidden and of lower priority than all the other expectations. Most people just expect the applications and service providers to handle their personal data in a secure and privacy respecting way. Only when the security is breached or the privacy disturbed do users realise that this is not always the case. People who are more aware of the risks either avoid them by not performing potentially risky actions, or by applying appropriate security and privacy-protecting technologies. This is, however, neither an easy nor convenient task and requires a lot of knowledge and patience. Often, the bigger security breaches are just as much the result of human error as they are about technical vulnerability. The question is, how to minimise the security risks associated with human oversight?

17.4.1.2   Ease of use and acceptance – the human factor

The use of security mechanisms has a price. This might be monetary or might be in terms of an impact on usability or functionality. If it is too complicated to use secu-rity mechanisms, people invent tricks, like writing passwords into their address book under 's' for secret. Many people are just frustrated because of the number of pass-words and PINs they have to remember. Simple, integrated, but still secure solutions are necessary, such as, for instance, biometric procedures. The use of biometrics, however, only authenticates a user, but the problem of authorisation still remains, because people belong to many groups, each with special access rights. After the development of a technical solution for a security problem, we face two further problems: does it match the social behaviour of the potential users, and is the usage legal in the international world we have today? People tend to be suspicious about sending their credit card number over the Internet, but are prepared to pay with the very same card in every shop in foreign countries. Some countries allow strong encryption; in other countries people have to stick to weak keys.

A critical consideration for the successful evolution of new wireless communications into the default communication medium is the acceptance of the user. Besides user friendliness of security, the issue of confidentiality is important here. Can the users be sure that their communication is not held against them later?

Right now, standards and infrastructures are being laid down without proper consideration of the real user requirements. We therefore stress that future research activities should also take into account the 'human factor' in security, for example, basic security standards, user-friendliness of systems. Moreover, the ways and conditions to establish trust in new electronic information relationships between government and citizens in different countries and legal areas, facilitated by various privacy-related ICT trends, should be analysed.

17.4.1.3   User perspective – privacy

The mobile terminal inhabits a far more intimate space in our daily lives than, for example, television or personal computer. Many mobile users are, therefore, generally more concerned about threats to their privacy and pay careful attention to the incoming stream of calls, content and messages.

The real user requirement for privacy is flexibility, and, in particular, choice in the level of risk he/she is prepared to enter into. Real privacy would allow the user to have control over the expiry date of personal information he/she is making available to a requester. Real flexibility would allow the user to have the means to control the amount of personal information he/she is giving to a requester.

It is obvious that the user should be in control over the release of her information, and, therefore, the ownership of the information should belong to the user. Moreover, most users do not want complete automaticity of private data exchange, but instead want to be able to grant or deny any transfer of such data.

However, tools to support such a controlled release of private information are scarcely available, difficult to handle or are achieved at the cost of loss of functionality. Further research in the field of secure user profile management tools, including user studies and user-driven development, is therefore required.

17.4.1.4   Wireless gets personal

The ultimate goal of wireless services is to drive the creation of new applications that take advantage of the unique characteristics of the mobile world – one that is immediate, constant and personalised. The latter aspect implies that information required for personalisation of content delivery is available at any time and any place.

Moreover, due to the user interface limitations and the inherent mobility of the user, mobile services are most beneficial for the user if they adapt to the current context of the user. 'Context' to this end may include the user's location, terminal capabilities, user preferences, user situation (work versus leisure, etc.), sensor data and many other kinds of information. In order to make applications truly context sensitive – beyond what location-aware services offer today – fragments of user context profiles have to be collected, stored and interpreted at different places and by different players, and they will need to be communicated and matched against each other.

For effective personalisation, the content or service provider needs information from the user. Users are willing to give this information in return for value-adding services. Consequently, large amounts of personal information (e.g. the user's name, location, e-mail address, passwords, encryption keys, etc.) are sent over the network. Users should be very careful to prevent unintentional leakage of this information to other sources. Convenient tools and interfaces for the user to control dissemination of such information for their desired customised content provision, however, are currently not available.[2] The way(s) in which the protection of personal data can effectively be guaranteed in new customisation approaches arising from the evolution, inherent characteristics and exploitation of mobile services should be analysed further.

17.4.1.5 Personal security technology

With the rise of mobile communications, the threats to our personal security and privacy promise to get only more complex. Advances in technology not only allow the government but also creditors, employers, advertisers, insurance companies and many others to gain a tremendous amount of information about all of us. The convergence of wireless and wired technologies also makes this information thoroughly fungible and exchangeable – and thus more easily manipulated. Greater emphasis on personal security technologies is therefore required.

The use of personal firewall software and hardware appliances to assist with the protection of the mobile terminal continues to expand. Personal data protection technologies such as file/disk encryption products are becoming common security measures.

Although the need for security is understood and appreciated by many users, the current PIN-based approach is under-utilized and can, therefore, be considered to provide inadequate protection in many cases. Many users respond positively towards alternative methods of authentication, such as fingerprint scanning and voice verification. Further research is needed to investigate the possibilities for non-intrusive, and possibly hybrid, methods for authentication (using a combination of techniques) that would best satisfy the needs of future subscribers.

Furthermore, no protection is provided for user-related information once the content or service provider acquires them for performing personalisation or adaptation services. The development of trust models or DRM schemes seem appropriate tools to tackle this issue. The notion of privacy rights management has partly been considered in DRM standardisation activities, for example, in the OMA DRM standardisation. Here, users can restrict the distribution of private data, such as images taken with a camera-phone. Further research in the area of personal security technologies, however, is required.

17.4.1.6 Case study – socio-cultural diversity within the European Union

Across Europe, an average of 42 per cent of all mobile phone users are interested in 3G.[3] Interestingly, mobile phone users in Eastern Europe show more interest in using 3G applications than their counterparts in Western Europe. Fifty-nine per cent of users in Turkey and 51 per cent in Poland said they are 'interested' compared to only 34 per cent in the United Kingdom or in Germany. Forty-eight per cent of men are interested in 3G compared to only 36 per cent of women. The desire for customised mobile applications and their impact on the user's privacy, however, has not been examined. Further assessment of the impact of the introduction of new customisation approaches in mobile applications and service delivery on the privacy experience of citizens in different EU states is needed.

17.4.2 Economic and business drivers for mobile privacy and security

This section analyses the economic incentives and hindrances to mobile security and the protection and disclosure of personal information from a business perspective.

Business demand for increased productivity and competitive advantage virtually guarantees that wireless mobility solutions will make their way into the core of enterprise IT infrastructure. Wireless mobility solutions promise a host of benefits both at the top and bottom lines of the balance sheet. However, outstanding security concerns about wireless technology have been one of the main reasons why these solutions have not gained greater acceptance to date.

The security of transactions and data has become essential for the supply of electronic services, including m-commerce and other public services like m-entertainment, m-health and m-voting. A low confidence in security could slow down the widespread introduction of such services.

Privacy is an area of considerable concern to many online consumers [4], and those companies, which provide adequate support for their customers' privacy – and particularly those which present this information in an effective manner – increase the likelihood of consumer loyalty. One main reason for companies to implement expensive privacy-enhancing tools is often legal pressure and the corresponding penalties involved.

17.4.2.1 When to take security and privacy measures?

Effective information security should be driven from a business perspective; that is, security measures should match the needs. This implies that next to the need for security not only the costs of security measures but also the added value of the implemented security measures should be evident. In other words, organisations must first know their needs, be well aware of what they are buying and know what the (hidden) associated costs are.

Figure 17.3   Business risks (from Reference 5)

The need for security is based on business risks. Business risks are the likelihood that a threat due to the existence of a matching vulnerability results in a security incident causing damage to the organisation. The significance of the damage depends on the business value of assets affected and the impact of the security incident (see Figure 17.3).

From Figure 17.3, we conclude that all security measures should focus on the mitigation of vulnerabilities and threats, and hence the mitigation of risk. Moreover, assets having a high business value and those that are exposed to significant risks impose the most significant business risk.

Not all information and IT services are of equal importance to an organisation or exposed to the same level of risk. The cost of information security should, therefore, be appropriate to the importance of the information and/or IT services and the level of risk exposed. Information security should be driven from a business perspective. Effective and efficient information security requires security measures to be an integral part of business processes. Since there is a billing relationship between operator and customer in the mobile environment, this link is even more precious than in the fixed environment, where most of the data exchanged does not cause immediate costs to the customer. Due to the restricted capabilities of the mobile terminals, personalisation is essential to provide a good user experience. This, on the other hand, implies that many personal attributes are stored and managed. The distribution, protection and management of these personal user data should be also viewed under the same principles as above, that is, how vulnerable are the data, what are the real threats and how expensive and efficient are different options of managing that data.

17.4.2.2   Business benefits of security and privacy measures

Security measures do cost money. On the other hand, it is obvious that security incidents resulting from not implementing security measures also cost money. Making a balanced choice between the costs of security implementation and the costs of incidents requires having accurate knowledge of the needs for security. Security needs should be based on accurate knowledge of the risks an organisation is exposed to, the costs of these risks and the costs of avoiding these risks.

In the case of privacy the business value is even more vague. To a user privacy is a very important asset, but she expects that her privacy will be protected by force of legal requirements. Some companies attempt to attract customers by offering highly sensitive and developed privacy policies to protect consumers. These companies assume that privacy is very important to their customers and that customer loyalty and uptake of services will be affected by the non-existence of privacy policies or the breach of customer privacy. Some companies, on the other hand, prefer to offer no user privacy so that they can build customer profiles and offer optimized services to potential customers. What is needed is the development of a context-and market-related categorisation of the strategic value of personal information, including location information.

If there is no solid business case for privacy-preserving applications, it is unlikely that a solution will ever be deployed. A good example is the case for anonymous payments. Numerous models for anonymous payment have been proposed but none of them has been widely adopted in commercial systems.

17.4.2.3   There is more than just security and privacy measures

Information system security comprises more then just the installation and/or implementation of security measures. Information system security is the application of managerial and administrative procedures and technical and physical safeguards to ensure not only the confidentiality, integrity and availability of information that is processed by an information system but also of the information system itself, together with its environment as well. Such procedures and safeguards not only need to deter and delay improper access to information systems, they must also ensure that any improper access is detected; that is, individuals have to be made accountable for their actions and must be made aware of the dangers.

Technical tools are an essential part of implementing real privacy, but without the corresponding guidelines on the usage of these systems, their value is quite low. Guidelines are needed to ensure that the actual goal of user protection can be reached. The procedures described in such a framework are needed to deal with storage, processing, collection and access to the personal data.

17.4.2.4   Technical support for advanced business models

Currently, it is not clear what business models will fit future open, heterogeneous, packet-oriented networks, with wireless multi-hop scenarios. It is envisaged that in such an environment, the role of subscriber and network provider, as well as the role of subscriber and content provider, may not be clearly distinguishable. Although it is not predictable which business models will gain acceptance we will need to provide the appropriate security-related support.

There is a need to evaluate various schemes and protocols to authenticate or register a mobile device and, if necessary, its resource usage. This is especially true when applying mobility protocols over different administrative domains. The suitability of an optimal authentication protocol depends on the assumed mobility scenario, its privacy issues and also strongly depends on the purposes of the assumed accounting scenario. Thus, the requirements for an integrated authentication, authorisation and accounting framework may differ significantly with different business models and charging schemes. For example, pre- or post-paid usage-based schemes have completely different requirements for an authentication scheme/protocol than fixed contract models have. The greatest challenge here may arise from the possible multiplicity of roles of involved parties, resulting in multi-directional flow of charges.

17.4.2.5 Summary

Summarising, the business drivers for security are as follows.

- *Responsibility*. Professional security management is a must for every company in order to be perceived as trustworthy. One could say it is related to 'responsible fatherhood'.
- *Market requirements*. A company must be able to guarantee a certain level of secure service delivery and availability. Users do not want to be bothered with security but expect the appropriate measures to be in place.
- *Legislation*. Laws and supervising authorities force businesses to take the appropriate security measures.
- *Security creates value*. Customers are likely to return when they have experienced a solid and secure first encounter. Security costs, but insecurity costs more money. It prevents bad media publicity and the consequent erosion of brand.

17.4.3 Legal drivers for mobile privacy and security

Effective regulation plays an important part both in the initial and ongoing shaping of liberalised telecommunication markets. The specific processes and structures for implementing interconnection, access, pricing and other regulations will shape the market, and thereby shape the extent and significance of universal service problems.

Law/self-regulation challenges are:

- appropriate for all types of Information Society services and applications;
- protect key interests such as consumer rights and privacy;
- involve all actors;
- promote self-regulation and co-regulation;
- simple, flexible and technology-neutral.

Policy measures can be more effective if they are part of an international approach, respect the effective functioning of the internal market, build on increased cooperation between individual countries and states, and support innovation and the ability of enterprises to compete at the global level.

A very important subject with any new technology is deciding how current laws deal with new issues that might be raised by a new generation of products. In the following sections we shall address several drivers for mobile privacy and security.

Issues that cannot be tackled technically or are hampered by social or economical forces may perhaps only be solved via legislation. For instance, governments usually push the introduction of smartcards for identification of citizens since there is little economical or social support for it; even though it is technically possible.

17.4.3.1 Trusted computing platforms

A truly secure computing platform might reduce the threat of private data exposure to unscrupulous data processors. A secure computing platform could make enforcement of the legal use of data possible. For instance, a data controller might require data processors to use a secure computing platform to access personal data.

For example, the EU directive on a Community framework for electronic signatures has some requirements on the 'secure signature-creation device', a device that creates an electronic signature in a secure way. The private key has to be under exclusive control of the owner of an electronic signature. The secure signature-creation device is often a smartcard, but it could be a mobile device as well if it complies with the requirements of the EU directive.

17.4.3.2   Cyber crime

An issue of growing concern is the challenge posed by the emergence of cyber crime, such as electronic money laundering, illegal money gambling, malicious hacking or copyright infringement. International cooperation is already well advanced in a number of key areas, such as the fight against organised trans-national crime on new communication networks. Faced with new forms of high technology and computer crime on global networks (reported criminal hacking cases are doubling every year), governments have responded vigorously [6].

In Europe (Europol), as well as in the wider international environment (P8), specialised task forces have been set up, and trans-border operational cooperation reinforced in such key areas as the real-time 'trap and trace' of online criminals and 'search and seize' of digital evidence. Efforts are similarly being made to harmonise the criminalisation of computer offences and avoid digital havens. A high-level group, set up following the Dublin Council, is finalising an Action Plan to fight cyber crime. These efforts are crucial to reinforce trust and confidence in trans-national electronic commerce.

17.4.3.3   Privacy of personal data and location information

Personal data play an important role in many areas, such as public administration, health care, social security, telecommunication, financial services and direct marketing. Slowly but surely almost everyone is becoming a celebrity. However, personal data also need to be protected in order to safeguard the rights and freedom of citizens.

In Europe, the fundamental right to privacy is recognised in the European Convention of Human Rights and Fundamental Freedoms, which states that everyone has the right to respect for his private and family life, his home and his correspondence. With the development of information technology and its potential for controlling data, the mere recognition of a fundamental and constitutional principle of privacy in general appeared to be too insufficient to effectively safeguard the growing need of privacy protection with regard to processing of personal data. This resulted in the EU Data Protection Directives: a general Data Protection Directive that constitutes a general framework regulation on the collection, use and transfer of personal data and the Directive on Privacy and Electronic Communications that regulates issues such as confidentiality of communications, calling and connected line identification, unsolicited communications.

Pursuant to the European Directives and Personal Data Protection Acts, independent institutions must be available to check that the personal data are used carefully and are protected and that citizens' privacy continues to be sufficiently guaranteed, both now and in the future. Governments should be advisors in this, codes of conduct should be tested, technological developments must be studied, information given, complaints handled, processing of personal data evaluated and, if necessary, enforcement action should be taken.

Moreover, the responsibility of citizens and organisations for adequate privacy protection should be promoted and self-regulation within the statutory frameworks should be supported.

New technologies should respect the fundamental right to privacy; for example, DRM systems should incorporate privacy safeguards in accordance with the EU Data Protection Directive. A large number of national initiatives and human rights groups have also addressed the issue of location privacy: the privacy of location information has to be protected thoroughly in mobile networks.

A detailed analysis of privacy regulations can be found in the Roadmap for European legal research in privacy and identity management of the RAPID project [7].

17.4.3.4   Mobile spam

Spamming, or unsolicited messages sent by text to the mobile phone, is becoming ever more prevalent. These mobile spam messages are undoubtedly irritating and can also be costly. Despite this, however, text messaging has become a new and effective way to market a company, brand or product. Despite the fact that the Wireless Marketing Association has established a code of best practice for marketers, which includes an opt-out from receiving messages, the European Union is tightening up the laws on m-commerce. The EU Directive on Privacy and Electronic Communications states that individuals must be protected from spamming for direct marketing purposes. As such, automated direct marketing activities (e.g. e-mail, SMS messages) may only be allowed in respect of subscribers who have given their prior consent. The Directive, however, does not specify what technical measures are to be put in place to effect this measure. Moreover, it also appears that spamming for purposes other than direct marketing, including 'unsolicited commercial messages' might not be caught by the clause. It is unclear whether spam of, for example, political, religious or philosophical nature, or unsolicited invitations to sign up for free services is covered as well.

A point of controversy in the EU Directive on Privacy and Electronic Communications is that it allows member states to decide whether consent should be on an 'opt-in' or 'opt-out' basis. This contradicts with existing laws in several individual EU states. For instance, in the Netherlands, legislation concerning the processing of personal data unequivocally requires prior consent for receiving spam.

The EU Directive also does not include any measures if only a single unsolicited commercial message is sent.

Legislation in the United States prohibits advertisers to cause nuisance to citizens via their mobile phone. However, text-oriented services for mobile phones are allowed. This loophole in the US legislation opens the door for mobile spam. In some cases, the mobile user also has to pay for the undesired messages.

The measures related to privacy of stored personal data are mostly covered by legislation already.

The fact that the legislation will be difficult to enforce does not, of course, detract from its validity. The majority of computer users will probably welcome the clear condemnation of unreasonable spamming, even if it cannot be easily backed up by action.

17.4.3.5 Roaming

With wireless networks it is not always easy to determine where we are connected. While moving around our own property it might be possible to inadvertently connect to the access point of the neighbour's wireless network because its signal is more powerful. It might be possible to unintentionally damage or overload the network by using bandwidth or other resources. Who is going to pay the bill if this neighbour gets billed for exceeding his bandwidth usage from his ISP? What does the law say about this?

To address this issue, the FBI stated in a memo to its agencies: 'Identifying the presence of a wireless network may not be a criminal violation, however, there may be criminal violations if the network is actually accessed including theft of services, interception of communications, misuse of computing resources, up to and including violations of the Federal Computer Fraud and Abuse Statute, Theft of Trade Secrets, and other federal violations'.

17.4.3.6 Mobile payments

There should be a legal framework around mobile payment systems. Questions that are raised include: what is the legal value of electronic money? Is electronic money legal tender? Will merchants be obliged to accept electronic money?

To encourage innovation and make European-wide solutions easier, the European Commission proposed a directive on electronic money institutions: a directive relating to the taking up and pursuit of the business of credit institutions and a directive on the taking up, pursuit of and prudential supervision of the business of electronic money institutions. In the past, only banks could issue electronic money. As this reduces competition and slows down innovation, the EU directive introduces a new institution, the 'Electronic Money Institution' (EMI). An EMI can issue electronic money under a lighter regulatory framework than a credit institution. This should be more attractive for non-banking institutions. The directive also intends to define the concept of electronic money. It is, however, not clear if all schemes fit into the current definition (e.g. server-based schemes, bonus points and loyalty schemes, software-based products).

An EU Recommendation concerning transactions carried out by electronic payment deals in particular with the legal aspects of the relationship between holders and issuers within an electronic payment system. It gives a recommendation about the rights and obligations of holders and issuers, for example, liability issues, notification obligations and burden of proof.

17.4.3.7   Mobile voting

Mobile voting should conform to the basic election rights laid down in international and regional convents and in national constitutions: non-discrimination, freedom and secrecy of the vote, one-person one-vote, and security, reliability, verifiability and confidentiality of the voting system. While the principles of non-discrimination and of democratic elections are fundamental, the legal system, however, has to be adapted to allow a specific technical implementation of mobile voting.

A detailed analysis of the legal issues of Internet voting can be found in D4 Volume 2 of the Cybervote project [8].

17.4.3.8   Mobile entertainment

Multimedia will be key in the future, both in its own right and as the future base for both the entertainment and professional industries and services.

The Internet has the potential to develop into the link between the current networks and the digital delivery systems of the future. The future delivery systems of the Information Society could be combinations of broadband cable and Internet-type telecom backbones and integrated satellite and mobile systems. Customer terminals could become some sort of hybrid between television, PC and mobile phone/terminal. The growing multimedia sector will be a major challenge for regulators and competition and anti-trust policy since no sector specific regulation is fully capable of considering the impact of alliances implying sector convergence.

17.4.3.9   Challenges

A sound and flexible regulatory framework, which generates confidence for both business and consumers and ensures full and unlimited access to a single market, is an essential key to Europe's success. Such a regulatory framework will be a major competitive advantage in itself. Steps must also be taken to improve the business environment: to exchange best practices, facilitate access to venture capital and stimulate training. Ultimately, global solutions must be found. The Community should be in the lead in exploring and offering solutions at an international level.

Europe, the United States and the Asian Pacific area share a strong interest in the development of the global telecoms system. Market opening is key. The aim of regulators must be to open markets and ensure pro-competitive market structures for the choice and benefit of the users and as a base for the development of our countries. All regions must have the opportunity to contribute their best technology. The result should be systems reflecting best global practice and experience.

A substantial body of legislation relevant to network and information security is already in place, notably as part of the EU's legal framework for telecommunications, electronic commerce and electronic signatures. However, the rise of new mobile services that fully exploit the inherent characteristics of mobility (e.g. location and context awareness) impose new legal requirements on providers of such telecommunications services to take appropriate technical and organisational measures to safeguard the security of their services. Besides further research on the development of technologies and organisational measures in order to provide an appropriate level of security, a corresponding and appropriate legal framework must be defined as well.

Large-scale introduction of wireless networks (e.g. wireless local loop, wireless local area networks, third-generation mobile) will bring the challenge of effectively encrypting data transmitted over radio signals. It will, therefore, be increasingly problematic to require by law weak encryption of those signals.

There is a scope for the government to further review developments in convergence between telecommunication and broadcasting, in particular, as regards merging legal frameworks to ensure that carriage regulation comes within the scope of a single regulatory framework.

17.5   SWOT analysis

An evaluation of the internal and external environment is an important part of any strategic planning process. Internal environmental factors usually can be classified as strengths (S) or weaknesses (W), and those external can be classified as opportunities (O) or threats (T). Such an analysis of the strategic environment is referred to as a SWOT analysis.

The SWOT analysis provides information that is helpful in matching resources and capabilities to the competitive environment in which they operate. As such, it is instrumental in strategy formulation and selection. In this section, we provide a brief SWOT analysis of the European position with respect to mobile telecommunications, focussing, in particular, on security and privacy aspects.

17.5.1   Strength and weaknesses of the European Union compared to the United States and Far East

The state-of-the-art of world-wide wireless communications today is that of incompatibility. Each of the three major regions – North America, Asia and Europe – took different technology routes when moving from first-generation analogue to second-generation digital mobile communications. The focus for third-generation systems and beyond is motivated by a single global world-wide standard for mobile communications. However, deriving a single solution will be extremely difficult, if not impossible. Each region of the world has its own economic, political, regulatory and social issues, which affect decisions of this magnitude. Moreover, each region already makes use of existing own communication infrastructures. Any attempt to merge these into a single infrastructure will prove to be difficult. Consequently, the new mobile framework must allow for differentiation of services and the product but must be open for global roaming, service portability and multimedia.

The European telecommunications market is a dynamically growing industry with the potential to rival that of the United States in the next few years. The strengthening of the European Commission and European Union, combined with the Commission's eEurope Action Plan and the passing and implementation by Member States of new telecommunications directives, positions the European telecommunications industry well for the future.

Since 1998, Europe has taken steps to substantially liberalise many segments of the telecommunications services and infrastructure markets. As a result, the industry was the fastest growing segment of the European market in 2001. With further progress in competition and local loop unbundling, broadband deployment, Internet access, content development, information and network security, as well as other areas, individual EU member countries and the European Union, in general, will be engines of telecom growth in the future.

The European telecommunications market has been through a transition phase during the last 1–2 years, much of it due to the world wide economic downturn. As a result the market now consists of strong players that are focusing on core strengths. The indicators for success are:

1. traditional indicators of telecom reform – market and service development, consumer pricing and competition;
2. information infrastructure and network society development – investment and access technologies; Internet market development; e-commerce and e-government readiness and activity.

Revenue per capita is highest in the United States where there has been an historic tendency for people to use fixed network services more intensively than people in the European Union and Asia.

The mobile market penetration is still growing rapidly in the European Union. The United Kingdom and The Netherlands have caught up with the Nordic countries as the leaders in Europe. Europe is still a long way ahead of the United States and Canada. However, it must be noted that mobile penetration rates in Korea and some other Asian countries are approaching the world's highest levels, and there are many indications that world leadership in mobile market development is shifting to Asia.

Considering the prices for mobile services we see that the North European countries enjoy the lowest prices for residential mobile services. Mobile business is a more mixed picture, and prices are continuing to decline in all countries. The pricing policies differ a lot in different countries. In the United States, the charges for the representative call baskets for business and residential are about the same, a 1/1 ratio. France is 1.4/1; United Kingdom and Canada, almost 2/1; the Scandinavian countries between 3 and 4/1. The United States serves the business market best; the EU countries serve the residential market best. One reason for the higher charges to business in Europe compared to the United States is the application of very high mobile call termination prices, especially for roaming calls between countries, which are made possible by the monopoly on the termination call number [9].

Comparative data on mobile call termination prices is unavailable. However, there is increasing evidence that these prices are extremely high, not regulated effectively, if at all, and are providing a bottleneck for continued mobile service development, in particular international roaming. Studies have estimated call termination charges to

Table 17.2 ITU mobile/Internet index rankings

Country	Mobile/Internet score (/100)	Ranking
Hong Kong	65.88	1
Denmark	65.61	2
Sweden	65.42	3
United States	65.04	5
Norway	64.67	6
United Kingdom	63.00	8
Netherlands	62.25	9
Canada	61.97	11
Finland	61.22	12
Germany	55.53	17

be 40–70 per cent higher than the cost. This could have a major constraining effect on future growth rates.

The ITU has developed a measure for the preparedness of countries to take advantage of mobile Internet integration. It has developed an Index that measures how each economy is performing in terms of information and communication technologies (ICTs) while also capturing how poised it is to take advantage of future ICT advancements. The Index covers 26 variables sorted into three groups: infrastructure, usage and market structure. These three components combine for a score between a low of 0 and a high of 100. The rankings of the countries being compared in this report are provided in Table 17.2 [10].

17.5.1.1   Strengths

Europe has a strong base in technology and infrastructure. It has powerful telecommunication operators (incumbents as well as new players), highly reliable basic infrastructure and early deployment of advanced digital networks. Commitment to standardisation, exemplified by the success of industry-driven standards such as the GSM and DVD, is another crucial asset. So is Europe's commercial advance in key electronic commerce technologies such as smartcards and intelligent agents. Content development is another of Europe's greatest strengths. Content – computer software, business information, video entertainment – is the very essence of immaterial electronic commerce. European companies, in particular publishing and multimedia industries, are harnessing their considerable resources and know-how in the global electronic information markets. Similarly, highly innovative SMEs are positioning themselves successfully in specialised markets such as multimedia production and multilingual content localisation. Europe also has a competitive retail sector, with adapted product ranges and an in-depth knowledge of the various consumer tastes around the continent, a strength, which can be leveraged too.

Furthermore, the ability to trade electronically in a single currency – the Euro – across the world's largest single market will give European businesses considerable competitive advantages. Cross-border price transparency resulting from the Euro will stimulate the use of electronic commerce; conversely, electronic commerce will facilitate the transition to the Euro.

The market penetration for mobile services in Europe is about three times what it is in the United States; and for Internet services, market penetration in the United States is about three times what it is in Europe. As Europe consolidates its lead over the United States in mobile services, developments in mobile security technology are driving the roll-out of fully transactional m-banking and m-brokerage services in Europe.

A surge in demand for network infrastructure will dominate growth in the Asia-Pacific markets through 2005, predicts the Telecommunications Industry Association (TIA).

Summarising Europe's strengths:

- cultural and language diversity;
- leadership in technology, for example, mobile communications, smartcards, cryptography, watermarking and protocol verification;
- rapidly rising PC-based Internet access, rapid adoption of broadband access;
- highest mobile phone penetration;
- rapid growth of digital TV;
- governments' commitment;
- basic skills level (although not enough ICT);
- European single currency (€).

17.5.1.2   Weaknesses

There are also a number of important weaknesses that characterise the European mobile telecommunications scene.

Europe lags the United States in taking up new technologies, notably the Internet (this does not have too much impact on mobile services). These new technologies will carry the lion's share of growth in the new knowledge-based economy. Even if Europe has undisputed leadership in mobile telecommunication, it is not leading the race along the information highway.

The regulator tends to be thorough in following consultation procedures, but these procedures are often too slow in a market area that, because it is characterised by rapid market and technological change, requires speed in decision-making (this may be indeed a weakness on the short term, but may lead to large benefits on the long term; it acts more or less in the same way that open standardisation does). One reason for this may be insufficient resources in the European Union on the telecommunication side.

Another weakness with regards to single market penetration is that of cultural and language diversity. Language and cultural adaptations are necessary for each localised market within each of the European states; this adds costs and reduces coherency.

Traditionally, mobile telephony operators have obtained licenses based on the geographic boundaries of member states, the radio frequency allocation being in the sovereignty of each member state. This has often required duplication of infrastructures and local customisations even for neighbouring states with a common language. This, however, is changing with consolidation of mobile operators by the taking over and re-branding of existing operators by a few global players.

A further weakness that must be considered is the degree to which telecom operators in Europe have become debt-laden because of the costs of acquiring 3G licences and deploying new 3G-infrastructure. Operators need to see a quick uptake of 3G amongst customers, but so far there is a lack of compelling services to promote that development. It may be that consumers have not forgiven operators for the disappointments of WAP, or that the right mix of handsets, interfaces and services has not yet reached the marketplace. A related weakness particular to 3G is the emergence of wireless hotspots as an alternative solution to 3G for mobile Internet access: since they are cheap and fast to deploy, they may quickly take up market share in this important revenue generating application.

17.5.2 Opportunities for mobile security and privacy research

Ultimately, the main beneficiaries of improved mobile privacy and security mobile communications are citizens and companies: they can enjoy a better organised handling of their legitimate security and privacy expectations in communication, transactions, etc. This should ultimately result in improved economic performance.

IT companies trying to realise the security infrastructure needed for these expectations will benefit from focussed research in this area, in collaboration with academia.

Moreover, secure mobile communications opens the way to applications such as mobile payment and finance, mobile ticketing, mobile voting and location-based services (ambient intelligence) with increased convenience and confidence. This will lead to further development of an electronic marketplace with higher efficiency and lower prices for consumers. It is also clear that for mobile to thrive in the areas of public safety and healthcare, security is a prerequisite.

Following on from the success of GSM, European collaboration has played a central role in ensuring a single set of standards in 3G systems to the benefit of users and of suppliers of equipment and services worldwide.

The development of world-leading European expertise in security and privacy for mobile telecommunications has been enormously aided by EU funding in the past. Successful projects that have directly contributed to the development of relevant standards and technologies include ASPeCT, USECA and SHAMAN. These projects have helped to establish a well-connected network consisting of a critical mass of key researchers and technologists in industry and academia.

We have identified an expanding need to give special attention to security and privacy measures to accompany the development and use of mobile/wireless communications technologies and services. In addition, the interaction with existing and evolving fixed network standards and technologies needs to be assured to avoid fragmentation of communications technology. The European Union is well placed to take advantage of its current position of strength in this area: a significant opportunity for the European Union is represented by the combination of the availability of a highly skilled pool of human resources coupled with a set of challenging research problems in the area of mobile security and privacy.

17.5.3   Threats to mobile security and privacy research

The advantages gained in Europe by the early adoption of second- and third-generation systems are under threat more as adoption rates increase significantly in other regions, especially in the Far East. One other possibility is that if the security and privacy challenges that we have identified are not met in Europe, then they will certainly be addressed elsewhere, though perhaps more slowly. The community of mobile users, especially early adopters of mobile technology, are increasingly aware of the security and privacy issues associated with mobile technology and the press are not slow to publicise any failings in this respect. Because of this, a lack of progress towards solving our identified research challenges is likely to significantly delay the uptake of third and future generations of mobile technology.

The impact of this delay or even failure on the European telecommunications industry is potentially significant. The risk of not maintaining a unity of approach and standards is that users will adopt fragmented, proprietary solutions which may be divergent, flawed, incompatible and of limited lifetime.

17.5.3.1   UMTS–WLAN competition

Competition to well-defined regulatory standards also comes from less highly regulated business sectors. This is well illustrated by the case of the competition to 3G systems by WLANs based on the IEEE 802.11 standards for wireless data/Internet access [11]. It is generally agreed that it will be a significant challenge to integrate the two technologies such as to provide seamless mobile/wireless data/Internet service for consumers.

The separate development tracks of 3G and WLAN have exposed two different approaches to bringing mobile/wireless data/Internet services to market. On the one hand, the mobile communications track for 3G services and, on the other, the Internet track for the WLAN services. Traditionally, it is the Internet track that has led to faster market introduction at the expense of a higher risk of insular solutions and standard defects; however, it is the 'first-to-market' solutions with their higher volumes and lower unit costs that tend to dominate the consumer market in the end.

A threat to European research lies in siding with either of these tracks and depending on the success of a single technology.

In the case of 3G, success depends upon: what its (end-) user community is perceived to be by the different players involved; how to define interfaces to the infrastructure and how to develop communications services and information content. The manufacturers of user equipment are waiting for the content services to be available, while the service providers need to adapt their services to suitable handsets.

In the case of WLAN, success instead depends upon the usability of the interface to the network and the 'roaming' capabilities. Other critical issues are security along with organisational solutions and contractual agreements for cost sharing by accounting and billing procedures.

The development and diffusion of WLAN as a data transport infrastructure has resulted in an ongoing incremental evolution of the technology during the course of its deployment. WLAN has proved very successful when used in selected market segments. But WLAN still lacks important features necessary to become a widespread, enabling and open service infrastructure. The price of implementing a minimal standard is paid for by difficulties in creating seamless interoperability, enabling access to 'hot-spots' without complicated end-user configuration.

On the other hand, 3G is intended to provide a service infrastructure built on a transport infrastructure. The transport infrastructure of 3G is believed to be well designed, and provides roaming, billing and security without user awareness. At the same time, to provide a broad range of services for the end-users, third parties must provide a range of value-added services as well as user-friendly handsets suited to present the services in a proper manner. The current situation is that while we have technical standards that promise to create a general infrastructure for multi-purpose mobile, broadband communications, there are not enough implementations, available user terminal equipment as well as the digital content that can make it potentially economically sustainable.

It may seem a paradox that the evolutionary, market-driven development and diffusion of WLAN solutions in a rather competitive environment has proven a success. This is due to a certain level of coordination amongst the various competing technical solutions. At the same time, the emphasis on competition in the telecommunication market at the expense of cooperation and cost sharing may have as a consequence that the building of a new mobile infrastructure will be further delayed. Analysis shows that one of the main challenges related to 3G is to create new market segments where companies may compete. Furthermore, it is increasingly likely that each network will develop specific, proprietary services not accessible from other networks.

The competition between 3G and WLAN services for mobile/wireless data/Internet service will ultimately benefit the consumer, as access prices will drop. However, the consequence must be the erosion of available investment for the development of future 4G services, effecting privacy and security research in mobile systems.

17.6   Recommendations for further research

Based on the register of requirements [2] and current state-of-the-art of technologies and architectures relevant to mobile communication security and privacy [3], we could formulate an extensive menu of open research issues seen as necessary to address in order to ensure the success of future mobile/wireless communication systems [3]. Given the non-technical aspects of mobile privacy and security and the strengths, weaknesses, opportunities and threats in this area we have reviewed and refined the research challenges. The full list of research issues and their ranking can be found in the long version of the Final Roadmap of the PAMPAS project [12].

In this section, we will describe the prioritisation procedure of the identified research issues, present the high-priority research issues and conclude with a general trend analysis. Finally, we will discuss the necessary standardisation activities that are lacking in several areas.

17.6.1   Ranking of research issues

Having compiled a long list of potential security research challenges the question arises as to which ones have the highest priority or are most essential for further research. The prioritisation of research challenges for mobile security and privacy can be done by many criterias, e.g. technological importance, economical impact, consumer expectations and many others, as illustrated in Figure 17.4.

In order to come to a prioritisation of the numerous research issues, we have chosen to rank each open research issue with respect to the following question: 'How important is this open research issue for the benefit and success of mobile communications in the European Union?'.

Of course, many of the conclusions will probably apply outside Europe, but this chapter was written within a project that had the specific goal of examining European research priorities, and hence the authors make no specific claims about whether the conclusions apply more generally.

The criteria that were taken into account during the ranking procedure were:

- technological feasibilities and time horizon ('likely to be solved within the next 4 years');
- socio-cultural and economical drivers;
- the human factor;
- legal boundaries;
- the SWOT analysis of Section 17.5;
- future expectations of mobile communication (based on an ambitious and imaginative set of scenarios for mobile security and privacy in the near future (<2010) that challenge both application and enabling technology);
- the mobile-specificity of the issue and its impact on the mobile world.

Given the background of the authors and the fact that work was largely technology driven we admit that the technological criteria may have tended to carry most weight.

17.6.2   High-priority research issues

We have identified the following areas with respect to advanced mobile privacy and security research:[4]

- trusted platforms for mobile security and privacy;
- mobile network/transport security and privacy;
- mobile application security and privacy;
- mobile privacy and identity management;
- basic security and privacy technologies for mobile environments.

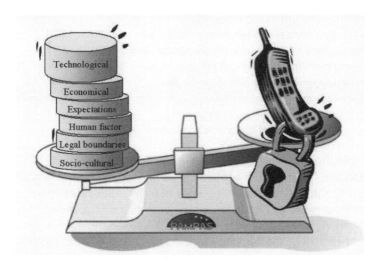

Figure 17.4   Prioritising mobile security and privacy issues

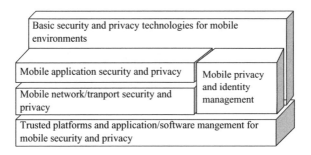

Figure 17.5   Schematic representation of the different areas related to mobile privacy and security

The areas have overlaps and relationships to each other and are illustrated in Figure 17.5.

The topics of needed and beneficial research in the area of mobile privacy and security that were identified by the ranking process as being the most crucial are as follows.

1. Trusted platforms and application/software management for mobile security and privacy
   - trusted mobile devices – secure and user friendly;
   - full exploitation of the SIM card;
   - secure reconfiguration of mobile devices;
   - DRM security architecture and protocols.
2. Mobile network/transport security and privacy
   - protection of the core network against attacks;
   - heterogeneous network access control security;

- seamless security handover at network level;
- security architectures for service networks;
- protection of the service network against attacks.
3. Mobile application security and privacy
   - mobile application security framework;
   - user-centric mechanisms allowing controlled release of personal information.
4. Mobile privacy and identity management
   - single sign-on based on mobile authentication;
   - authorisation privacy;
   - authentication via security tokens using mobile devices;
   - location-based services versus location privacy;
   - privacy-preserving mobile applications with tuneable anonymity.
5. Basic security and privacy technologies for mobile environments
   - lightweight stream ciphers;
   - truly practical cryptographic mechanisms in constrained environments;
   - delegation of cryptographic operations;
   - lightweight key management infrastructures;
   - conference and group keying.

17.6.3   Research issues in detail

The following subsections describe the different research issues in more detail.

17.6.3.1   Trusted mobile devices – secure and user-friendly

End-to-end security protocols can only be meaningful between properly secured endpoints. At the client side, mobile devices with a secure default configuration need to be provided. Trusted paths from the mobile device to the user need to be foreseen ('what you see is what is happening'). A good balance should be found between allowing potentially dangerous tools (such as cookies and mobile code) and enforcing the appropriate usage of them (i.e. in security applications).

Furthermore, substantial research and development efforts concerning trusted execution environments (operating system and hardware platform) for mobile devices remain of paramount importance. There seems to be a fundamental conflict between, on the one hand, the desire for user-friendliness, functionality, openness and freedom, including the user's right (in a non-organisational context) to disable/enable and configure the security features of this platform and, on the other hand, the level of security that such a platform can offer. If users have options to choose from, it will always be possible to lure them into choosing the least secure configuration, and whatever security the platform could offer will be circumvented. This might not be a problem for the few experienced users, but it is a real threat for all the inexperienced end-users.

Thus, a secure, personal and trusted device is a core ingredient for secure and mobile solutions. We need more research into and development of mobile devices that provide just enough user-friendly functionality and still remain authentic and trustworthy.

17.6.3.2   Full exploitation of the SIM card

The SIM (in combination with the mobile phone) is an important asset for the mobile operator. An extended SIM could provide the necessary platform for many different security functions. Therefore, it has to be researched, what – besides pure authentication – can be the role of an extended SIM card in a future mobile terminal, what are the technological challenges and what measures are required to establish trust. Note that Europe is also leading in the smartcard area, a lead that needs to be maintained.

17.6.3.3   Secure reconfiguration of mobile devices

Security issues concerning the reconfiguration of mobile equipment heavily depend on the kind of data to be transferred to the mobile device (such as pure application software, software updating the mobile equipment's operating system, software and parameter data modifying the radio properties of the mobile equipment, data used to reconfigure FPGAs) and on the entities involved in the reconfiguration process (such as software providers, network operators, device manufacturers, the mobile device, its user). The conceivable reconfiguration scenarios can be rather complicated, and for such scenarios, it will probably not be possible to provide a reliable security architecture. Current reconfiguration scenarios that serve the needs of the different entities involved and that simultaneously allow for trustworthy security solutions cannot be seen as providing a complete picture, and must, therefore, be addressed by future research.

With respect to a given, concrete reconfiguration scenario, security issues concerning authorisation (who is permitted to initiate and execute the download), data origin and integrity, privacy (not everybody should know which software is on my mobile equipment) and conformance with regulatory requirements have to be investigated. In addition, protection mechanisms have to be in place to meet the requirements of network operators and service providers. The main issues are the split of authority for reconfiguration in decentralised scenarios, and approaches to dealing with failures. Furthermore, some reconfiguration scenarios may require investigation of digital rights management, confidentiality and non-repudiation issues.

17.6.3.4   DRM security architectures and protocols

Today's DRM standards and proprietary DRM solutions still have many shortcomings and security weaknesses. Many of them are based on obfuscation and security by obscurity. A good DRM system should be secure even if all technical details are published. This requires underlying security technology components like tamper-resistant memory, tamper-resistant execution environments, tamper-resistant network interfaces, secure clocks, device-specific keys and certificates. Using such secure components and secure environments, secure DRM architectures can be developed by specifying suitable data formats, and authentication and data exchange protocols. Such secure DRM architectures have yet to be developed to maturity, and to be tested and scrutinised.

17.6.3.5 Protection of the core network against attacks

Mobile networks are opening up more and more, leading to an increased vulnerability to both internal and external attacks (and failures). Mechanisms need to be researched that reduce the risk of attacks against the core network.

Of particular importance is protection against Denial-of-Service attacks, which becomes increasingly important as IP becomes ever more pervasive. Trust assumptions valid today that allow hop-by-hop security solutions in the core network may be no longer valid, and may necessitate alternative solutions for core network security in the future. The partial sharing of network resources (e.g. radio access networks) among operators raises many security issues. A means to counter Denial-of-Service attacks in the core network is the use of intrusion detection tools. Such tools need to be tailored to the requirements of the mobile telecommunications environment to be effective. The range of Internet solutions for IDS is quite broad, but investigation of adaptation and enhancements of existing solutions and standards is required for the special case of a mobile infrastructure.

Other forms of attacks against core networks, for example, eavesdropping, impersonation or viruses also require research.

17.6.3.6 Heterogeneous network access control security

The increasing heterogeneity of the networking environment is one of the long-term trends, which requires new security approaches. The main challenges include the following.

- The investigation and development of unified, secure and convenient authentication mechanisms that can be used in different access networks and for different services (single sign-on). Authentication and key agreement is the central component of secure access procedures. Currently, there are no protocols available that are light-weight, can be carried over arbitrary access networks and are flexible enough to be re-used in many different contexts in future mobile systems. The development of new authentication protocols may be required here.
- In addition to the key agreement, the efficient negotiation of security contexts is not available in some settings. The secure negotiation of configuration information is also an open issue.
- Access procedures need to be hardened against Denial-of-Service attacks.
- Today's access procedures do not provide non-repudiation.
- Architectures for the flexible use of different types of access credentials (e.g. payment tokens) need much further research and standardisation.
- Efficient key distribution for multicast remains a research topic.

17.6.3.7 Seamless security handover at the network level

Seamless handover means that a user can switch between points of attachment to networks with performance guarantees and without experiencing degradation in the quality of the service. Currently, this is only possible between networks of the same or similar type. Security is an issue here, as re-authentication often causes prohibitive delays, so efficient security context transfer schemes are needed. Security solutions still need to be developed for seamless handover between different network types.

17.6.3.8   Security architectures for service networks

Service networks, that is, the parts of mobile operator networks hosting services and applications, become increasingly important. The internal and inter-domain security architecture for those service networks is still in its infancy and needs to be researched in detail.

17.6.3.9   Protection of service networks against attacks

Service networks offer numerous internal and external interfaces. Some of the communication partners might change frequently; for instance, external service providers or end-users. Thus, it is essential to protect service networks against any kind of internal or external attack. Protection could, for instance, be improved by conducting systematic threat analyses and by designing a catalogue of countermeasures against, for example, Denial of Service and Distributed Denial of Service attacks. These could include cryptographic measures (cookies, client puzzles, etc.) as well as networking measures (intrusion detection systems, active networking technology). Research in service network protection is of utmost importance.

17.6.3.10   Mobile application security framework

Given the growing heterogeneity of the user's communication environment, there is a need to develop and validate a coherent framework for security and trust for mobile applications. The traditional approach to security is to adopt a layered architecture to solve specific aspects of an overall problem. However, what is needed is a holistic, comprehensive approach to security for mobile systems, as the security of a system is only as strong as the weakest element. Key elements of the framework are a suitable architecture for mobile applications and a methodology guiding the use of the framework. Security architectures need to ensure that the user can access a multitude of applications with only one or a small number of access credentials, using unified, easy to use procedures.

17.6.3.11   User-centric mechanisms allowing controlled release of personal information

There is a need to develop user-centric mechanisms to allow controlled release of personal, preference-related and location-based information, and to deliver assurances to owners about how personal information will be used by second and third parties and in compliance with any local data protection regulations.

17.6.3.12   Single sign-on based on mobile authentication

Single sign-on is currently driven by the (non-European) IT giants. There is, however, a chance that mobile operators, with their authentication mechanisms already deployed, could play an important role in single sign-on solutions. Investigations should include not only technical aspects but also end-user experience.

17.6.3.13   Authorisation privacy

Authorisation plays a particular role on privacy, since too often a lot of personal information is distributed to enable access control. In particular, the client–server model, on which most Internet applications are developed, can be considered as privacy-intrusive: generally, the server grants or denies access to a client according to the identity claimed by the client.

In many cases, the customer should be allowed to access services by providing, instead of his/her identity, a proof that he/she is authorised, for example, an attribute certificate proving his/her association membership, a credit card ownership, that he/she has a driving licence, etc. Different certification authorities could issue such attributed certificates, which is a way to implement multiple identities. Zero-knowledge techniques can be used to guarantee that the certificate belongs to the user. Of course, the management of multiple identities raises problems of ease-of-use, as well as unlinkability.

Another related approach is to use cryptographic functions to prove certain attributes of a certificate without disclosing other attributes.

More research is needed to develop similar solutions and experiments on large-scale operations should also be supported.

17.6.3.14   Authentication via security tokens using mobile devices

Mobile devices can be used as a security token (e.g. a user can open a door with his PDA). Specific solutions already exist (some cars can be started by the user only if he/she carries a smartcard), but more generalised techniques such as a challenge–response via a wireless PDA need further research. These solutions also need to be standardised so that it should be possible to use any conformant mobile device as a security token in authentication protocols.

17.6.3.15   Location-based services versus location privacy

The user may be provided with certain services having properties based on his/her location (e.g. finding the nearest specialty restaurant). Using these services may reveal the user's location information, but this is not always desirable. Sometimes, the user may wish to keep his location private. That is why there is a need for location-based services, which preserve location privacy. Location measurement control is a specific instance of this issue. Where is the user location measured (device, network, etc.) and what are the conditions for such a measurement and access to that data? If the location is measured in the device, only the user may be able to access it, and decide to give it away as proof. Is it possible to decouple the identity of the user from the location of the mobile device in order to preserve privacy?

17.6.3.16   Privacy-preserving mobile applications with tuneable anonymity

Technical tools exist, and are gaining popularity, that provide (data) anonymity and that can protect the user's privacy to some determined extent. In some cases, this is achieved at the cost of loss of functionality. Further research and evaluation of application-specific privacy-preserving solutions is needed.

More generally, anonymity is not a black-or-white issue. A model for measuring the degree of anonymity should be further developed and we should learn to correctly interpret this model. This would help in finding the balance between, on the one hand, privacy and personalisation (e.g. the banner application) and, on the other hand, privacy and performance (most current solutions for anonymous communication trade-off anonymity with bandwidth efficiency).

17.6.3.17 Lightweight stream ciphers

Research is needed to develop lightweight stream ciphers with a well-understood level of security for application in constrained environments. Current efforts such as NESSIE (www.cosic.esat.kuleuven.ac.be/nessie/) have not resulted in such primitives.

17.6.3.18 Truly practical cryptographic mechanisms in constrained environments

Making truly practical in constrained environments the cryptographic mechanisms associated with payment, digital rights management, privacy and anonymity protection is a challenge to cryptography from mobile applications. These mechanisms include special signature schemes, electronic cash schemes, and key exchange and authentication protocols.

17.6.3.19 Delegation of cryptographic operations

Further research is needed to develop mechanisms that support delegation of cryptographic operations from constrained mobile devices to more powerful but less trusted devices.

17.6.3.20 Lightweight key management infrastructures

There is a need to develop and standardise lightweight key management infrastructures supporting the deployment of public key technology in mobile networks, and to ensure the interoperability of infrastructures from different mobile standards.

17.6.3.21 Conference and group keying

Many scenarios involving multi-party mobile communications assume the existence of efficient cryptographic protocols for dynamic conferencing and group keying. While 'theoretical' protocols already exist in the cryptographic literature, work is needed to ensure these are sufficiently lightweight and reliable in the mobile setting.

17.6.4 Analysis of the research topics

Taking into account all relevant criteria for mobile security and privacy we can summarise the results as follows.

To establish a coherent future for mobile trust and security in the developing mobile service world, it is of paramount importance to consider security as a fundamental pillar from the beginning of the design. Traditional approaches are not holistic enough to cover all the approaching technology interactions. Networks of services can only be built on a trustworthy foundation of solid interworking security.

To provide valuable and interesting content to users, content providers need to get involved, since an operator of a mobile network will not be able to provide all the content itself. For content providers it is of utmost importance that their content is paid for, else they will not provide it. The Digital Rights Management security that is available today is often proprietary, has shortcomings in security and is not tuned to the mobile environment. Hence, the development of the trustworthy DRM security architecture and the investigation of suitable protocols are of utmost importance.

Establishing high trust in the mobile device, particularly taking into account the existence of the future SIM as security token, is a prerequisite for private and secure mobile communication. Europe is leading in the smartcard area, and should exploit this advantage by leveraging existing technologies and knowledge. The user must be able to perform sensitive actions in a trusted execution environment, authenticate securely and to have trustworthy communication. To reach a secure and trusted mobile environment the SIM should play a central role and the opportunities that could be provided by this secure token should be investigated.

The increasing heterogeneity of radio technologies like UMTS, WLAN and Bluetooth in the new mobile environment is one of the most visible trends that can be observed. Flexible authentication and authorisation procedures have to be in place to provide convenient and user-friendly access to all these networks. Different types of credentials provided at different layers and handover of secure sessions when changing the network should be possible. Single sign-on (SSO), which is currently driven by non-European IT companies, can provide a unique opportunity for mobile network operators who can integrate SSO into their systems and provide a comfortable usage environment for the user.

Future service networks and core networks offer numerous interfaces toward users and service providers. Some communication partners might change frequently. To provide services without intensive consideration of the security issues would pose a high risk not only to the infrastructure, but also to the services provided. The list of threats range from Denial of Service attacks, injection of malicious code, viruses, worms, hijacking of sessions and servers, to theft of confidential information. To protect the service networks and core networks, intrusion detection systems, proxies, firewalls and other technologies that are suited to the mobile environment have to be investigated. If these concerns are not addressed the whole deployment of new services is at risk.

In order to provide user-friendly and personalised services on mobile devices with their limited display capabilities, personal user data need to be stored and processed. This is a highly sensitive issue for the user, especially if the user location is involved. Suitable mechanisms that provide the user with control over her data and that offer various degrees of anonymity suitable for every situation to protect the user's privacy in a practical manner is essential to providing trust in the new services. Authorisation should only require the really necessary information and not a complete personal profile. If users do not trust the mobile environment to handle their data in a privacy-preserving manner they will simply not use it. Therefore, research in this area is important so as to establish market awareness of economic privacy solutions before products and services with somewhat shaky privacy provision are allowed to establish a de facto situation.

Numerous security and privacy-enhancing tools have already been investigated (e.g. signature schemes, DRM, electronic cash, authentication protocols, etc.), but many of these are impractical in constrained environments. The actual practical requirements of security and privacy tools need to be investigated and their efficiency (round trip time, usage of memory, etc.) and performance to be tested. Other required functionalities have not been investigated from a lightweight perspective at all, for example, stream ciphers, key management infrastructures and conference and group keying. Mobile profiling of security and privacy functions (e.g. outsourcing expensive operations to servers) could build a cornerstone of the new security approach for the mobile environment.

17.6.5   Assessment of standardisation

The success of much of the research work, which will address the research issues listed in this document, is inherently dependent on the consensual adoption of its results and recommendations by industry actors. This is true, in particular, for all results concerning protocols between entities in a public environment. There, standards are a must because otherwise interoperability of entities produced by different manufacturers or under the control of different operators cannot be guaranteed. The lack of standards may cause market fragmentation and prevent adequate growth of the market. One important reason for the success of European industry in the mobile arena is the successful establishment of the GSM standard, which has now by far the largest footprint of any mobile standard. In contrast, North America alone has three different digital radio standards. Therefore, there will need to be a well-managed relationship between the projects addressing the standards-relevant research and the appropriate standards bodies and industry forums. But it has to be borne in mind that much of the research addressed in this document will be on achieving medium- to long-term research objectives. The standards bodies currently seen as most relevant are 3GPP and the IETF for network security, OMA and W3C for application security, IEEE 802 (WLAN) and the Bluetooth SIG for short-range wireless security, the Liberty Alliance Project, OASIS and others for web security and identity management, and ISO and IEEE for standards in cryptography. Privacy has been discussed in the Location Interoperability Forum (LIF), a joint effort from telecom vendors, service providers and operators to come to a common location interchange format. The LIF has now been integrated into the OMA. The same is true for the Location Positioning Workgroup of the WAP Forum. The IETF Geographical Location/Privacy Workgroup (GeoPriv) is currently working on standards and recommendations for location-based services. In addition, there are numerous industrial forums, especially

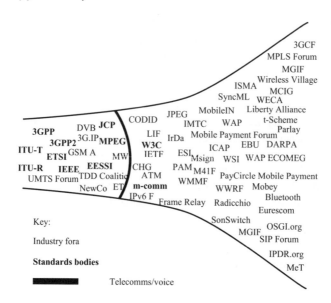

Figure 17.6   The explosion of industry forums and standards bodies

in the area of applications, for example, m-commerce, whose relevance for standards may become clearer over time.

Due to the growing complexity of mobile networks we observe an explosion of industry forums and standards bodies. An overview of the standard bodies and industry forums is shown in Figure 17.6. Drivers for the explosive growth are the increased separation of network and application standards, a growing number of specifications for mobile communication and participation of a larger number of stakeholders (broadcasting and media also come together). As it is one of the objectives of OMA to establish a single mobile industry standards forum to function as the driving force responsible for creating service-level interoperability, many of the micro-forums are integrated in OMA.

National Data Protection organisations are actively striving for safeguards that preserve the integrity of the user's privacy. Emerging context and location-aware systems force them to broaden their working ground.

Legislative initiatives to control sensitive information, such as location data, are under way both within the European Union and in the United States. In short, they specify that users of mobile location services must be protected by privacy safeguards, must be fully informed of the purposes of the usage of the mobile location services, and must have the right to determine the use of their personal information. The European Telecommunication Directives discuss the idea of a user's right to choose whether to grant (opt-in) or deny (opt-out) any use of personal information at any time. In some cases, the European Directives conflict with national legislation, that is, national laws are either more or less strict than those described in the Directives. These conflicts should be resolved.

Acknowledgements

The work described here was performed in the PAMPAS (Pioneering Advanced Mobile Security and Privacy) project. PAMPAS was an accompanying measure funded within the IST Programme of the European Union's Fifth Framework Programme (FP5) under contract number IST-2001–37763. The object of PAMPAS was to prepare the ground for research initiatives in the upcoming Sixth Framework Program (FP6). PAMPAS focussed on the area of privacy and security for beyond-3G mobile systems and applications. Participants of the PAMPAS project were: Ericsson Eurolab Deutschland GmbH, Vodafone Group Services Ltd, Siemens AG, Katholieke Universiteit Leuven Research & Development, Royal Holloway and Bedford New College and the Telematica Instituut.

The authors acknowledge the contributions of the PAMPAS project members, especially Vaia Sdralia, Sorin Iacob, Andreas Fuchsberger, Dries Schellekens, Dave Singelee, Robert Maier, Rolf Blom, Axel Busboom and Ricarda Weber for their contributions to the mobile privacy and security research.

Without the involvement and expertise input of the constituency of stakeholders, the work described here would be less valuable.

Notes

1   Identity may not be the only attribute to be authenticated; other attributes such as location, functionality or capability may be more significant in certain circumstances [1].
2   The Platform for Privacy Preferences Project (P3P), developed by the World Wide Web Consortium, is emerging as an industry standard providing a simple, automated way for users to gain more control over the use of personal information on Web sites they visit (http://www.w3.org/P3P).
3   According to a report by market analysts Taylor Nelson Sofres on 3G roll-out; www.tnsofres.com/press/pressstory.cfm?storyID=643.
4   This is a revision and clustering of the open research issues in the area of mobile privacy and identity management as identified in the PAMPAS project and described in the Refined Roadmap [3].

References

1   ZAKIUDDIN, S., CREESE, S., ROSCOE, B. and GOLDSMITH, M.: 'Authentication in pervasive computing', PAMPAS Workshop #1, Leuven, 16/17 September 2002, position paper, http://PAMPAS.eu.org/Position_Papers/ QinetiQ.pdf
2   PAMPAS, Pioneering Advanced Mobile Privacy and Security: Deliverable D02, 'Preliminary roadmap', December 2002, available at http://www.pampas.eu.org/
3   PAMPAS, Pioneering Advanced Mobile Privacy and Security: Deliverable D03, 'Refined roadmap', February 2003, available at http://www.pampas.eu.org/

4 Privacy & American Business: 'Privacy on and off the Internet: what consumers want' (Privacy & American Business, Hackensack, NJ, 2002)

5 PIEPERS, E.: 'Cost-effective information security', SANS Institute Information Security Reading Room, 6, June 2001, http://rr.sans.org/audit/cost-effective.php

6 See the EU Information Technologies Programme, http://www.cordis.lu/esprit/

7 RAPID: 'Roadmap for European legal research in privacy and identity management', http://www.law.kuleuven.ac.be/icri/publications/421rapid.pdf

8 CyberVote: D4 Volume 2, 'Report on electronic democracy projects, legal issues of Internet voting and users (i.e. voters and authorities representatives) requirements analysis', http://www.eucybervote.org/KUL-WP2-D4V2-v1.0.pdf

9 MELODY, W. H.: 'Trends in European telecommunication: 2002 status report of Denmark's progress in telecom reform and information infrastructure development', October 2002, Denmark, LIRNE.NET, http://www.lirne.net/resources/denmark_2002.pdf

10 ITU 2002: 'Internet for a mobile generation report', Statistical Annex ITU, Geneva, http://www.itu.int/osg/spu/publications/mobileinternet/

11 A general comparison between 3G and WiFi is described in LEHR, W. and MCKNIGHT, L. W.: 'Wireless Internet access: 3G vs. WiFi', *Telecommunications Policy*, 2003, **27**, pp. 351–70

12 PAMPAS, Pioneering Advanced Mobile Privacy and Security: Deliverable D04, 'Final roadmap – extended version', May 2003, available at http://www.pampas.eu.org/

Index